AF358665

Intelligent Image Processing and Sensing for Drones

Intelligent Image Processing and Sensing for Drones

Editor

Seokwon Yeom

Basel • Beijing • Wuhan • Barcelona • Belgrade • Novi Sad • Cluj • Manchester

Editor
Seokwon Yeom
AI
Daegu University
Gyeongsan
South Korea

Editorial Office
MDPI
St. Alban-Anlage 66
4052 Basel, Switzerland

This is a reprint of articles from the Special Issue published online in the open access journal *Drones* (ISSN 2504-446X) (available at: https://www.mdpi.com/journal/drones/special_issues/Drone_Image).

For citation purposes, cite each article independently as indicated on the article page online and as indicated below:

Lastname, A.A.; Lastname, B.B. Article Title. *Journal Name* **Year**, *Volume Number*, Page Range.

ISBN 978-3-7258-0751-2 (Hbk)
ISBN 978-3-7258-0752-9 (PDF)
doi.org/10.3390/books978-3-7258-0752-9

Contents

 *drones*

Special Issue on Intelligent Image Processing and Sensing for Drones

Seokwon Yeom

Department of AI, Daegu University, Gyeongsan 38453, Republic of Korea; yeom@daegu.ac.kr;
Tel.: +82-53-850-6643

1. Introduction

Recently, the use of drones or unmanned aerial vehicles (UAVs) for various purposes has been increasing [1]. Drones can be remotely controlled or programmed to capture scenes from a distance. This capture method is cost-effective and does not require highly trained personnel. Drones are widely used in various applications such as industrial and infrastructure inspections [2,3], agricultural and environmental monitoring [4,5], geographical surveying [6], search and rescue missions [7], security and surveillance [8], and so on.

Multiple sensors can be mounted on a drone. In addition to visible cameras, infrared thermal imaging and multispectral imaging equipment can be mounted on a drone [9,10]. LiDAR and SAR are active sensors that can be used in drones [11–13]. These mobile aerial imaging sensors provide a new perspective on research and development for a variety of applications. However, more challenges are often posed than fixed and ground sensors because of the unique sensing environments and limited resources of drones. Certainly, information acquired by a drone is of tremendous value; thus, intelligent analysis of the data is necessary to make the best use of them.

This Special Issue focuses on a wide range of intelligent processing of images and sensor data acquired by drones. The objectives of intelligent processing range from the refinement of raw data to the extraction and processing of featured attributes and the symbolic representation or visualization of the real world. This can be achieved through image/signal processing or deep/machine learning algorithms. The latest technological developments will be shared through this Special Issue. Researchers and investigators are invited to contribute original research or review articles to this Special Issue.

Eight research papers and two review papers were verified through a thorough review process. Many valuable and recent technologies have been provided in the selected papers to solve real problems. The first volume of this Special Issue on the topic is closed; more in-depth research on the same topic is expected in the second volume of this Special Issue. It is anticipated that the scope of intelligent processing will be even broader in the future.

2. Overview of Published Articles

This Special Issue was introduced to collect the latest research on relevant topics, and more importantly, to address the current practical and theoretical challenges. In the following, papers are categorized into several subtopics: industrial applications, positioning and tracking, visualization of the real world, and advances in computer vision.

2.1. Industrial Applications

In the first contribution entitled 'Method for Complex Road Cracks Collected by UAV Based on HC-Unet++', Cao, H.; Gao, Y.; Cai, W.; Xu, Z.; and Li, L. proposed a new deep learning network model called HC-Unet++ for the detection and segmentation of road cracks. Their method is based on convolutional neural networks, and they show that UAV aerial photography plays an important role in road maintenance and traffic safety. The new

Citation: Yeom, S. Special Issue on Intelligent Image Processing and Sensing for Drones. *Drones* **2024**, *8*, 87. https://doi.org/10.3390/drones8030087

Received: 28 February 2024
Accepted: 29 February 2024
Published: 4 March 2024

network model can eliminate complex backgrounds, effectively detect cracks of various irregular shapes, and reduce crack discontinuity.

Detection of power lines and transmission towers is critical for the safety of power grid operations. The second contribution entitled 'Transmission Line Segmentation Solutions for UAV Aerial Photography Based on Improved UNet' by He, M.; Qin, L.; Deng, X.; Zhou, S.; Liu, H.; and Liu, K. enhances the UNet algorithm, which is a deep learning segmentation model. The authors improved the UNet algorithm by extracting features using a lightweight backbone structure and then reconstructing them into contextual information features.

Ali, M.A.H.; Baggash, M.; Rustamov, J.; Abdulghafor, R.; Abdo, N.A.-D.N.; Abdo, M.H.G.; Mohammed, T.S.; Hasan, A.A.; Abdo, A.N.; Turaev, S.; et al., in the third contribution entitled by 'An Automatic Visual Inspection of Oil Tanks Exterior Surface Using Unmanned Aerial Vehicle with Image Processing and Cascading Fuzzy Logic Algorithms', proposed a method for visual inspection of external surface defects on oil tanks. Two cascade fuzzy logic algorithms were developed to detect defects and remove noise.

The fourth contribution is a review paper entitled 'An Overview of Drone Applications in the Construction Industry'. Choi, H.-W.; Kim, H.-J.; Kim, S.-K.; and Na, W.S. presented a comprehensive overview of the applications of drones in the construction industry. The introduction of drones into the construction industry has brought about revolutionary advancements in all phases of construction projects. For example, drones equipped with high-resolution cameras in the design phases have revolutionized field surveys and aerial mapping. During the construction phase, drones play an important role in monitoring and inspecting the construction progress and ensuring safety. Drones can also efficiently detect and identify damage, enabling proactive maintenance, cost reductions, and extended asset lives.

2.2. Positioning, Detection, and Tracking

UAV landing technology is a critical tool that allows drones to land without human intervention, improving safety, efficiency, and operability in remote or challenging environments. However, high-precision autonomous landing is still a major challenge. Xu, Y.; Zhong, D.; Zhou, J.; Jiang, Z.; Zhai, Y.; and Ying, Z described, in the fifth contribution entitled 'A Novel UAV Visual Positioning Algorithm Based on A-YOLOX', a UAV positioning algorithm called attention-based YOLOX, which improves the accuracy of automatic landings for UAVs.

While the use of drones offers significant benefits for a variety of applications, it also presents a number of hazards. These hazards range from concerns about privacy and security to risks of physical injury and environmental impacts. In the sixth contribution entitled 'Non-Linear Signal Processing Methods for UAV Detections from a Multi-Function X-Band Radar', Kumar, M. and Kelly, P.K. developed a nonlinear processing technique for UAV detection using a portable radar system.

Drones have become an invaluable tool in search and rescue (SAR) missions. In the seventh contribution entitled 'Thermal Image Tracking for Search and Rescue Missions with a Drone', Yeom, S. developed an effective thermal image tracking method. His method shows promising results for handling challenging environments such as complex backgrounds, heavy occlusions, and complex maneuvering of drones.

2.3. Visualization of the Real World

Amala Arokia Nathan, R.J.; Kurmi, I.; and Bimber, O., in the eighth contribution entitled 'Inverse Airborne Optical Sectioning', presented inverse airborne optical sectioning, an optical analogy to inverse synthetic aperture radar. Moving targets, such as walking people, that are heavily occluded by vegetation can be made visible and tracked using cameras on drones hovering over forests. They introduced the principles of inverse synthetic aperture imaging and suppressed the signal of occluders by filtering the Radon transform of the image integral.

The development of UAVs has significantly increased the type and number of datasets available for image synthesis. An improved image synthesis model, SYGAN, was proposed in the ninth contribution entitled 'Improved Image Synthesis with Attention Mechanism for Virtual Scenes via UAV Imagery' by Mo, L.; Zhu, Y.; Wang, G.; Yi, X.; Wu, X.; and Wu, P. A spatial adaptive normalization module and a sparse attention mechanism were introduced on the basis of a generative adversarial network (GAN) for image synthesis.

2.4. Potentials of Drones in Computer Vision

Vision transformers can be used for various computer vision applications, including image classification, object detection, image segmentation, image compression, image super-resolution, image denoising, anomaly detection, and drone imagery. In the tenth contribution entitled 'A Comprehensive Survey of Transformers for Computer Vision', Jamil, S.; Jalil Piran, M.; and Kwon, O.-J. reviewed the state of the art and compiled a list of available models and discussed the pros and cons of each vision transformer model.

3. Conclusions

This Special Issue covers various applications of images and signals acquired by drones. It also shows a wide range of potential as well as the versatility of drones in the near future, encompassing a richness of research fields. This is reflected in the wide range of methodologies adopted in the studies, including deep learning, traditional machine learning, signal and image processing, and estimation theory. From a methodological perspective, deep learning appears in five of the eight research papers (Contributions 1, 2, 5, 7, 9). Among them, a deep learning method was combined with estimation theory in Contribution 7. Two papers (Contributions 6 and 8) take advantage of methods using the signal processing regime, and one paper (Contribution 3) combines image processing with machine learning algorithms. It is anticipated that research in the field of deep learning will increase further, while different approaches can potentially be combined with one another in this era of rapid change and development.

As a final note, I would like to thank all the authors contributing to this Special Issue. I also appreciate the dedicated reviewers and editors for their efforts and expertise. I sincerely hope that readers will find great inspiration from current and future technologies related to drones in this Special Issue.

Conflicts of Interest: The authors declare no conflict of interest.

List of Contributions:

1. Cao, H.; Gao, Y.; Cai, W.; Xu, Z.; Li, L. Segmentation Detection Method for Complex Road Cracks Collected by UAV Based on HC-Unet++. *Drones* **2023**, *7*, 189. https://doi.org/10.3390/drones7030189.
2. He, M.; Qin, L.; Deng, X.; Zhou, S.; Liu, H.; Liu, K. Transmission Line Segmentation Solutions for UAV Aerial Photography Based on Improved UNet. *Drones* **2023**, *7*, 274. https://doi.org/10.3390/drones7040274.
3. Ali, M.A.H.; Baggash, M.; Rustamov, J.; Abdulghafor, R.; Abdo, N.A.-D.N.; Abdo, M.H.G.; Mohammed, T.S.; Hasan, A.A.; Abdo, A.N.; Turaev, S.; et al. An Automatic Visual Inspection of Oil Tanks Exterior Surface Using Unmanned Aerial Vehicle with Image Processing and Cascading Fuzzy Logic Algorithms. *Drones* **2023**, *7*, 133. https://doi.org/10.3390/drones7020133.
4. Choi, H.-W.; Kim, H.-J.; Kim, S.-K.; Na, W.S. An Overview of Drone Applications in the Construction Industry. *Drones* **2023**, *7*, 515. https://doi.org/10.3390/drones7080515.
5. Xu, Y.; Zhong, D.; Zhou, J.; Jiang, Z.; Zhai, Y.; Ying, Z. A Novel UAV Visual Positioning Algorithm Based on A-YOLOX. *Drones* **2022**, *6*, 362. https://doi.org/10.3390/drones6110362.
6. Kumar, M.; Kelly, P.K. Non-Linear Signal Processing Methods for UAV Detections from a Multi-Function X-Band Radar. *Drones* **2023**, *7*, 251. https://doi.org/10.3390/drones7040251.
7. Yeom, S. Thermal Image Tracking for Search and Rescue Missions with a Drone. *Drones* **2024**, *8*, 53. https://doi.org/10.3390/drones8020053.
8. Amala Arokia Nathan, R.J.; Kurmi, I.; Bimber, O. Inverse Airborne Optical Sectioning. *Drones* **2022**, *6*, 231. https://doi.org/10.3390/drones6090231.

9. Mo, L.; Zhu, Y.; Wang, G.; Yi, X.; Wu, X.; Wu, P. Improved Image Synthesis with Attention Mechanism for Virtual Scenes via UAV Imagery. *Drones* **2023**, *7*, 160. https://doi.org/10.3390/drones7030160.
10. Jamil, S.; Jalil Piran, M.; Kwon, O.-J. A Comprehensive Survey of Transformers for Computer Vision. *Drones* **2023**, *7*, 287. https://doi.org/10.3390/drones7050287.

References

1. Alzahrani, B.; Oubbati, O.S.; Barnawi, A.; Atiquzzaman, M.; Alghazzawi, D. UAV assistance paradigm: State-of-the-art in applications and challenges. *J. Netw. Comput. Appl.* **2020**, *166*, 102706. [CrossRef]
2. Shanti, M.Z.; Cho, C.S.; de Soto, B.G.; Byon, Y.J.; Yeun, C.Y.; Kim, T.Y. Real-time monitoring of work-at-height safety hazards in construction sites using drones and deep learning. *J. Saf. Res.* **2022**, *83*, 364–370. [CrossRef] [PubMed]
3. Fan, J.; Saadeghvaziri, M.A. Applications of drones in infrastructures: Challenges and opportunities. *Int. J. Mech. Mechatron. Eng.* **2019**, *13*, 649–655.
4. Rao Mogili, U.M.; Deepak, B.B.V.L. Review on Application of Drone Systems in Precision Agriculture. *Procedia Comput. Sci.* **2018**, *133*, 502–509. [CrossRef]
5. Hodgson, J.; Baylis, S.; Mott, R.; Herrod, A.; Clarke, R.H. Precision wildlife monitoring using unmanned aerial vehicles. *Sci. Rep.* **2016**, *6*, 22574. [CrossRef] [PubMed]
6. Rossi, G.; Tanteri, L.; Tofani, V.; Vannocci, P.; Moretti, S.; Casagli, N. Multitemporal UAV surveys for landslide mapping and characterization. *Landslides* **2018**, *15*, 1045–1052. [CrossRef]
7. Schedl, D.C.; Kurmi, I.; Bimber, O. An autonomous drone for search and rescue in forests using airborne optical sectioning. *Sci. Robot.* **2021**, *6*, eabg1188. [CrossRef] [PubMed]
8. Dilshad, N.; Hwang, J.; Song, J.; Sung, N. Applications and challenges in video surveillance via drone: A brief survey. In Proceedings of the 2020 International Conference on Information and Communication Technology Convergence (ICTC), Jeju Islan, Republic of Korea, 21–23 October 2020; IEEE: New York, NY, USA, 2020.
9. Beaver, J.T.; Baldwin, R.W.; Messinger, M.; Newbolt, C.H.; Ditchkoff, S.S.; Silman, M.R. Evaluating the use of drones equipped with thermal sensors as an effective method for estimating wildlife. *Wildl. Soc. Bull.* **2020**, *44*, 434–443. [CrossRef]
10. Carrasco-Escobar, G.; Manrique, E.; Ruiz-Cabrejos, J.; Saavedra, M.; Alava, F.; Bickersmith, S.; Prussing, C.; Vinetz, J.M.; Conn, J.E.; Moreno, M.; et al. High-accuracy detection of malaria vector larval habitats using drone-based multispectral imagery. *PLoS Neglected Trop. Dis.* **2019**, *13*, e0007105. [CrossRef] [PubMed]
11. Risbøl, O.; Gustavsen, L. LiDAR from drones employed for mapping archaeology–Potential, benefits and challenges. *Archaeol. Prospect.* **2018**, *25*, 329–338. [CrossRef]
12. Schreiber, E.; Heinzel, A.; Peichl, M.; Engel, M.; Wiesbeck, W. Advanced buried object detection by multichannel, UAV/drone carried synthetic aperture radar. In Proceedings of the 2019 13th European Conference on Antennas and Propagation (EuCAP), Krakow, Poland, 31 March–5 April 2019; IEEE: New York, NY, USA, 2019.
13. Li, C.J.; Ling, H. Synthetic aperture radar imaging using a small consumer drone. In Proceedings of the 2015 IEEE International Symposium on Antennas and Propagation & USNC/URSI National Radio Science Meeting, Vancouver, BC, Canada, 19–24 July 2015; IEEE: New York, NY, USA, 2015.

Article

Segmentation Detection Method for Complex Road Cracks Collected by UAV Based on HC-Unet++

Hongbin Cao [1], Yuxi Gao [1], Weiwei Cai [2], Zhuonong Xu [1,*] and Liujun Li [3]

[1] College of Computer & Information Engineering, Central South University of Forestry and Technology, Changsha 410004, China
[2] School of Artificial Intelligence and Computer Science, Jiangnan University, Wuxi 214122, China
[3] Department of Soil and Water Systems, University of Idaho, Moscow, ID 83844, USA
* Correspondence: t19980580@csuft.edu.cn

Abstract: Road cracks are one of the external manifestations of safety hazards in transportation. At present, the detection and segmentation of road cracks is still an intensively researched issue. With the development of image segmentation technology of the convolutional neural network, the identification of road cracks has also ushered in new opportunities. However, the traditional road crack segmentation method has these three problems: 1. It is susceptible to the influence of complex background noise information. 2. Road cracks usually appear in irregular shapes, which increases the difficulty of model segmentation. 3. The cracks appear discontinuous in the segmentation results. Aiming at these problems, a network segmentation model of HC-Unet++ road crack detection is proposed in this paper. In this network model, a deep parallel feature fusion module is first proposed, one which can effectively detect various irregular shape cracks. Secondly, the SEnet attention mechanism is used to eliminate complex backgrounds to correctly extract crack information. Finally, the Blurpool pooling operation is used to replace the original maximum pooling in order to solve the crack discontinuity of the segmentation results. Through the comparison with some advanced network models, it is found that the HC-Unet++ network model is more precise for the segmentation of road cracks. The experimental results show that the method proposed in this paper has achieved 76.32% mIOU, 82.39% mPA, 85.51% mPrecision, 70.26% dice and Hd95 of 5.05 on the self-made 1040 road crack dataset. Compared with the advanced network model, the HC-Unet++ network model has stronger generalization ability and higher segmentation accuracy, which is more suitable for the segmentation detection of road cracks. Therefore, the HC-Unet++ network model proposed in this paper plays an important role in road maintenance and traffic safety.

Keywords: road cracks; drone; feature fusion; SEnet; Blurpool; HC-Unet++

1. Introduction

As an important part of the transportation hub, roads provide the most basic facilities for the entire transportation network. Road cracks are a common phenomenon of road surface damage. After the road is completed, due to natural damage factors such as long-term exposure to the sun [1], rain erosion, and human factors such as frequent rolling of vehicles, construction materials, and construction quality, the road appears to incur varying degrees of crack damage. With the development of cracks, the overall structure of the road will gradually change [2], which will affect the service life and safety of the road to a certain extent, and even cause traffic accidents in severe cases. In order to reduce the cost [3] and prolong the service life of the road, it is necessary to survey and repair the road as soon as possible. Therefore, finding an accurate and effective method for identifying road cracks and repairing cracks in time is of great significance to the structural safety of roads and traffic safety.

Citation: Cao, H.; Gao, Y.; Cai, W.; Xu, Z.; Li, L. Segmentation Detection Method for Complex Road Cracks Collected by UAV Based on HC-Unet++. *Drones* **2023**, *7*, 189. https://doi.org/10.3390/drones7030189

Academic Editor: Seokwon Yeom

Received: 16 February 2023
Revised: 5 March 2023
Accepted: 7 March 2023
Published: 10 March 2023

In the early days, many scholars proposed the use of traditional image processing methods to solve the task of crack identification, such as the use of typical digital image processing algorithms to extract crack features to identify cracks [4] and using multi-threshold image segmentation methods to reduce computational cost and improve segmentation accuracy [5]. For example, Jin et al. [6] proposed a detection and segmentation algorithm for mixed cracks, using the threshold method of histogram to obtain the approximate location of cracks. Li et al. [7] used the improved Otsu threshold segmentation algorithm and adaptive iterative threshold segmentation method to identify airport runway cracks containing runway markings. Li et al. [8] proposed a new unsupervised multi-scale fusion crack detection algorithm that extracts candidate cracks in the image through the minimum intensity path. Additionally, a multivariate statistical hypothesis-testing crack assessment model was developed to detect pavement cracks. Xu [9] and others used the second-order differential operator edge detection algorithm to identify cracks in infrared images at different times. Zhao et al. [10] combined the improved Canny edge detection algorithm with the edge filter of road surface edge detection to effectively detect edge cracks in road surface images while eliminating noise interference. Liang et al. [11] used wavelet technology to detect crack edges and eliminated background noise interference by searching and analyzing the maximum value of wavelet coefficients. At the same time, a threshold method was used to judge whether there were cracks on the pavement. Peggy et al. [12] performed a separable 2D continuous wavelet transform for several scales based on the continuous wavelet transform. The propagation between scales was analyzed by searching for the maximum value of the wavelet coefficients in order to determine whether there were cracks. Cheng et al. [13] built an end-to-end diagnostic mechanism based on continuous wavelet transform, one which adaptively captured features through automatic feature extraction. Although traditional image processing methods can identify and extract cracks to a certain extent, this method has poor generalization performance and can only be applied to specific contexts. Due to the complexity of road cracks, in the course of the crack identification process, whether or not the feature extraction is feasible, there is no guarantee that a crack will be detected.

With the development of deep learning, many scholars have proposed using convolutional neural network image processing methods to identify cracks. In addition, road crack images collected by UAV [14] can be more challenging due to lighting conditions, viewpoints, and scales. Researchers used a Gaussian noise residual network to extract crack features [15] and tried to use MVMNet's multiple detection method to identify cracks [16]. Zhu et al. [17] proposed a feature fusion enhancement module coupled with a convolutional network attention mechanism. It was used to improve the interaction between feature maps and strengthened the dependence between feature channels, so as to achieve the identification of cracks. Cha et al. [18] utilized sliding windows to divide the image into blocks and used a CNN convolutional network to predict whether cracks existed in the block. Fan et al. [19] proposed a new threshold method based on a CNN convolutional network to extract cracks in classified color images. Sadrawi et al. [20] proposed to use the lenet network model to identify cracks, and the model finally classified the cracks in the image into horizontal, vertical and massive cracks. Huang et al. [21] used FCN to extract feature hierarchy to detect cracks in subway shield tunnels. Zou et al. [22] utilized an end-to-end trainable deep neural network to fuse the multi-scale deep convolutional features learned in the layered convolution stage to capture the line structure to better detect cracks. However, these convolutional neural networks cannot solve the following three problems well in the segmentation of cracks: 1. As shown in Table 1a, there are many complex background interferences such as zebra crossings of different colors, stains, manhole covers, etc. During the segmentation process, this interference information will affect the extraction and segmentation of fracture features by the network. 2. As shown in Table 1b, road cracks usually appear in irregular shapes, and the conventional convolutional neural network cannot fully capture the crack feature information in the feature extraction stage. 3. As shown in Table 1c, in the process of fracture segmentation, the down-sampling of each layer will cause information loss, resulting in the phenomenon of fracture discontinuity in the segmentation result.

Table 1. Traditional network segmentation results.

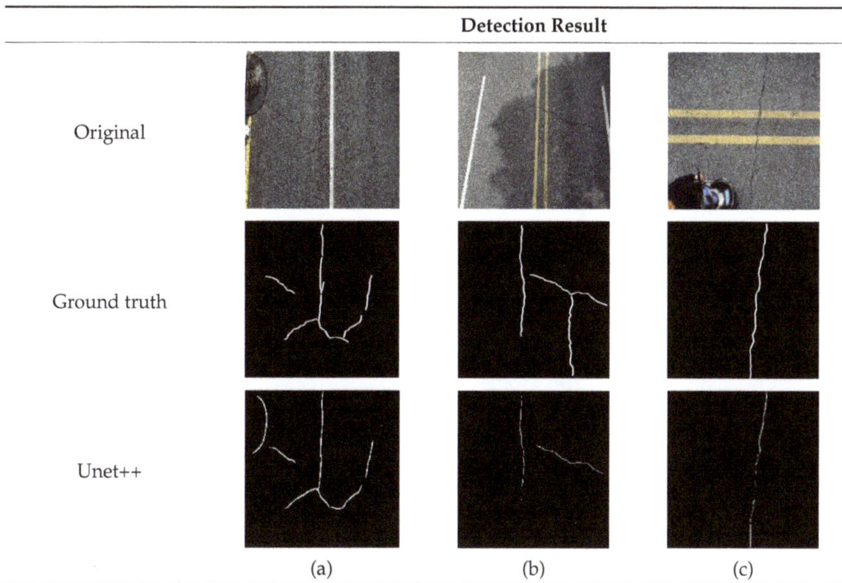

In order to be able to completely identify irregularly shaped cracks, Yuan et al. [23] designed a RDA detail-attention module for crack detection, one which enhanced the segmentation of irregular cracks by accurately locating the spatial location of cracks. Billah et al. [24] proposed an encoder-based deep network architecture to strengthen the extraction of clear features of cracks to effectively find the exact location of cracks and identify cracks with irregular shapes. Li et al. [25] designed a multi-scale feature fusion method. A multi-scale parallel structure was obtained through various sampling rates and pooling methods. This structure can obtain more receptive fields to improve the ability of the network to identify disordered cracks. Considering the particularity of road crack segmentation, a deep parallel feature fusion module is proposed in this paper. This module is located in the deep layer of the network. The purpose is to obtain more receptive fields so that the network can effectively locate cracks and judge their shapes, so as to enhance the ability of network segmentation to identify irregular cracks.

In order to enhance the extraction of important information and suppress the interference of useless information, many scholars have tried different methods. Yang et al. [26] proposed the AFB attention fusion block, which was used to replace the original skip connection to enhance feature extraction. Zhang et al. [27] combined MobileNets with the convolutional attention module of CBAM. Firstly, the residual structure of MobileNetV2 was introduced to eliminate the accuracy drop caused by separable convolution in the depth direction, and then CBAM was embedded into the convolutional layer to enhance the effect of important information. Yang et al. [28] proposed a UAV-supported edge computing method that was able to integrate different levels of feature map information into low-level features. This allowed the network to remove the complexity of the background and the inhomogeneity of the crack intensity. Qiao et al. [29] proposed the scSE attention mechanism. This module was divided into upper and lower branches. Each branch obtained a different matrix and multiplied the input image and finally stitched it to achieve the recalibration of the feature map. Hu et al. [30] proposed the SEnet attention mechanism. Through the compressed weight matrix, different weights are assigned to different positions in the channel to help the network obtain important information. Based on the complexity of the background when detecting road cracks, this study uses the SEnet attention mechanism and uses it in the decoding and encoding stages deep in the network structure. The feature map is corrected by the obtained one-dimensional weight vector to achieve the purpose of eliminating the interference of non-fracture information and enhancing the feature extraction of fracture information during feature extraction.

In order to solve the fracture discontinuity phenomenon that occurs during the fracture segmentation process, Xiang et al. [31] adopted a pyramid module to divide the feature map into different sub-regions. Crack information was extracted from the global view by aggregating contextual information in different regions to enhance the continuity of pavement crack detection. Han et al. [32] proposed a jump-level round-trip sampling structure to solve the problem of interruption of continuous cracks in the segmentation process by improving the ability of different receptive fields to perceive information at different scales. Jiang et al. [33] proposed a segmentation framework with an enhanced graph network branch to improve crack segmentation continuity by adding a new feature extraction branch to enrich feature map information. In this paper, due to the loss of part of the information caused by the maximum pooling operation in the Unet++ network downsampling, the segmentation cracks are not continuous in the segmentation results. Therefore, the Blurpool pooling operation is introduced, which can alleviate the shift-equivariance to the greatest extent, greatly reduce the loss of some crack features when the feature map is down-sampled, and solve the crack discontinuity in the segmentation.

Based on the problems of complex background, irregular crack shape and discontinuous crack segmentation in the process of road crack segmentation. We propose a road crack network segmentation model based on HC-Unet++. The contributions of this paper are as follows:

1. A UAV-based road crack dataset is constructed, one which contains 1040 images of road cracks with complex backgrounds and irregular crack shapes. These images are precisely annotated and used to train a network model to solve the problem of road crack segmentation.

2. We propose a deep parallel feature fusion module. This module operates on the network's deepest layer and can obtain a larger receptive field, making global feature extraction possible. This module is partitioned into two parallel branches; each branch extracts crack features using Conv, BN, and Relu operations and stitches the extracted feature maps. The spliced feature map has more comprehensive irregular crack features, thereby enhancing the network's capacity to segment irregular cracks.

3. The SEnet attention mechanism is introduced and used in the deep encoding and decoding stages of the network. First, the input feature map is compressed into a one-dimensional vector through the spatial dimension. Then after the 1*1 convolution, channel feature learning is performed to weight the one-dimensional vector. Finally, the one-dimensional weight of channel attention information is learned, which is corrected in combination with the input character map. The redundant irrelevant information contained in the output feature map is reduced, and the important crack feature information is increased.

4. Refers to the Blurpool pooling procedure. This pooling operation can eliminate aliasing to the greatest extent, return the original incorrect output to its correct location, and achieve translation invariance to a significant degree. The Blurpool pooling operation replaces the maximum pooling in the original network, making the output robust to small input translations and minimizing the loss of crack features during downsampling. The problem of crack discontinuity in the segmentation result has been resolved.

5. Experimental results show that compared to other advanced methods, the network model proposed in this article on the home-made road cracks dataset has stronger generalization capabilities and higher accuracy. In generalization experiments, its multiple indicators have also achieved excellent results. Therefore, the HC-Unet++ network

has been more versatile in the segmentation of road cracks, making it more efficient and cost-effective.

2. Materials and Methods

2.1. Data Acquisition

The experimental dataset of this experiment was taken by the team researchers on Shaoshan South Road and Furong South Road in Changsha City with a DJI Mini3Pro drone (as shown in Figure 1), with a resolution of 5472 × 3468. When dataset is collected, time synchronization and control of the drone is very important [34,35]. During the shooting process, we hovered the UAV 3 to 5 m above the ground and set the viewing angle to 87 degrees in order to capture close-up images of cracks. This type of crack image includes more individual crack targets. When photographing distant cracks, we set the flight altitude of the drone between 8 and 10 m. Currently, the UAV has a two-lane field of vision, and the obtained image depicts a crack scene that is relatively complete. From the acquired photos, a total of 1040 images with complex backgrounds and irregular crack shapes were selected as the route crack dataset for research. Among them, 734 pictures contained complex backgrounds such as zebra crossings, manhole covers, stains, scratches, etc., and 813 pictures had different crack shapes. Then the resolution of these images was adjusted to 512 × 512 as input images, and these images were manually labeled using the Labelme tool. Afterwards, the background of the marked picture was black, the crack was white, and it was stored in the form of json. In training, we divided the dataset into training set, validation set and test set in a ratio of 8:1:1.

Figure 1. Collection of crack images.

2.2. Methods

2.2.1. HC-Unet++

In order to narrow the semantic gap between the feature maps from the encoder and decoder networks in the Unet network model, Zhou et al. [36] proposed a network segmentation model for Unet++. In this network model structure, the hollow Unet network was filled. It has connections at every point in the horizontal direction, so that features at different levels can be captured. Due to the different sensitivities of the receptive field at different depths, the shallow receptive field is more sensitive to small targets, while the deep receptive field is more sensitive to large targets. Additionally, splicing together through the feature concat can integrate the advantages of the two. The Unet++ network adds dense skip connections to reduce the semantic gap between feature maps, making it achieve good results in the general segmentation field.

However, in the field of road crack segmentation, because of the complex and changeable situation, a higher performance detection method is needed. In addition, in the actual segmentation, it is found that the cracks always appear in various irregular shapes. Additionally, it is accompanied by different complex backgrounds that interfere with the segmentation process, such as: zebra crossing backgrounds of different colors, water stains, stains, scratches, etc. In addition, due to the loss of information caused by the traditional down-sampling method of the network, the segmentation results show discontinuity of cracks, that is, the interruption of continuous crack segmentation. Based on these three questions, this paper uses Unet ++ as the backbone network to propose a HC-Unet++ network model (as shown in Figure 2a) to perform semantic segmentation on road cracks. In this model, we propose a deep parallel feature fusion module, as shown in Figure 2b, which can solve the problem of difficult segmentation of disordered fractures. The SEnet attention mechanism is introduced, as shown in Figure 2c, which can eliminate the interference of complex background in the road crack image, thereby improving the accuracy of segmentation. Replace the maximum pooling operation in the original network with the Blurpool pooling operation, as shown in Figure 2d. This operation reduces the loss of some fracture features during down-sampling, thereby reducing fracture discontinuity in the segmentation results. The following sections will provide more details.

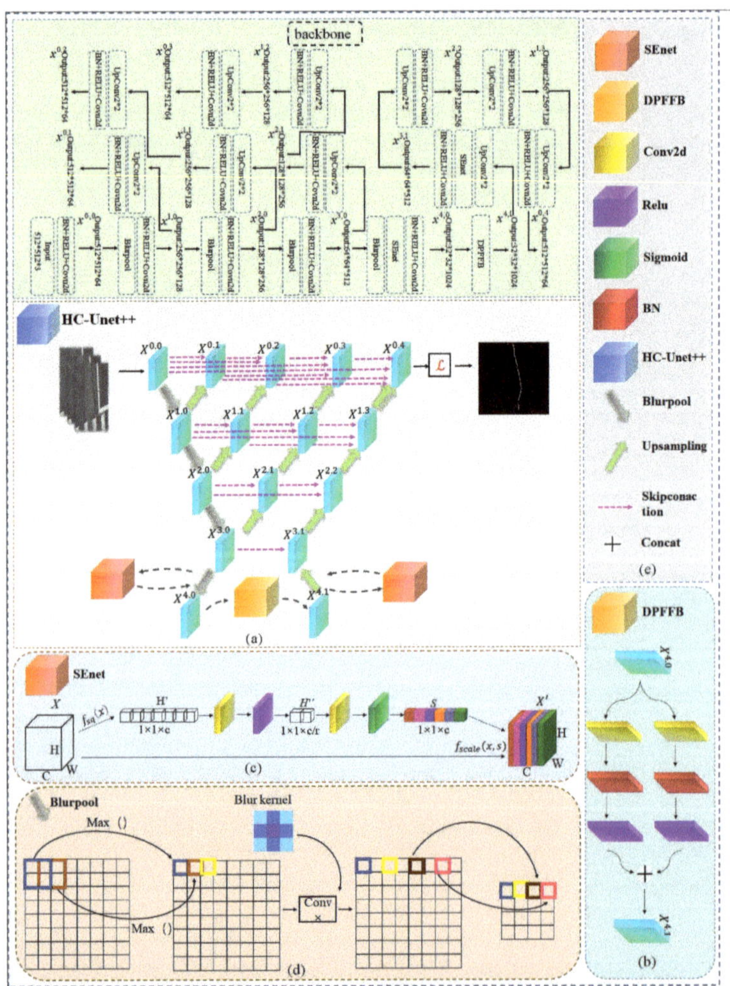

Figure 2. Network structure diagram: (**a**) is the network structure of HC-Unet++. (**b**) is a flow chart of the DPFFB module. (**c**) is a flowchart of the SEnet attention mechanism. (**d**) is a workflow diagram of Blurpool pooling. (**e**) is the meaning of the corresponding modules in (**a–c**).

2.2.2. Deep Parallel Feature Fusion Module

In road crack segmentation, there will be various cracks with irregular shapes, which makes it difficult for the network to extract complete crack features, which brings certain difficulties to the actual segmentation. In order to improve the ability of the network to segment irregular cracks, it is necessary to add a special feature map processing module deep in the network. This module is used to obtain a larger receptive field and more deep semantic information to enhance the sensitivity of the network to the characteristics of road cracks. Therefore, the shape of the crack can be effectively judged, and the irregularly-shaped road crack can be identified.

However, in the Unet++ network, there is a lack of special processing of feature maps deep in the network, resulting in unsatisfactory segmentation capabilities for irregular cracks. In order to solve this problem, we propose a deep parallel feature fusion module (as shown in Figure 2b) to be placed into the bottom layer of the network (as shown in Figure 2a). It can be seen that the module placed at the bottom layer of the network can have a larger receptive field and be more sensitive to crack features. Therefore, the position of the road crack can be more accurately located and the shape of the crack can be judged. When the input feature map enters the module, the module will divide the feature map into upper and lower branches and perform independent and identical operations to obtain more complete crack features. It makes the network judge the shape of the crack more clearly.

The detailed operation steps of the deep parallel feature fusion module are as follows:

(1) First, the input feature map is divided into upper and lower branches, and 1×1 convolution is performed on the upper and lower branches, respectively.

(2) Subsequently, the two branches perform BN normalization processing on the feature map, respectively. This will normalize the distribution of the data to the standard normal distribution, so that the input value of the subsequent activation function is in the sensitive area. The formula for BN processing is as follows:

$$\mu = \frac{1}{m} \sum_{i=1}^{m} x_i \tag{1}$$

$$\sigma^2 = \frac{1}{m} \sum_{i=1}^{m} (x_i - \mu)^2 \tag{2}$$

$$y(i) = \gamma \frac{x_i - \mu}{\sqrt{\sigma^2 + \varepsilon}} + \beta \tag{3}$$

Among them, μ is the mean value in σ^2 a batch, is the variance in a batch, x_i is the input, $y(i)$ is the output after the BN layer, γ and β is the learnable parameters, which will change with the gradient during the training process.

(3) After that, the upper branch and the lower branch, respectively, pass through the relu activation function to increase the nonlinear factor. The expression describing it is as follows:

$$y(i)' = \begin{cases} y(i) & y(i) > 0 \\ 0 & y(i) \leq 0 \end{cases} \tag{4}$$

Among them, $y(i)'$ represents the output and $y(i)$ represents the input.

The last two branches concatenate the extracted fracture features to strengthen the segmentation of irregular fracture shapes.

2.2.3. SEnet

In order to enhance the extraction of important features and reduce the interference of useless information, HU et al. [30] proposed the SEnet attention mechanism, which obtained a one-dimensional vector by global pooling the input feature map, and then sent the one-dimensional vector to the fully connected layer to make the one-dimensional vector have weight. Finally, the obtained one-dimensional weight was multiplied by the original input feature map to achieve the purpose of correcting the feature map.

One of the difficulties in road crack detection is the elimination of complex backgrounds such as: zebra crossings of different colors, water stains, stains, shadows, etc. In order to effectively solve this problem, we introduced the SEnet attention mechanism, and replaced the original fully connected layer with a 1×1 convolutional layer to reduce the amount of calculation. At the same time, in the upsampling, in order to protect the integrity of the features, we set the scaling factor r to 2. Subsequently, the attention mechanism was placed in the deep layer of the network: $x^{3.0}$ between to $x^{4.0}$ and $x^{4.1}$ to $x^{3.1}$. In the deep layers of the network, the obtained 1D weight vector has a large receptive field. The one-dimensional weight is used to correct the input feature map to enhance the extraction of important features and suppress the interference of useless information, so as to achieve the purpose of eliminating complex backgrounds.

The specific implementation steps of the attention mechanism in this article are as follows:

(1) First, the input feature map X is globally pooled, and the one-dimensional feature vector H' ($1 \times 1 \times C$) is generated by compressing X in the spatial dimension. This allows it to have per-channel global information. The H' calculation formula for is as follows:

$$H' = f_{sq}(X) = \frac{1}{H \times W} \sum_{i=1}^{H} \sum_{j=1}^{W} X(i,j) \tag{5}$$

Among them, f_{sq} represents average pooling.

(2) The one-dimensional vector $H\prime$ passes through the 1×1 convolution kernel so that the number of channels becomes $1/r$. Then the activation function relu is entered to get the compressed one-dimensional vector H'' ($1 \times 1 \times C/r$). H'' ($1 \times 1 \times C/r$). The H'' calculation formula for is as follows:

$$H'' = \delta(w_1 a) \tag{6}$$

Among them, δ represents the activation function relu, and w_1 represents the dimensionality reduction parameter.

(3) The compressed one-dimensional vector is restored to the original number of channels H'' through a 1×1 convolution kernel, and it then enters the activation function sigmoid to obtain a weighted one-dimensional vector s ($1 \times 1 \times C$). Through (2), (3) two steps realize the interaction between channels, thus improving the computational efficiency of the network. The s calculation formula is as follows:

$$s = \sigma(w_2 a) \tag{7}$$

Among them, σ represents the activation function sigmoid, and w_2 represents the dimension-raising parameter.

(4) Finally, multiply the obtained one-dimensional weight s with the input feature map X to achieve the purpose of correcting the input feature map. This results in the output feature map X'. The X' calculation formula is as follows:

$$X' = f_{sacle}(s, X) = s_c X_c \tag{8}$$

2.2.4. Blurpool

Zhang et al. [37] found that modern convolutional networks were not displacement invariant. Commonly used downsampling methods such as maximum pooling and average pooling ignore the sampling theorem and do not insert low-pass filtering before downsampling. This will cause, when using these downsampling methods, small changes in the input that will cause the output value to fluctuate violently. This is not the desired result. In order to reduce the phenomenon wherein the output changes violently with the small displacement of the input, that is, to achieve displacement invariance to the greatest extent, the Blurpool pooling method is used to replace the maximum pooling in the original network. Its structure diagram is shown in Figure 2d, and its position in the network is shown in Figure 2a.

In the actual road crack identification process, the traditional maximum pooling operation will cause the use of each layer of down-sampling to cause the loss of some crack features, resulting in continuous crack interruptions in the segmentation results. In order to solve this problem, we introduce the Blurpool pooling method and use this operation to replace the original maximum pooling operation. The introduction of this operation can maximize translation invariance during downsampling to reduce the loss of some fracture features, thereby solving the phenomenon of fracture discontinuity in the segmentation results.

The specific implementation steps are as follows:

(1) Max operation with stride = 1. In this process, the operation has translation invariance and will not cause aliasing to information.

(2) Subsequently, a low-pass filter Blur is added before downsampling. Its function is to eliminate aliasing to the greatest extent and return the wrong output to the original position as much as possible. That is, the translation invariance is obtained to the greatest extent. The formula for translation invariance is as follows:

$$\tilde{F}(x) = \tilde{F}(shift_{\Delta h, \Delta w(x)}) \quad \forall (\Delta h, \Delta w) \tag{9}$$

Note, though that this formula is valid only when the translation amount is an integer multiple of N.

(3) Finally, downsample the module after low-pass filtering and output the result.

It can be seen that the aliasing of the downsampling has been greatly improved after adding the low-pass filter Blur. This strengthens the extraction of fracture features by the network during downsampling, and greatly reduces the discontinuity of fractures in the segmentation results.

2.3. Experimental Environment and Settings

2.3.1. Data Preparation

All tests in this study were performed on the same hardware and software platform. Table 2 is the hardware environment and software environment of this experiment.

Table 2. Experimental environment.

	CPU	AMD EPYC 7543 32-Core Processor
Hardware environment	ARM	80GB
	Video memory	48GB
	GPU	A40
	OS	windows 11
Software Environment	PyTorch	1.11.0
	Python	3.8
	Cuda	11.3

2.3.2. Training Methods

In order to avoid the mismatch between height and width, we adjust the height and width of the image to equal size, and the image input size is unified to 512 × 512 during training. The batch size is set to 2, the momentum parameter is 0.9, and the Adam optimizer is selected. The optimizer combines the advantages of two optimization algorithms, momentum and RMSProp, and can overcome the problem of sharp decrease of Adgrad gradient while automatically adapting to different learning rates for different parameters. This enables the network to avoid frequently updated parameters from being affected by a single outlier sample. Additionally, using cross-entropy loss as the loss function, a total of 300 epochs are trained. In the experiment, we divide the dataset of road crack detection into training set, verification set and test set with a ratio of 8:1:1. Table 3 contains the experimental parameters and settings.

Table 3. Parameter Settings.

Size of Image	Batch_Size	Momentum	Initial lr	Optimizer	Iterations
512×512	2	0.9	e^{-4}	Adam	300

Figure 3 is the change of train loss and val loss with epoch. It can be seen from the figure that the network model tends to be stable at about 10 Epoch, and the network model converges quickly.

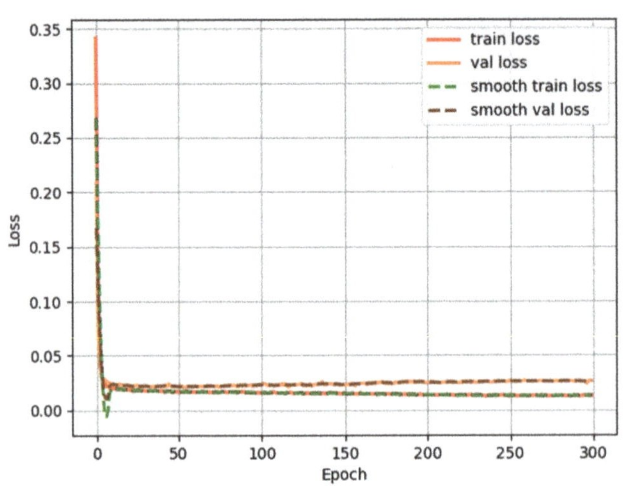

Figure 3. Train loss and val loss.

3. Experimental Results and Analysis

3.1. Experimental Evaluation Criteria

The evaluation index used in this experiment is composed of these five coefficients: *mIOU*, *mPA*, *mPrecision*, *dice*, and Hd95. In addition, in the following formula, *TP* is true (predicted result is a crack, and the actual result is a crack), *FP* false positive (predicted result is a crack, and the actual result is not cracked), and *FN* is false negative (the predicted result is non-cracked, and the actual result is a crack), *TN* is true negative (predicted result is non-crack, and actually is non-crack).

mIOU is the average intersection ratio, which represents the average intersection ratio of each class in this dataset, and its expression is as follows:

$$mIOU = \frac{1}{k+1}\sum_{i=0}^{k}\frac{TP}{FN + FP + TP} \tag{10}$$

mPA is the category average pixel accuracy rate, which represents the proportion of correctly classified pixels for each category, and then accumulates the average value. The expression about it is as follows:

$$mPA = \frac{1}{k+1}\sum_{i=0}^{k}\frac{TP + TN}{FN + TP + TN + FP} \tag{11}$$

mPrecision indicates the proportion of correct predictions in the samples that are predicted to be positive, and then accumulate and average, the expression is as follows:

$$mPrecision = \frac{1}{k+1}\sum_{i=0}^{k}\frac{TP}{FP + TP} \tag{12}$$

The value *dice* indicates the ratio of the intersection area between the predicted value and the real value to the total area, and is sensitive to internal filling. Its expression is as follows:

$$dice = \frac{2 * TP}{TP + FN + TP + FP} \tag{13}$$

Hausdorff_distance is the maximum distance from one set to the nearest point of another set, and is sensitive to the segmented boundary. Its expression is as follows:

$$h(A, B) = \max_{a \in A} \{ \min_{b \in B} \{ d(a, b) \} \} \tag{14}$$

In order to eliminate the unreasonable distance caused by some outliers, we refer to the Hd95 index. It can arrange the distances of these closest points in descending order, and select the distance ranked as 5% as the final value of Hd95.

3.2. Module Performance Analysis

This section tests the performance of each module we built in detail, divided into Section 3.2.1, Section 3.2.2 and Section 3.2.3. They respectively test the effectiveness of the deep parallel feature fusion module, the effectiveness of the SEnet module, and the effectiveness of the Blurpool module.

3.2.1. Effectiveness of Deep Parallel Feature Fusion Module

In this section, we use the self-made road crack dataset to verify the effectiveness of the deep parallel feature fusion module. We will use the SPP [38], ASPP [39] modules in comparison with the deep parallel feature fusion module, duly embedding them into the Unet++ network according to the position of DPFFB in the HC-Unet++ network. Table 4 is the comparison result of these feature extraction modules. It can be seen from the table that when the feature extraction block is embedded in the network, mIOU is improved and Hd95 is reduced by a certain value. Compared with SPP and ASPP, the mIOU obtained by embedding the DPFFB module is higher, and the Hd95 value is smaller. This shows that in the segmentation of road cracks, the performance of DPFFB is better than that of SPP and ASPP modules, which shows that this module is more suitable for the road crack segmentation task of this research. However, the embedding of the DPFFB module also brings a certain number of parameters, which inevitably increases the training time of the network.

Table 4. Comparison of feature extraction blocks.

Networks	Unet++	SPP+Unet++	ASPP+Unet++	DPFFB+Unet++
mIOU	70.39%	71.26%	71.39%	72.44%
Params	47.19 M	47.49 M	47.71 M	48.24 M
Hd95	12.62	10.16	10.03	8.16

3.2.2. Effectiveness of SEnet

To evaluate the effectiveness of the SEnet, we validate the SEnet module with the self-made road crack dataset. We will choose CA [40], CBAM [41] attention mechanism and SE attention mechanism for comparison. Additionally, we will imitate the location of SEnet in the HC-Unet++ network and embed them in the Unet++ network. Table 5 is the comparison result of these attention mechanisms. It can be seen from the table that after embedding the attention mechanism, the mIOU of the network model has been improved and Hd95 has been reduced. Compared with CBAM, although SEnet lacks the feature map correction of spatial attention, its data is slightly better than CBAM, and the overall parameter amount is also less than CBAM. Compared with CA, SEnet not only has better data than CA, but also has 0.78 M less parameters than it. This proves that the SEnet mechanism is more suitable for the road crack segmentation in this experiment.

Table 5. Comparison of Attention Mechanisms.

Networks	Unet++	CBAM+Unet++	CA+Unet++	SE+Unet++
mIOU	70.39%	72.98%	72.84%	73.14%
Params	47.19 M	47.85 M	48.13 M	47.35 M
Hd95	12.62	8.31	9.03	7.96

3.2.3. Effectiveness of Blurpool

To evaluate the effectiveness of Blurpool, we used the self-made road crack dataset to validate the Blurpool module. We conducted a comparative experiment between the Blurpool and the traditional Maxpool, and Figure 4 shows the loss training curve and accuracy curve of the two. It can be seen from the figure that when Blurpool is used instead of Maxpool, the convergence speed of network model training is accelerated and the accuracy is also improved, which shows that compared with Maxpool, Blurpool is more suitable for the network model in this paper.

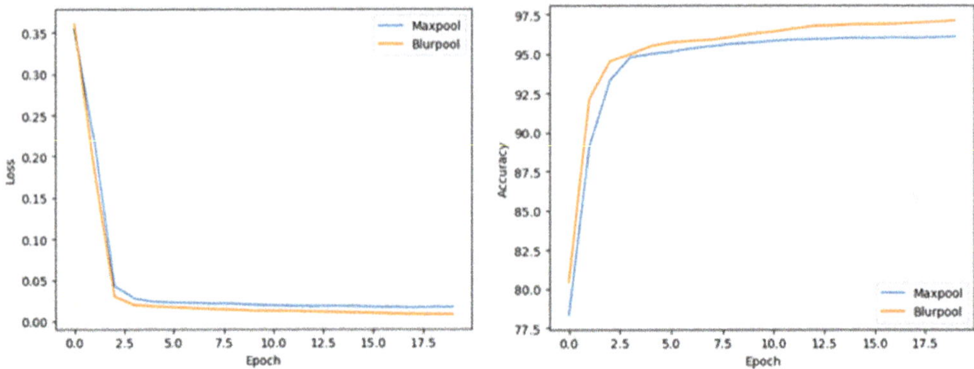

Figure 4. Comparison of Maxpool and Blurpool.

3.3. Ablation Experiments

To evaluate the performance of the method proposed in this paper, we conduct ablation experiments. In the ablation experiment, we use a self-made road crack dataset for experiments and use Unet++ as the backbone structure of the network. On this basis, one or more methods proposed in this paper is added to compare and form ablation experiments to further prove the effectiveness of the deep parallel feature fusion module, SEnet module, and Blurpool module. The experimental results are shown in Table 6.

Table 6. Ablation Experiment Results.

Number	Method	mIOU (%)	mPA (%)	mPrecision (%)	Hd95	Param
1	HC-Unet++	76.32	82.39	85.51	5.05	48.40 M
2	DPFFB+SE+Maxpool	75.12	81.12	84.69	5.83	48.40 M
3	SE+Blur	74.86	80.33	83.82	6.71	47.35 M
4	DPFFB+Blur	73.69	78.69	82.47	7.23	48.24 M
5	DPFFB+Maxpool	71.82	77.41	81.57	8.16	48.24 M
6	SE+Maxpool	72.54	76.20	82.05	7.96	47.35 M
7	Blur	71.16	75.21	81.94	11.26	47.19 M
8	Unet++	70.39	73.50	80.73	12.62	47.19 M

From Table 6 it can be seen that: DPFFB, Blurpool, and SEnet can all improve the accuracy in the process of road crack identification, and the HC-Unet++ model that combines these three methods performs the best. From the comparison between the fourth group and the seventh group, it can be seen that, after adding the DPFFB module, the Hd95 index changes significantly. This shows that after adding the DPPFB module, the model's ability to locate fractures is enhanced, and relatively complete fracture characteristics can be obtained, which improves the network's ability to identify irregular fractures. At the same time, the embedding of this module also brings some parameters. From the comparison between the third group and the seventh group, it can be seen that after adding the SEnet attention mechanism, mIOU, mPrecision, and mPA are significantly improved, and Hd95 is also significantly reduced. It shows that this module effectively eliminates the interference of complex background in the depth of the network, but it inevitably increases the number of parameters while improving the segmentation ability of the model. From the comparison between the fourth group and the fifth group, it can be seen that, after replacing Maxpool with Blurpool pooling, mIOU, mPrecision, and mPA have a slight increase and Hd95 has a slight decrease. Ultimately, the replacement of this module does not increase the computational load of the network.

These eight sets of experiments fully demonstrate the contribution of deep parallel feature fusion, SEnet, and Blurpool to the model's accuracy. It also shows that the HC-Unet++ proposed by us is more suitable for the detection of road cracks than Unet++.

In addition, we also compared the effects of DPFFB+SE+Maxpool, SE+Blur, and DPFFB+ Blur to further analyze the performance of this method, as shown in Table 7. In Figure a of Table 7, the Unet++ network is affected by the zebra crossing background, and some zebra crossings are misjudged as road cracks, while the other four networks perform relatively well. In Figure c of Table 7, Unet++ not only misses the judgment of road cracks, but also misjudges the manhole cover in the picture as a crack. After adding the DPFFB module and the SE module, this phenomenon of missed judgment and misjudgment disappears. This is because the network is more sensitive to fracture characteristics, can identify irregular fractures, and has improved anti-interference ability against complex backgrounds. However, the information is partially lost during downsampling, which makes the segmentation cracks appear discontinuous. After replacing Maxpool with Blurpool, the fracture interruption phenomenon of model segmentation is reduced, and the identified fractures are more fine and complete in comparison. This proves that the Blurpool module can alleviate the displacement variability of output information, reduce the loss of information during downsampling, and solve the problem of discontinuous crack segmentation. In the SE+Blur network in Figure d of Table 7, since the DPFFB module is erased, the network loses the special operation module for feature extraction. This leads to the reduction of the ability to divide irregular cracks in the network, so there is a serious phenomenon of missed judgments. When the DPFFB module is added, this missed judgment phenomenon disappears, which further proves the ability of the DPFFB module to identify irregular cracks. In Figure b of Table 7, when the SE module is erased from the HC-Unet++ network, the model misjudges it as a crack due to the black scratch background. After adding SEnet attention, the one-dimensional weight recalibrates the feature map to reduce the extraction of interference information. That is, the influence of the black scratch background is eliminated, which proves the ability of the SENet module to eliminate complex backgrounds.

Table 7. Visual comparison of the test results.

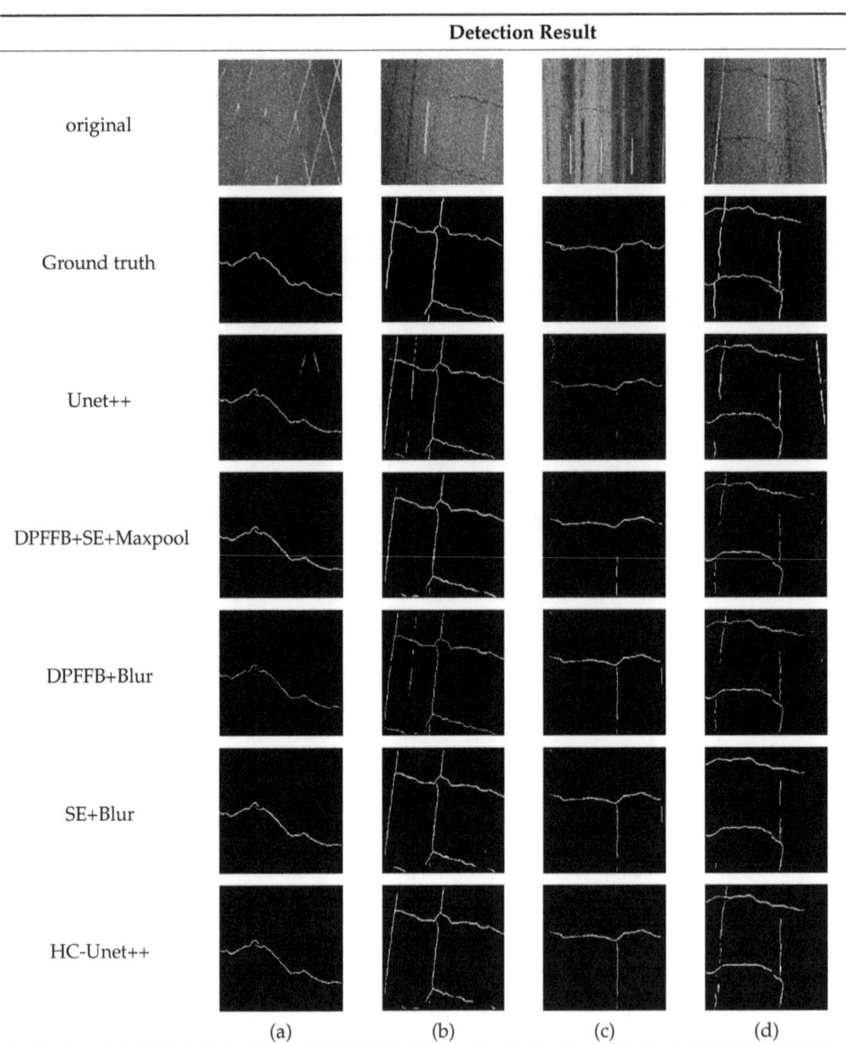

3.4. Comparsion of HC-Unet++ with Other Methods

To further analyze the performance of HC-Unet++, we compared it with state-of-the-art network models. These network models are: BC-Dunet [42], U2-Net [43], CS2-Net [44]. Extremec3net [45], DCNet [46]; the experimental results about them are shown in Table 8. In the table, we calculated the mIOU, mPA, mPrecision, dice, and Hd95 of different models for detecting road cracks. These five methods performed well on our self-made dataset, and their average mPA exceeded 80% and reached 80.39%. Among them, the mPrecision value of U2-Net surpassed 85.51% of HC-Unet++ network and reached 85.64%. However, compared with these five advanced network models, HC-Unet++ performed better, especially since the Hd95 value was only 5.05, which shows that the edge detection ability of the HC-Unet++ model in this paper is very good. In addition, the mIOU, mPA, and dice values of HC-Unet++ are all better than the other five networks, and the mPrecision value is only 0.13% lower than the mPrecision value in U2-Net. Overall, experimental data show that our proposed HC-Unet++ network model is more suitable for road crack segmentation than some advanced network models.

Table 8. Comparison with advanced networks.

Number	Method	mIOU (%)	mPA (%)	mPrecision (%)	Dice (%)	Hd95
1	HC-Unet++	76.32	82.39	85.51	70.26	5.05
2	BC-Dunet [42]	72.41	78.59	79.38	61.19	9.82
3	U2-Net [43]	73.28	80.63	85.64	63.74	11.68
4	CS2-Net [44]	73.19	79.50	82.73	64.51	7.34
5	Extremec3net [45]	74.76	81.99	81.98	67.84	9.57
6	DCNet [46]	72.53	81.24	83.75	63.49	12.58

3.5. Generalization Experiments

In order to verify that our network model has good generalization ability, we used [47] concrete crack conglomerate dataset and Crack500 crack dataset, respectively, and used them in FCN [48], Unet [49], Unet++, HC-Unet++ models for training. Among them, the Concrete Crack Conglomerate Dataset has a total of 10,993 images and 1000 crack images are selected for training, and Crack500 has a total of 476 images and is used for training. They are divided into a training set, a verification set and a test set, in a ratio of 8:1:1; the experimental environment for training is the same as the experimental configuration. The experimental results are shown in Table 9. From the perspective of data, the performance of our HC-Unet++ model on these two different datasets is better than the other three comparative network models. Additionally, some of the indicators of the experimental results are even slightly better than the experimental data on the self-made dataset.

Table 9. Generalization experiments.

Dataset	Method	mIOU (%)	mPA (%)	mPrecision (%)	Dice (%)	Hd95
Concrete Crack Conglomerate	HC-Unet++	77.23	86.45	85.91	71.38	3.17
	FCN [48]	69.38	80.13	79.19	60.64	14.89
	Unet [49]	71.06	82.54	83.64	62.52	11.68
	Unet++	73.71	81.67	82.91	67.98	9.98
Crack 500	HC-Unet++	76.91	84.04	87.49	69.54	4.68
	FCN [48]	70.41	79.65	78.91	59.94	13.29
	Unet [49]	73.95	83.24	81.03	65.23	10.68
	Unet++	73.83	82.94	84.57	64.26	9.35

In order to better demonstrate the good generalization performance of the HC-Unet++ model, we selected some renderings of the experiments in this section for comparison, as shown in Table 10. It can be seen from the figure that the performance of the FCN network is the worst. As shown in Figure b–f of Table 10, there are varying degrees of misjudgment and missed judgment which are caused by the lack of global context information in the network. The overall performance of the Unet and Unet++ networks is roughly the same, and both of them have slightly missed or misjudged. The HC-Unet++ network performs the best. It can be seen that most of the prediction maps are basically consistent with the label maps. This further demonstrates the excellent capabilities of our model on these two datasets. It shows that our network has good generalization ability on other datasets.

Table 10. Generalization experiment effect diagram.

	Crack 500			Concrete Crack Conglomerate Dataset		
original						
Ground truth						
HC-Unet++						
Unet++						
Unet [49]						
FCN [48]						
	a	b	c	d	e	f

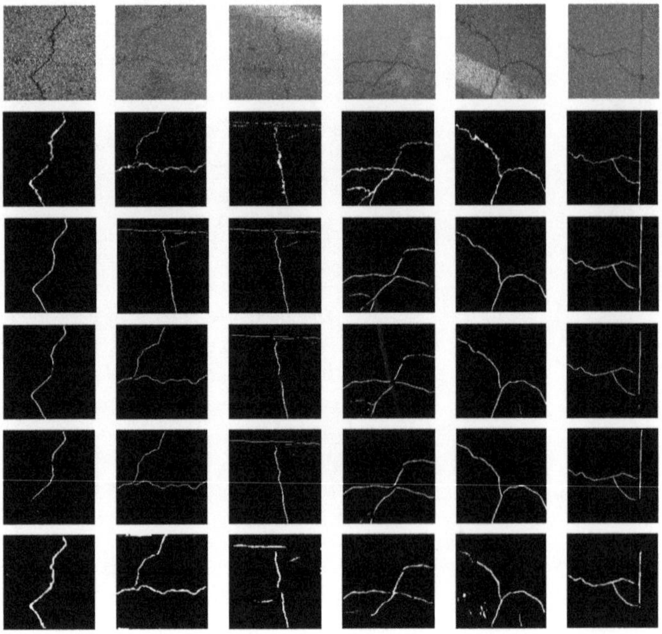

4. Discussion

To segment road cracks, in this paper we proposed the HC-Unet++ network model. In this study, we construct a new road crack dataset and use HC-Unet++ to train the network. Experiments show that our proposed HC-Unet++ network model is effective for segmenting road cracks. To a certain extent, it can solve the complex background of road cracks, irregular cracks and other problems. Nonetheless, we still need more research:

(1) HC-Unet++ networks are relatively large, so the training takes a relatively long time. How to reduce the parameters of the network without affecting the accuracy is a problem we need to solve in the future.

(2) As shown in Figure 5, in the process of HC-Unet++ network segmentation: When the UAV is flying at a high altitude, cracks will appear in relatively small forms in the image, and the result of model segmentation at this time is poor. In the future, we will need to add an efficient module to effectively segment out small cracks.

(3) There is a lack of quantification of cracks, and physical parameters such as the length, width, and area of road cracks cannot be obtained. In order to quantify cracks, we need to develop an effective method to quantify cracks in the future.

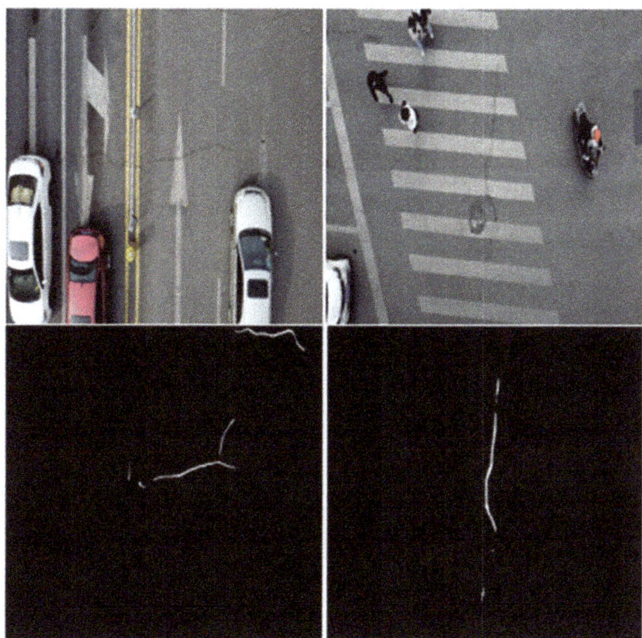

Figure 5. Segmentation of fine cracks.

5. Conclusions

This paper proposes HC-Unet++ road crack segmentation recognition technology. First of all, the road-crack dataset of this experiment was constructed by using camera shooting and UAV aerial photography, and the label processing was performed using the labelme tool. Then it was input into the HC-Unet++ network model for training. HC-Unet++ used Unet++ as the basic network structure, and embedded a deep parallel feature fusion module to improve the sensitivity of the network to crack features. This enabled the network to obtain relatively complete fracture information, thereby identifying irregular fractures. After adding the SEnet attention mechanism to eliminate the complex background interference in the road cracks, and replacing the Maxpool with the Blurpool module, the variability of displacement was greatly alleviated. This increased the network's extraction of some fracture features in the down-sampling of each layer, making the fractures obtained by network segmentation more continuous. The experimental results show that the HC-Unet++ network model achieves an average intersection ratio of 76.32%, an average pixel accuracy rate of 82.39%, an average precision rate of 85.51%, and Hausdorff 95 is 5.05. As well, in the generalization experiment, the performance of our network model is still stable, which shows that the HC-Unet++ network has good adaptability and provides data value for road maintenance and traffic safety.

Road crack segmentation is still an important research direction in the engineering field of image recognition technology, a task which is of great significance for prolonging the service life of roads and reducing traffic safety hazards. Although the segmentation and detection of road cracks based on convolutional neural network has achieved outstanding results, further improvement is still needed. The next step of this research will be to think about how to compress the network scale without affecting the segmentation accuracy and the physical quantification of the road cracks.

Author Contributions: H.C.: Methodology, Writing—original draft, Conceptualization. Y.G.: Software, Data acquisition, Investigation. W.C.: Model guidance. Z.X.: Validation, Project administration. L.L.: Visualization, Writing—Review and Editing. All authors have read and agreed to the published version of the manuscript.

Funding: This work was supported by National Natural Science Foundation in China (Grant No. 61703441); in part by the Key Project of Education Department of Hunan Province (Grant No. 21A0179); in part by the Changsha Municipal Natural Science Foundation (Grant No. kq2014160); in part by Hunan Key Laboratory of Intelligent Logistics Technology (2019TP1015).

Data Availability Statement: All self-made datasets for this study are available by contacting the authors.

Conflicts of Interest: The authors declare no conflict of interest.

References

1. Hu, G.X.; Hu, B.L.; Yang, Z.; Huang, L.; Li, P. Pavement crack detection method based on deep learning models. *Wirel. Commun. Mob. Comput.* **2021**, *2021*, 5573590. [CrossRef]
2. Ren, J.; Zhao, G.; Ma, Y.; Zhao, D.; Liu, T.; Yan, J. Automatic Pavement Crack Detection Fusing Attention Mechanism. *Electronics* **2022**, *11*, 3622. [CrossRef]
3. Johnson, A.M. *Best Practices Handbook on Asphalt Pavement Maintenance*; Minnesota Technology Transfer/LTAP Program, Center for Transportation Studies: Minneapolis, MN, USA, 2000.
4. Lu, J.; Liu, A.; Dong, F.; Gu, F.; Gama, J.; Zhang, G. Learning under concept drift: A review. *IEEE Trans. Knowl. Data Eng.* **2018**, *31*, 2346–2363. [CrossRef]
5. Xing, Z. An improved emperor penguin optimization based multilevel thresholding for color image segmentation. *Knowl.-Based Syst.* **2020**, *194*, 105570. [CrossRef]
6. Tang, J.; Gu, Y. Automatic crack detection and segmentation using a hybrid algorithm for road distress analysis. In Proceedings of the 2013 IEEE International Conference on Systems, Man, and Cybernetics, Manchester, UK, 13–16 October 2013; pp. 3026–3030.
7. Peng, L.; Chao, W.; Shuangmiao, L.; Baocai, F. Research on crack detection method of airport runway based on twice-threshold segmentation. In Proceedings of the 2015 Fifth International Conference on Instrumentation and Measurement, Computer, Communication and Control (IMCCC), Qinhuangdao, China, 18–20 September 2015; pp. 1716–1720.
8. Li, H.; Song, D.; Liu, Y.; Li, B. Automatic pavement crack detection by multi-scale image fusion. *IEEE Trans. Intell. Transp. Syst.* **2018**, *20*, 2025–2036. [CrossRef]
9. Xu, D.; Zhao, Y.; Jiang, Y.; Zhang, C.; Sun, B.; He, X. Using improved edge detection method to detect mining-induced ground fissures identified by unmanned aerial vehicle remote sensing. *Remote Sens.* **2021**, *13*, 3652. [CrossRef]
10. Zhao, H.; Qin, G.; Wang, X. Improvement of canny algorithm based on pavement edge detection. In Proceedings of the 2010 3rd international congress on image and signal processing, Yantai, China, 16–18 October 2010; pp. 964–967.
11. Liang, S.; Sun, B. Using wavelet technology for pavement crack detection. In *ICLEM 2010: Logistics for Sustained Economic Development: Infrastructure, Information, Integration, Proceedings of the International Conference of Logistics Engineering and Management (ICLEM) 2010, Chengdu, China, 8–10 October 2010*; Zhang, J., Xu, L., Zhang, X., Yi, P., Jian, M., Eds.; American Society of Civil Engineers: Reston, VA, USA, 2010; pp. 2479–2484.
12. Subirats, P.; Dumoulin, J.; Legeay, V.; Barba, D. Automation of pavement surface crack detection using the continuous wavelet transform. In Proceedings of the 2006 International Conference on Image Processing, Atlanta, GA, USA, 8–11 October 2006; pp. 3037–3040.
13. Cheng, Y.; Lin, M.; Wu, J.; Zhu, H.; Shao, X. Intelligent fault diagnosis of rotating machinery based on continuous wavelet transform-local binary convolutional neural network. *Knowl.-Based Syst.* **2021**, *216*, 106796. [CrossRef]
14. Liu, W.; Quijano, K.; Crawford, M.M. YOLOv5-Tassel: Detecting tassels in RGB UAV imagery with improved YOLOv5 based on transfer learning. *IEEE J. Sel. Top. Appl. Earth Obs. Remote Sens.* **2022**, *15*, 8085–8094. [CrossRef]
15. Wang, B.; Yan, Z.; Lu, J.; Zhang, G.; Li, T. Explore uncertainty in residual networks for crowds flow prediction. In Proceedings of the 2018 International Joint Conference on Neural Networks (IJCNN), Rio de Janeiro, Brazil, 8–13 July 2018; pp. 1–7.
16. Hu, Y.; Zhan, J.; Zhou, G.; Chen, A.; Cai, W.; Guo, K.; Hu, Y.; Li, L. Fast forest fire smoke detection using MVMNet. *Knowl.-Based Syst.* **2022**, *241*, 108219. [CrossRef]
17. Zhu, W.; Zhang, H.; Eastwood, J.; Qi, X.; Jia, J.; Cao, Y. Concrete crack detection using lightweight attention feature fusion single shot multibox detector. *Knowl.-Based Syst.* **2023**, *261*, 110216. [CrossRef]
18. Cha, Y.J.; Choi, W.; Büyüköztürk, O. Deep learning-based crack damage detection using convolutional neural networks. *Comput.-Aided Civ. Infrastruct. Eng.* **2017**, *32*, 361–378. [CrossRef]
19. Fan, R.; Bocus, M.J.; Zhu, Y.; Jiao, J.; Wang, L.; Ma, F.; Cheng, S.; Liu, M. Road crack detection using deep convolutional neural network and adaptive thresholding. In Proceedings of the 2019 IEEE Intelligent Vehicles Symposium (IV), Paris, France, 9–12 June 2019; pp. 474–479.

20. Sadrawi, M.; Yunus, J.; Abbod, M.F.; Shieh, J.-S. Higher Resolution Input Image of Convolutional Neural Network of Reinforced Concrete Earthquake-Generated Crack Classification and Localization. *IOP Conf. Ser. Mater. Sci. Eng.* **2020**, *931*, 012005. [CrossRef]

21. Huang, H.-W.; Li, Q.-T.; Zhang, D.-M. Deep learning based image recognition for crack and leakage defects of metro shield tunnel. *Tunn. Undergr. Space Technol.* **2018**, *77*, 166–176. [CrossRef]

22. Zou, Q.; Zhang, Z.; Li, Q.; Qi, X.; Wang, Q.; Wang, S. Deepcrack: Learning hierarchical convolutional features for crack detection. *IEEE Trans. Image Process.* **2018**, *28*, 1498–1512. [CrossRef]

23. Yuan, G.; Li, J.; Meng, X.; Li, Y. CurSeg: A pavement crack detector based on a deep hierarchical feature learning segmentation framework. *IET Intell. Transp. Syst.* **2022**, *16*, 782–799. [CrossRef]

24. Billah, U.H.; Tavakkoli, A.; La, H.M. Concrete crack pixel classification using an encoder decoder based deep learning architecture. In Proceedings of the Advances in Visual Computing: 14th International Symposium on Visual Computing, ISVC 2019, Lake Tahoe, NV, USA, 7–9 October 2019; Proceedings, Part I 14. pp. 593–604.

25. Li, C.; Wen, Y.; Shi, Q.; Yang, F.; Ma, H.; Tian, X. A pavement crack detection method based on multiscale Attention and HFS. *Comput. Intell. Neurosci.* **2022**, *2022*, 1822585. [CrossRef]

26. Yang, L.; Fan, J.; Huo, B.; Li, E.; Liu, Y. A nondestructive automatic defect detection method with pixelwise segmentation. *Knowl.-Based Syst.* **2022**, *242*, 108338. [CrossRef]

27. Zhang, Y.; Huang, J.; Cai, F. On bridge surface crack detection based on an improved YOLO v3 algorithm. *IFAC-Pap.* **2020**, *53*, 8205–8210. [CrossRef]

28. Yang, J.; Li, H.; Zou, J.; Jiang, S.; Li, R.; Liu, X. Concrete crack segmentation based on UAV-enabled edge computing. *Neurocomputing* **2022**, *485*, 233–241. [CrossRef]

29. Qiao, W.; Liu, Q.; Wu, X.; Ma, B.; Li, G. Automatic pixel-level pavement crack recognition using a deep feature aggregation segmentation network with a scse attention mechanism module. *Sensors* **2021**, *21*, 2902. [CrossRef]

30. Hu, J.; Shen, L.; Sun, G. Squeeze-and-excitation networks. In Proceedings of the IEEE Conference on Computer Vision and Pattern Recognition, Salt Lake City, UT, USA, 18–23 June 2018; pp. 7132–7141.

31. Xiang, X.; Zhang, Y.; El Saddik, A. Pavement crack detection network based on pyramid structure and attention mechanism. *IET Image Process.* **2020**, *14*, 1580–1586. [CrossRef]

32. Han, C.; Ma, T.; Huyan, J.; Huang, X.; Zhang, Y. CrackW-Net: A novel pavement crack image segmentation convolutional neural network. *IEEE Trans. Intell. Transp. Syst.* **2021**, *23*, 22135–22144. [CrossRef]

33. Chen, J.; Yuan, Y.; Lang, H.; Ding, S.; Lu, J.J. The Improvement of Automated Crack Segmentation on Concrete Pavement with Graph Network. *J. Adv. Transp.* **2022**, *2022*, 2238095. [CrossRef]

34. Liu, W.; Xia, X.; Xiong, L.; Lu, Y.; Gao, L.; Yu, Z. Automated vehicle sideslip angle estimation considering signal measurement characteristic. *IEEE Sens. J.* **2021**, *21*, 21675–21687. [CrossRef]

35. Rehak, M.; Skaloud, J. Time synchronization of consumer cameras on Micro Aerial Vehicles. *ISPRS J. Photogramm. Remote Sens.* **2017**, *123*, 114–123. [CrossRef]

36. Zhou, Z.; Rahman Siddiquee, M.M.; Tajbakhsh, N.; Liang, J. Unet++: A nested u-net architecture for medical image segmentation. In Proceedings of the Deep Learning in Medical Image Analysis and Multimodal Learning for Clinical Decision Support: 4th International Workshop, DLMIA 2018, and 8th International Workshop, ML-CDS 2018, Held in Conjunction with MICCAI 2018, Granada, Spain, 20 September 2018; Proceedings 4. pp. 3–11.

37. Zhang, R. Making convolutional networks shift-invariant again. In Proceedings of the International Conference on Machine Learning, Long Beach, CA, USA, 10–15 June 2019; pp. 7324–7334.

38. He, K.; Zhang, X.; Ren, S.; Sun, J. Spatial pyramid pooling in deep convolutional networks for visual recognition. *IEEE Trans. Pattern Anal. Mach. Intell.* **2015**, *37*, 1904–1916. [CrossRef]

39. Chen, L.-C.; Papandreou, G.; Kokkinos, I.; Murphy, K.; Yuille, A.L. Deeplab: Semantic image segmentation with deep convolutional nets, atrous convolution, and fully connected crfs. *IEEE Trans. Pattern Anal. Mach. Intell.* **2017**, *40*, 834–848. [CrossRef]

40. Hou, Q.; Zhou, D.; Feng, J. Coordinate attention for efficient mobile network design. In Proceedings of the IEEE/CVF Conference on Computer Vision and Pattern Recognition, Nashville, TN, USA, 20–25 June 2021; pp. 13713–13722.

41. Woo, S.; Park, J.; Lee, J.-Y.; Kweon, I.S. Cbam: Convolutional block attention module. In Proceedings of the European Conference on Computer Vision (ECCV), Munich, Germany, 8–14 September 2018; pp. 3–19.

42. Liu, T.; Zhang, L.; Zhou, G.; Cai, W.; Cai, C.; Li, L. BC-DUnet-based segmentation of fine cracks in bridges under a complex background. *PLoS ONE* **2022**, *17*, e0265258. [CrossRef]

43. Qin, X.; Zhang, Z.; Huang, C.; Dehghan, M.; Zaiane, O.R.; Jagersand, M. U2-Net: Going deeper with nested U-structure for salient object detection. *Pattern Recognit.* **2020**, *106*, 107404. [CrossRef]

44. Mou, L.; Zhao, Y.; Fu, H.; Liu, Y.; Cheng, J.; Zheng, Y.; Su, P.; Yang, J.; Chen, L.; Frangi, A.F. CS2-Net: Deep learning segmentation of curvilinear structures in medical imaging. *Med. Image Anal.* **2021**, *67*, 101874. [CrossRef]

45. Park, H.; Sjösund, L.L.; Yoo, Y.; Bang, J.; Kwak, N. Extremec3net: Extreme lightweight portrait segmentation networks using advanced c3-modules. *arXiv* **2019**, arXiv:1908.03093.

46. Li, F.; Li, W.; Gao, X.; Liu, R.; Xiao, B. DCNet: Diversity convolution network for ventricle segmentation on short-axis cardiac magnetic resonance images. *Knowl.-Based Syst.* **2022**, *258*, 110033. [CrossRef]

47. Bianchi, E.; Hebdon, M. *Concrete Crack Conglomerate Dataset*; University Libraries, Virginia Tech: Blacksburg, VA, USA, 2021.

48. Long, J.; Shelhamer, E.; Darrell, T. Fully convolutional networks for semantic segmentation. In Proceedings of the IEEE Conference on Computer Vision and Pattern Recognition, Boston, MA, USA, 7–12 June 2015; pp. 3431–3440.
49. Ronneberger, O.; Fischer, P.; Brox, T. U-net: Convolutional networks for biomedical image segmentation. In Proceedings of the Medical Image Computing and Computer-Assisted Intervention–MICCAI 2015: 18th International Conference, Munich, Germany, 5–9 October 2015; Proceedings, Part III 18. pp. 234–241.

 *drones*

Article

Transmission Line Segmentation Solutions for UAV Aerial Photography Based on Improved UNet

Min He [1,2], Liang Qin [1,2,*], Xinlan Deng [1,2], Sihan Zhou [1,2], Haofeng Liu [1,2] and Kaipei Liu [1,2]

1   School of Electrical and Automation, Wuhan University, Wuhan 430072, China; whuhemin@whu.edu.cn (M.H.); 2018302070241@whu.edu.cn (X.D.); warren_zsh@whu.edu.cn (S.Z.); 2018302070013@whu.edu.cn (H.L.); kpliu@whu.edu.cn (K.L.)
2   Hubei Key Laboratory of Power Equipment & System Security for Integrated Energy, Wuhan 430072, China
*   Correspondence: qinliang@whu.edu.cn; Tel.: +86-189-861-729-77

Abstract: The accurate and efficient detection of power lines and towers in aerial drone images with complex backgrounds is crucial for the safety of power grid operations and low-altitude drone flights. In this paper, we propose a new method that enhances the deep learning segmentation model UNet algorithm called TLSUNet. We enhance the UNet algorithm by using a lightweight backbone structure to extract the features and then reconstructing them with contextual information features. In this network model, to reduce its parameters and computational complexity, we adopt DFC-GhostNet (Dubbed Full Connected) as the backbone feature extraction network, which is composed of the DFC-GhostBottleneck structure and uses asymmetric convolution to capture long-distance targets in transmission lines, thus enhancing the model's extraction capability. Additionally, we design a hybrid feature extraction module based on convolution and a transformer to refine deep semantic features and improve the model's ability to locate towers and transmission lines in complex environments. Finally, we adopt the up-sampling operator CARAFE (Content-Aware Re-Assembly of FEature) to improve segmentation accuracy by enhancing target restoration using contextual neighborhood pixel information correlation under feature decoding. Our experiments on public aerial photography datasets demonstrate that the improved model requires only 8.3% of the original model's computational effort and has only 21.4% of the original model's parameters, while achieving a reduction in inference speed delay by 0.012 s. The segmentation metrics also showed significant improvements, with the mIOU improving from 79.75% to 86.46% and the mDice improving from 87.83% to 92.40%. These results confirm the effectiveness of our proposed method.

Keywords: transmission line segmentation; UAV; UNet; light-weighting model; ACmix; CARAFE

Citation: He, M.; Qin, L.; Deng, X.; Zhou, S.; Liu, H.; Liu, K. Transmission Line Segmentation Solutions for UAV Aerial Photography Based on Improved UNet. *Drones* **2023**, *7*, 274. https://doi.org/10.3390/drones7040274

Academic Editor: Diego González-Aguilera

Received: 23 March 2023
Revised: 14 April 2023
Accepted: 15 April 2023
Published: 17 April 2023

1. Introduction

Transmission lines are a critical part of the power system with broad coverage, long transmission distances, and high reliability requirements. However, these lines are often exposed to a complex external environment, where the presence of vegetation and buildings of varying heights on the ground poses a potential threat. The proximity of these objects to high-voltage transmission lines can lead to accidents such as line tripping. Furthermore, during the inspection process, the wings of the drone may collide or tangle with the power lines, which poses a significant risk to the safety of the drone's flight and the stable operation of power facilities. Therefore, it is crucial to effectively monitor and segment transmission lines to ensure the safety of the power grid and low-altitude drones.

The detection methods for transmission lines can be divided into traditional image-processing methods and deep learning-based methods. Among the two methods for detecting transmission lines, traditional image-processing methods have been used for transmission line extraction based on edge detection algorithms. For example, Zhou et al. [1] proposed a color space variable-based classification and feature extraction method for transmission line images. It classifies images based on the relationship between the values of variables in each color space of the transmission line image and its corresponding image features, specifically focusing on different light intensities. Then, the edge extraction of power line images is performed by the OTSU (NobuyukiOtsumethod) algorithm [2]. This method mainly considers the power line segmentation under different lighting conditions; although the power lines are effectively extracted, the application scenario is mainly the power lines under low- to high-altitude photography. That is, the background is mainly the sky. When the perspective shifts from high altitude to low altitude shooting, the background is mostly houses, mountains, and rivers. As such, this type of algorithm cannot effectively solve the power line segmentation in complex backgrounds. Zhou et al. [3] took an improved Ratio operator [4] with horizontal orientation to extract power lines and then group and fit the power line segments. Since the experimental object of this literature is mainly for power line segmentation in horizontal distribution scenarios, it has a certain specificity. At the same time, the Ratio operator is susceptible to complex backgrounds with large variations in pixel gray values, especially for areas with relatively flat gray values, which are not as effective as the normal operator. Zhou et al. [5] perform transmission line segmentation in complex backgrounds by proposing a detection operator based on local contextual information. Compared with the literature 4 comprehensive optimization of power lines under a variety of angle changes, mainly considering the horizontal, vertical, and diagonal distribution of transmission lines, with a certain degree of algorithmic stability. Some studies are also based on Hough transform [6] for transmission line detection [7,8]. Shan et al. [9] design auxiliary devices for the segmentation of transmission lines and the method does not have a universal. In summary, since this type of algorithm does not introduce any prior knowledge and the algorithm model does not need to prepare a large number of samples for training in advance, it has the advantage of a low sample size requirement due to its less restrictive way of data collection. In addition, the power lines extracted using the Hough transform tend to lose the width information [10]. Both edge detection-based power line extraction algorithms and power line extraction algorithms with a priori knowledge share a common challenge: it is difficult to adaptively adjust the model parameters to maintain the good performance of the algorithm in the scenario-changing test data. All these methods use artificially defined shallow features in constructing power line extraction models.

In this problem, deep learning models with their strong feature extraction capability are effectively used in the field of power systems. A variety of deep learning-based image classification, target recognition, image segmentation, and other power vision algorithms are rapidly developing [11–16]. Two power-line recognition algorithms based on VGG-19 [17] and ResNet-50 [18] in aerial images are proposed by Yetgin Ö E et al. [19]. However, this method can only determine whether the image contains power lines and does not achieve the segmentation and detection of power lines. The optimal model for detailed detection of power lines is the segmentation model and the classification model can only determine the presence or absence of power lines in the transmission line in the image. The target detection model can only be presented in the form of a rectangular box and when the transmission line spans the whole picture, the detection effect will occupy the whole picture with a rectangular box, which is not conducive to the accurate positioning of the transmission line and is likely to generate too much redundant information. In the segmentation of power lines for transmission lines. Yang et al. [20] designed to use VGG-16 as the feature extraction backbone of UNet and combine the attention mechanism based on global average pooling and global maximum pooling for UNet skipping connection layer information supplementation. Finally, the four decoding layers of UNet are fused as a whole to output the final segmentation features. Since the process involves the fusion of four different scale features, it can effectively reduce the loss of multi-scale feature information, but at the same time, it also brings an increase in computational effort. Han et al. [21] proposed a UNet segmentation model based on GhostNet as the backbone by optimizing and improving the model. For light-weighting, the model and the attention mechanism

Shuffle Attention further optimize the detection accuracy of the model, achieving effective segmentation of power lines. However, the actual test speed in the article is still low, mainly because the decoder part of the segmentation network still has a large amount of computation. In segmentation studies targeting complex backgrounds, Xu et al. [22] also use VGG-16 as the feature extraction backbone to construct an improved UNet-based powerline segmentation network. In this study, a multi-level feature aggregation module is adapted to detect power lines at different pixel widths and orientations. That is, the two features after neighboring convolution are fused again for output, and the output features are then combined with the attention mechanism for semantic information enrichment and background noise suppression. The SPP (Spatial Pyramid Pooling) module is also combined with the perceptual field enhancement, Finally, the four features of the multiscale output are then fused for the overall output. Gao et al. [23] take proposed an efficient parallel branching network for real-time segmentation of overhead power lines. However, the data in these two literature studies are mainly processed by cropping the images at large resolutions and turning them into small-resolution images. From a complex background at a large resolution to a single background at a small resolution, the whole picture consists only of power lines and some of the power lines' attached power equipment. Choi et al. [24] propose a fusion of visible and infrared images of power lines for UNet model-based segmentation detection. The fusion method based on the channel attention mechanism is adopted to achieve the fusion of infrared images and visible images. This achieves effective segmentation of transmission lines in complex contexts by aiding the segmentation of power lines with the help of data from homomodal heterogeneous sources. However, by the above directions of model optimization for complex backgrounds, more pre-processing must take place, and the models cannot be directly used for actual detection scenarios.

Based on the above problem analysis, this paper proposes a lightweight real-time semantic segmentation network. The network adopts a structure that combines local feature refinement and global receptive field enhancement to effectively solve the problem of power line segmentation in complex backgrounds. Considering the characteristics of power lines, an up-sampling algorithm that combines adjacent pixel information is designed to achieve detailed segmentation of power lines and power poles. In detail, the innovations of this paper are as follows:

(1) To effectively extract features from power lines and towers, a lightweight DFC-GhostNet feature extraction module has been designed and incorporated into the backbone feature extraction network. Since both power lines and towers have long-range target features, the design divides the symmetric convolutional kernel into two modules: horizontal-based perceptual field enhancement and vertical-based perceptual field enhancement, based on the lightweight GhostNet architecture. This is intended to enable effective target identification and feature extraction.

(2) The refinement module of features is designed from the perspective of power line and pole tower feature refinement, combining a convolutional module with local sensory fields and a transformer module with global sensory in deep semantic features to refine target feature areas in complex backgrounds.

(3) In the process of feature image restoration from low resolution to high resolution, i.e., the decoder part of the semantic segmentation model. Traditional upsampling only considers the information distribution of sub-pixel points and without consider the semantic information of the entire feature map, which leads to the loss of feature map information, and the deconvolution leads to the increase of computation or even Checkerboard Artifacts, which affects the performance of the model. Therefore, a content-aware feature recombination module is designed. This module is mainly to construct a learnable upsampling operator for each pixel in low resolution to learn the distribution of image features. This paper aims to realize upsampling based on input content to improve the feature decoding effect of the final output layer.

2. Materials and Methods

2.1. Dataset Introduction

The dataset used in this paper is derived from publicly available data, where this dataset recorded videos collected by a UAV, Parrot-ANAFI, in two different states in the USA to guarantee the varieties of the scenes. The locations are randomly selected without any intentions or treatments to avoid noisy backgrounds. The UAV contains a 4K HDR camera and up to 2.8× lossless zoom. Zooming is exploited when collecting the video data to guarantee the high resolution of the objects without manual cropping. The dataset is extracted from a set of 80 videos. All aerial videos have a resolution of 3840 × 2160 with 30 fps. Different views were taken during data acquisition, including the front view, top view, and side view. This was mainly used to ensure that the deep learning model can detect the target from any angle [25]. The specific data are shown in Figure 1.

Figure 1. A sample dataset from different perspectives.

The dataset has a total of 1240 sample images; each image contains two categories of power transmission lines and power towers, corresponding to the Cable and Tower in the label, respectively. Since the distribution of targets in the image is different, the number of targets cannot be calculated by the number of image sheets, so the distribution of the number of pixel points is taken for comparison. The distribution of various types of target pixel points in this dataset is shown in Table 1. The Labelme software was used to annotate the sample images to produce the dataset in COCO format [26].

Table 1. Distribution of segmentation targets.

Target	Background	Cable	Tower
Total target pixels/million	109,860	22,510	37,140

2.2. Methods

2.2.1. Introduction of the UNet Model

UNet is a convolutional neural network for image segmentation, proposed in a paper by Olaf Ronneberger et al. in 2015, whose main idea is to combine the contextual and local information of an image to improve the segmentation accuracy [27]. The overall structure is shown in Figure 2.

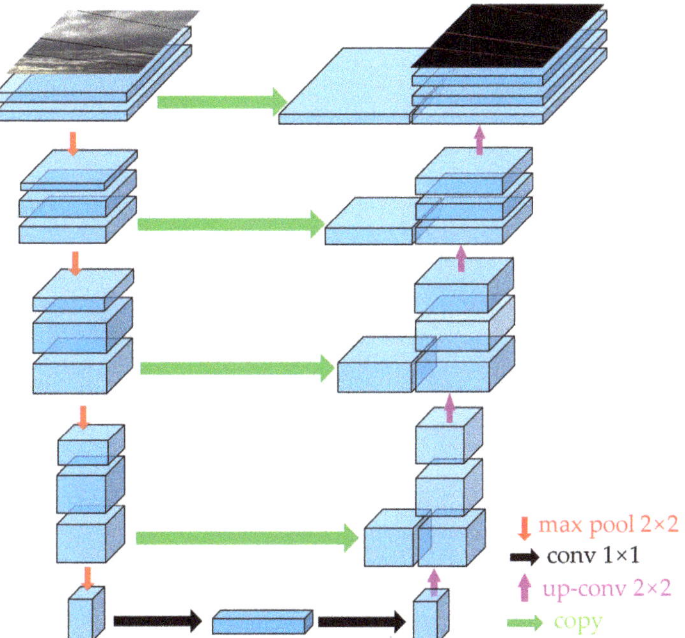

Figure 2. UNet structure diagram.

The structure of UNet is divided into two parts: the encoding path and the decoding path. The encoding path is used to extract the features and the decoding path is used to map the features back to the dimensions of the input image. There are cross-layer connections between the encoding and decoding paths, and these cross-layer connections help retain high-resolution feature information. Specifically:

(1) The encoding path consists of a series of convolution, maximum pooling, and ReLU activation functions. It starts from the input image, downsamples the feature map each time, and doubles the number of channels of the feature map. The result of this process is the generation of a series of high-level feature representations that preserve the global and semantic information of the image. The expression of the Relu function is as in Equation (1). This function is mainly used for the feature nonlinearization process in the model.

$$f(x) = max(0, x) \tag{1}$$

(2) The decoding path also consists of a series of convolutional and ReLU activation functions. It starts at the last layer of the encoding path, doubles the resolution of the feature map, and halves the number of channels each time. In the decoding path, each layer is stitched with the feature map of the corresponding layer in the encoding path to preserve the high-resolution feature information. The final output feature map size is equal to the input image size.

(3) Cross-layer connections are a key part of UNet. They connect the feature maps between the encoding path and the decoding path so that each layer in the decoding path can utilize the high-level feature representation of the corresponding layer in the encoding path. This helps to retain more contextual and semantic information, thus improving segmentation accuracy.

2.2.2. Comparison of Base Models

In this paper, after studying and researching the UNet model, we also compare the existing semantic segmentation algorithms with the UNet model, such as ResUnet [28],

ENet [29], PSPNet [30], Deeplabv3+ [31], HRNetv2 [32] and SegFormer [33] models, and designed multiple groups of baseline semantic segmentation detection models for comparison of segmentation experimental results; in the comparison of the segmentation results we took the following experimental index as the accuracy index of the model.

(1)   *mPrecision*

$$mPrecision = \frac{TP}{N \times (TP + FP)} \tag{2}$$

TP (True Positive) denotes positive samples correctly classified and *FP* (False Positive) denotes negative samples incorrectly classified. *N* denotes the number of categories of segmentation. *Precision* indicates the fraction of classes that are considered positive by the classifier and are indeed positive as a percentage of the classes considered positive by the classifier. Since it is a multi-category segmentation, the value of the mean Precision is taken as *mPrecision*.

(2)   *mRecall*

$$mRecall = \frac{TP}{N \times (TP + FN)} \tag{3}$$

FN (False Negative) denotes the misclassified positive sample. Recall indicates the fraction of classes that the classifier considers to be positive and are indeed positive as a percentage of all classes that are indeed positive. Since it is a multi-category segmentation, the value of the mean Recall is taken as *mRecall*.

(3)   *mPA*

$$mPA = \frac{1}{N} \sum_{i=0}^{N} \frac{p_{ii}}{\sum_{j=0}^{N} p_{ij}} \tag{4}$$

mPA is the category mean pixel accuracy. where i denotes the true value and j denotes the predicted value. p_{ii} denotes the prediction of i to i. p_{ij} denotes the prediction of i to j. The overall view is to calculate the proportion of correctly classified pixel points to all true categories, which is the same as the calculation principle of *mRecall*. In addition, the subsequent experimental results show that the values are the same for both.

(4)   *mIOU*

$$mIOU = \frac{TP}{N \times (FP + FN + TP)} \tag{5}$$

mIOU (mean Intersection over Union) is the classification of the mean intersection ratio. That is the mean value under the intersection of the total true label and the predicted value and the ratio.

(5)   *mDice*

$$mDice = \frac{2 \times Precision \times Recall}{N \times (Precision + Recall)} = \frac{2 \times TP}{N \times (FP + 2 \times TP + FN)} \tag{6}$$

mDice (mean Dice similarity coefficient) is another form of expression for the Intersection over Union. It represents the overlapping similarity between the segmentation result and the true result.

Each experimental metric is taken from the mean segmentation effect of the two segmentation objects of power lines and power poles. The experimental results are shown in Table 2:

Table 2. Comparison of common semantic segmentation algorithms.

Model	mIOU	mDice	mPA	mRecall	mPrecision	Parameters (M)	GFLOPs (G)	Runtime (s)
UNet	79.75	87.83	86.81	86.81	88.89	24.89	457.706	0.0516
ResUNet	70.43	80.85	80.18	80.18	81.59	13	647.93	0.0761
E-Net	61.31	72.88	69.82	69.82	81.46	0.349	4.432	0.0139
PSPNet	63.94	70.97	70.12	70.12	75.88	2.376	6.031	0.0238
Deeplabv3+	77.44	85.68	84.88	84.88	86.58	5.81	52.875	0.0337
HRNet	70.19	79.59	75.69	75.69	84.60	9.642	32.948	0.0282
SegFormer	74.64	83.55	82.80	82.80	84.38	13.720	26.696	0.0352

(1) From the perspective of detection accuracy, the UNet network has the highest detection accuracy among similar models, but still has poor segmentation of transmission lines in some complex scenes. As shown in Figure 3. Figure 3a shows the transmission line in the background of trees and houses. Figure 3c shows the segmentation experiment with the sky in the background. The experimental results show that there are obvious problems of incorrect segmentation and omitted segmentation in test Figure 3b (marked by the blue area in the figure). For simple background transmission, line segmentation presents better results, as shown in Figure 3d.

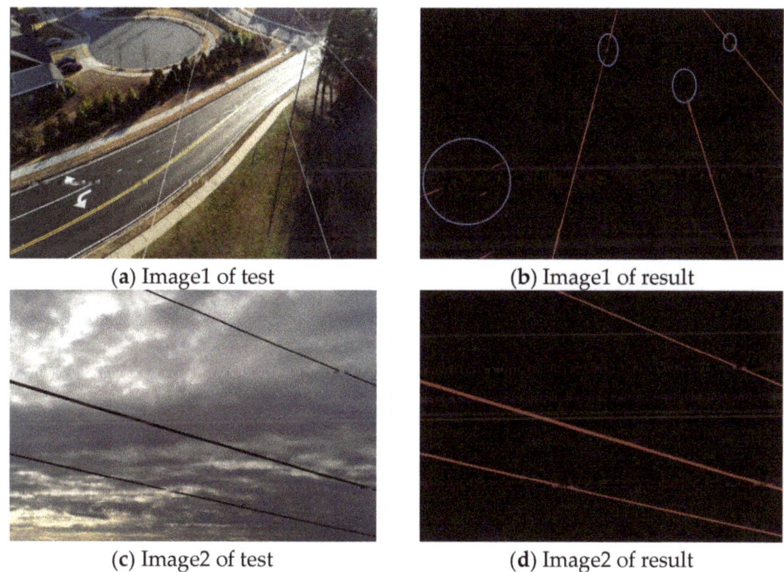

(a) Image1 of test (b) Image1 of result

(c) Image2 of test (d) Image2 of result

Figure 3. Test result graph of UNet model.

(2) From the perspective of detection speed: E-Net model has the smallest number of model parameters and computational effort. It can also be found that despite the small number of parameters of ResUNet, its GFLOPs value is very large. The main reason is that its structure has a large number of residual branches. The UNet model also has large parameters and GFLOPs. To better apply the model-to-edge applications, it is necessary to improve the light weighting of the UNet model and further improve the detection accuracy.

2.2.3. Feature Extraction Module Based on DFC-GhostNet

In computer vision, the architecture of deep neural networks plays a crucial role in various tasks. To better apply the model to mobile applications, it is necessary to consider not only the performance of the model, but also its computational efficiency, especially the actual inference speed. VGG-16 is mainly adopted as the backbone feature extraction network in more UNet model studies. While VGG-16 mainly takes 3×3 convolution for stacking, 3×3 convolution has a larger model computation. Based on this problem, another solution is to replace the original 3×3 convolution with a depth-separable convolution, such as MobileNetv3 [34], GhostNet [35], EfficientNet [36], ShuffleNet [37], etc. In the depth-separable convolution, the input feature maps ($H \times W \times C$) are first grouped by channel dimension, which is conducive to the dispersion of channels and the reduction of model computation. Here is a comparison of the computational effort with the original convolution.

Suppose the input feature map size is $H \times W \times c_1$. The size of the convolution kernel is $h_1 \times w_1 \times c_2$. The output feature size is $H \times W \times c_2$. If the depth-separable convolution is taken for feature extraction, then the number of parameters of the model will be reduced by a factor of g, where g is the number of groups split by the group convolution for the feature channel. The comparison of their computational effort is as follows:

$$p_{Conv} = (h_1 \times w_1 \times c_1) \times c_2 \tag{7}$$

$$\begin{aligned} p_{G\_Conv} &= (h_1 \times w_1 \times \tfrac{c_1}{g}) \times \tfrac{c_2}{g} \times g \\ &= (h_1 \times w_1 \times \tfrac{c_1}{g}) \times c_2 \end{aligned} \tag{8}$$

$$N = \frac{p_{Conv}}{p_{G\_Conv}} = \frac{(h_1 \times w_1 \times c_1) \times c_2}{(h_1 \times w_1 \times \tfrac{c_1}{g}) \times c_2} = g \tag{9}$$

The above equation mainly describes the change in the covariate size of the model after taking the depth-separable convolution. Based on the above ideas, this paper first conducted different backbone comparison experiments under depth-separable convolution based on the experimental results in Table 3.

Table 3. Comparison of Semantic Segmentation Models under Light-weighting Backbone.

Backbone	mIOU	mDice	mPA	mRecall	mPrecision	GFLOPs (G)	Parameters (M)	Runtime (s)
Original	79.75	87.83	86.81	86.81	88.89	457.706	24.89	0.0516
MobileNetv3	81.13	88.68	86.86	86.86	90.71	71.329	8.910	0.0205
GhostNet	83.46	90.40	89.70	89.70	91.83	70.841	9.584	0.0284
EfficientNet	82.49	89.65	89.08	89.08	90.24	102.392	26.692	0.0304
ShuffleNet	68.32	79.57	73.72	73.72	87.93	76.331	9.008	0.0194

From the above experiments:

(1) In terms of model size and inference speed, the backbone under depth-separable convolution has a significant increase in computation and inference speed compared to the original model. The GFLOPs based on GhostNet are only less than 16% of the original model. MobileNetv3 is only 35.8% of the original model in the calculation of the number of parameters. The optimized models are lower than the original model in terms of inference delay.

(2) In terms of detection accuracy indicators, the light-weighting network with the highest detection accuracy is based on the GhostNet backbone. An *mIOU* improvement of 3.71% is found compared to the original network segmentation, along with an *mDice* improvement of 2.57%, *mPA* improvement of 2.89%, *mPrecision* improvement of 2.94%, and an *mRecall* improvement of 2.89%.

(3) In general, The UNet model based on GhostNet backbone has a significant advantage in detection accuracy compared to similar Light-weighting models with depth-separable convolution. Therefore, the subsequent adoption of the design of a light-weighting backbone model based on GhostNet should be prepared. Since the above experiments only changed the backbone encoding structure to GhostNet structure, subsequent experiments will adopt the same structure of GhostConv for feature extraction for the whole decoder, which is used to further reduce the computation of the model.

One of the main GhostNet models taken by the Ghost module is shown in Figure 4. First, the input features are compressed by the 1 × 1 convolution of the channels. Then take the residuals to construct the feature extraction method, where the upper branch takes a constant mapping of the residual branch to retain the original information. The lower branch takes a 3 × 3 depth-separable convolution and convolves the features in groups based on the number of channels. Finally, the two are connected based on Concat to complete the feature extraction. The above structure forms a Ghost unit and the whole network relies on the Ghost unit to form.

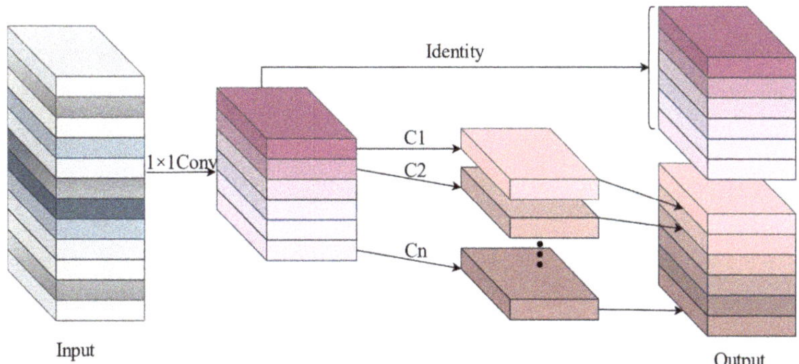

Figure 4. GhostNet convolutional structure.

Since the segmented objects in this paper are power lines and power towers, where the power lines are mainly thin and long targets, and the distribution area is not central in the image, but the data across the whole image, it is necessary to further optimize the backbone feature extraction by combining the characteristics of the segmented targets. Thus, Ghost convolution was considered for modeling based on long-range perceptual field enhancement.

In the study of attention mechanism, the channel-based attention mechanism can effectively explore the correlation between channels and at the same time adopt the fully connected way to combine the global feature map with the weight matrix, which can effectively enhance the weight of the target channels and suppress the proportion of invalid channels. The direct use of FC layers will bring about an increase in computation, so the design decomposes the original FC layers into feature extraction based on the horizontal and vertical directions so that the two FC layers model long-distance spatial information along their respective directions and finally combine the two FC layers to obtain the global perception field [38]. The structure is shown in Figure 5.

The overall calculation process is shown in Equation (10).

$$
\begin{aligned}
Y' &= X * F_{1\times1} \\
Y &= Concat([Y', Y' * F_{dp}]) \\
A &= UP(Conv_{1\times5}(Conv_{5\times1}(Down(X)))) \\
O &= Sigmiod(A) \odot Y
\end{aligned}
\tag{10}
$$

The formula is mainly described as the building block for the GhostNet convolution, where X represents the features of the input, $F_{1\times1}$ represents 1 × 1 convolution, Y' Represents the feature output after 1 × 1 convolution, F_{dp} is the depth-wise convolutional filter, and Y is the out feature of GhostNet. A is the attention mechanism branch, and the last O is the output weighted by the attention mechanism, where *Sigmoid* is the activation function.

The features modeled by the global perceptual field are fused with the original Ghost module features to effectively compensate for the loss of the global perceptual field of the original lightweight network, which results in less accurate transmission line segmentation.

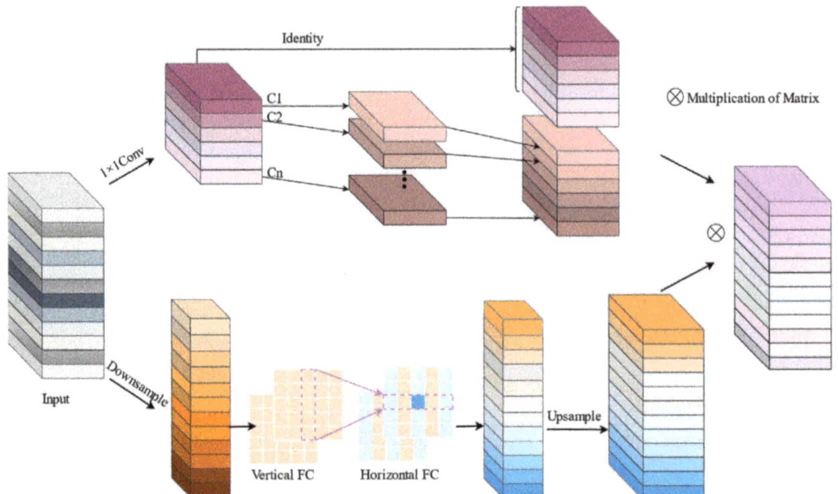

Figure 5. DFC-GhostNet convolutional structures.

2.2.4. Complex Background Feature Refinement Module Based on ACmix

Since power line targets in transmission lines are more subtle than tower targets, the model requires higher detail feature extraction in the recognition of power lines. A larger proportion of complex backgrounds are more likely to be blended into small target pixels at long distances. The reasonable use of attention mechanisms facilitates feature extraction networks to focus on target regions, learning the distribution pattern from the features, recalibrating them, and focusing on the position so that the segmentation model can capture the key information more efficiently and improve the segmentation ability of the model. The hybrid self-attentive and convolutional module ACMix(Mixed self-Attention Convolution) proposed in the literature [39] effectively combines the advantages of traditional convolutional and self-attentive modules. The Former leverages an aggregation function over a localized receptive field according to the convolution filter weights, which are shared in the whole feature map. The intrinsic characteristics impose crucial inductive biases for image processing. Comparably, the self-attention module applies a weighted average operation based on the context of input features, where the attention weights are computed dynamically via a similarity function between related pixel pairs. The flexibility enables the attention module to focus on different regions adaptively and capture more informative features and further distinguish the background from the detection target [39].

The ACmix structure is shown in Figure 6, which consists of two main phases:

(1) The feature learning stage. The input features are projected by three 1×1 convolutions, and N feature fragments are reconstructed separately to obtain a rich intermediate feature set containing $3 \times N$ feature maps to project features into a deeper space.

(2) The feature aggregation stage. The aggregation of information is performed according to both convolution and self-attentiveness and the features are aggregated mainly in terms of local and global information enhancement.

Specifically, for the convolutional path with a convolutional kernel size of k, a lightweight fully connected layer is first used to generate k^2 feature maps, and new features are generated by shifting and aggregating to process the input features in a convolutional manner to collect information from the local sensory field.

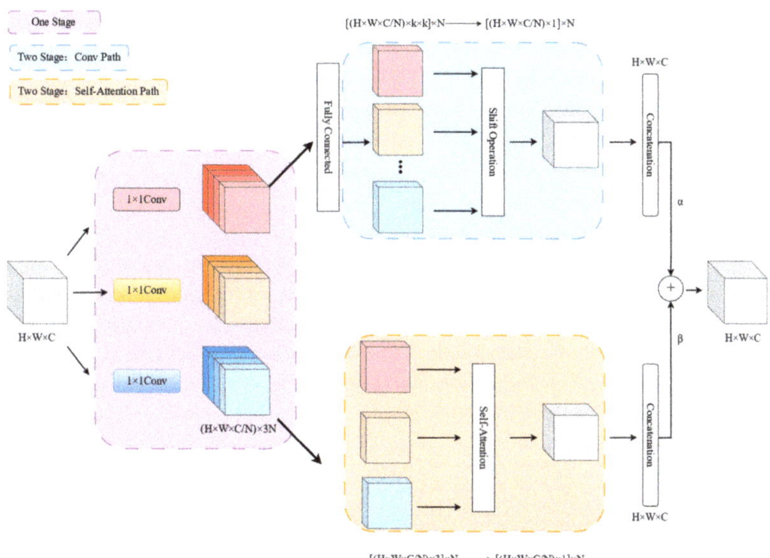

Figure 6. Structure of ACMix.

For the self-attention path, ACmix collects intermediate features into N groups, each group contains three features, corresponding to three feature mappings as Q (Query), K (Key), and V (Value), following the self-attentive module approach to collect information. Suppose f_{ij} and g_{ij} denote the tensor corresponding to the input and output of the pixel. $N_k(i, j)$ denotes the local pixel region with (i, j) as the center and spatial width k. Then, $A(W_q^{(l)} f_{ij}, W_k^{(l)} f_{ij})$ is the corresponding weight with respect to $N_k(i, j)$, as shown in Equation (11):

$$A(W_q^{(l)} f_{ij}, W_k^{(l)} f_{ab}) = \underset{N_k(i,j)}{softmax}\left(\frac{\left(W_q^{(l)} f_{ij}\right)^T \left(W_k^{(l)} f_{ab}\right)}{\sqrt{d}}\right) \tag{11}$$

Equation (11) mainly describes the expression of the Transformer attention mechanism, where $W_q^{(l)}$ and $W_k^{(l)}$ are the projection matrices of Q and K. d is the characteristic dimension of $W_q^{(l)} f_{ij}$. Meanwhile, in ACmix the multi-head self-attention is divided into two phases, as shown in Equations (12) and (13):

$$q_{ij}^{(l)} = W_q^{(l)} f_{ij}, k_{ij}^{(l)} = W_k^{(l)} f_{ij}, v_{ij}^{(l)} = W_v^{(l)} f_{ij} \tag{12}$$

$$g_{ij} = \overset{N}{\underset{l=1}{\|}} \left(\sum_{a,b \in N_k(i,j)} A(q_{ij}^{(l)}, k_{ij}^{(l)}, v_{ij}^{(l)}) \right) \tag{13}$$

where $W_v^{(l)}$ is the projection matrices of V, $q_{ij}^{(l)}$, $k_{ij}^{(l)}$, and $v_{ij}^{(l)}$ are query, key, and value matrices respectively. $\|$ is a cascade of N attention head outputs. The weight matrix is calculated by Equation (12) and the weighted characteristic information is obtained by Equation (13).

Finally, the final output F_{out} of ACmix is obtained by summing the outputs of the two-second stage pathways with additional learnable scalars α and β, as shown in Equation (14).

$$F_{out} = \alpha F_{att} + \beta F_{conv} \tag{14}$$

Equation (14) describes the final features after taking the weighted fusion of self-attentive mechanism feature extraction and convolutional feature extraction. F_{att} is the output of the self-attentive path and F_{conv} is the output of the convolutional path.

2.2.5. Feature Reconstruction Based on Contextual Neighborhood Pixel Information

Since the transmission line segmentation network also requires image restoration of the encoded features, the restored image is classified based on pixel points and the final set of classified pixel points is the final segmentation target. Therefore, the performance of the feature upsampling operator largely affects the continuity of pixel point segmentation. Currently, the nearest-neighbor sampling or bilinear interpolation sampling in the Up-Sample method is more commonly used. However, this method considers only sub-pixel domains. That is, image restoration by only some discrete feature pixels cannot capture the rich semantic information required for dense prediction tasks. Transmission lines in the power line object mainly across the characteristics of the map interval are large and the power tower is mainly a concentrated area. There is also a deconvolution operation, whose computational process mainly applies the same convolution kernel on the whole image for decoding the features without considering the correlation of the underlying feature content, which limits its responsiveness to local changes. Meanwhile, deconvolution is computationally intensive and prone to checkerboard artifacts. Based on the above two problems, this paper designs the CARAFE feature recombination operator for the restoration of features after depth encoding on transmission lines, and its structure is shown in Figure 7:

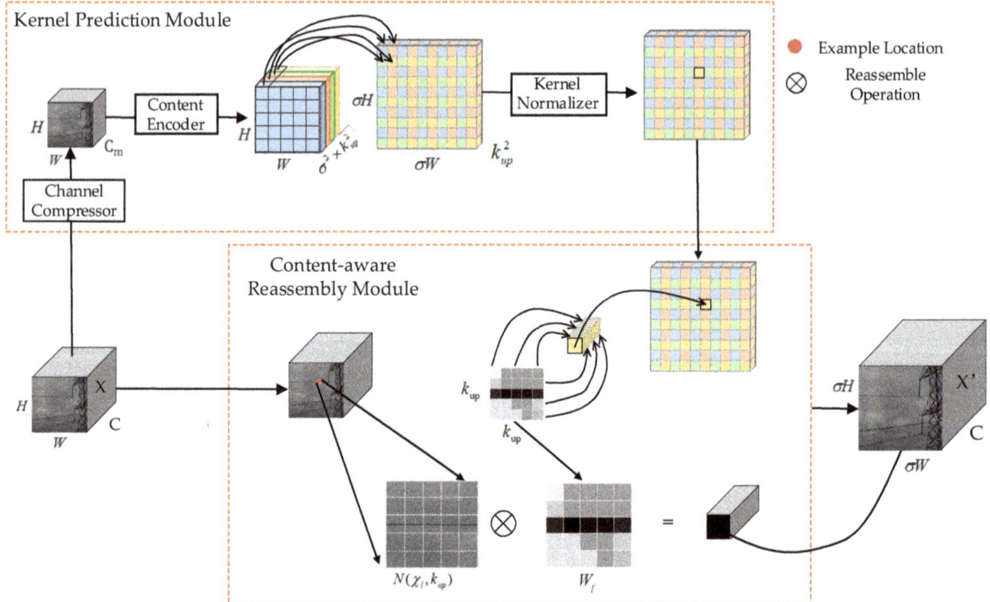

Figure 7. Structure of CARAFE.

In Figure 7, CARAFE consists of two main modules, namely the kernel prediction module and the content-aware reassembly module [40]. One of the kernel prediction modules is constructed as follows:

(1) In the kernel prediction module, the features are first compressed using 1×1 convolution for channel-based compression. The main purpose is to reduce the computational effort of the subsequent steps.

(2) Build a content-aware upsampling-based kernel. This kernel is mainly used to perform feature reduction for each pixel point in the original feature map. Since the size of

the original image is $H \times W \times C$, it takes $\delta \times \delta \times k \times k$ for each pixel point of any $H \times W$ (where δ is the upsampling multiplicity and k is the kernel size corresponding to each pixel point). Therefore, the total required prediction kernel size is $\delta \times H \times \delta \times W \times k \times k$. So, for the encoding of the content, the feature channel obtained in the first step is taken to be transformed by 3×3 convolution into $\delta \times \delta \times k \times k$. Finally, the upper sampling kernel of $\delta \times H \times \delta \times W \times k \times k$ is obtained.

(3) To improve the convergence of the model, the third step performs softmax-based normalization of the upsampling obtained in the second step so that the convolution kernel weights sum to 1.

Finally, a new feature recombination module is constructed by fusing the created upsampling kernel prediction module with the original features.

2.2.6. Improved UNet Algorithm Structure

In summary, the overall structure of the improved segmentation algorithm is shown in Figure 8. In this paper, we first propose to take the DFC-Ghost convolution block instead of the original convolution to extract the features of transmission line images, mainly by dividing the original FC layer into feature extraction based on the horizontal direction and vertical direction, it solves the problem of an excessive amount of model parameters on the one hand and improves the modeling ability of spatial information for long-distance semantic features on the other hand. Secondly, feature reuse is performed in the high semantic feature layer, and the hybrid attention structure built by the convolution module with local perceptual field enhancement and the Transformer module with global perceptual field enhancement is adopted for feature refinement to enhance model recognition in complex environments. Then, the decoder part also adopts DFC-Ghost convolutional block to extract features to ensure the light weighting of the model, while constructing a feature up-sampling structure based on convolutional kernel reorganization to effectively capture the neighboring pixel information for feature reduction. Finally, the output is segmented by 1×1 convolution for multi-category features. Meanwhile, the improved UNet model we have designed is used for transmission line segmentation. Therefore, we abbreviate the final fused improved model as TLSUNet.

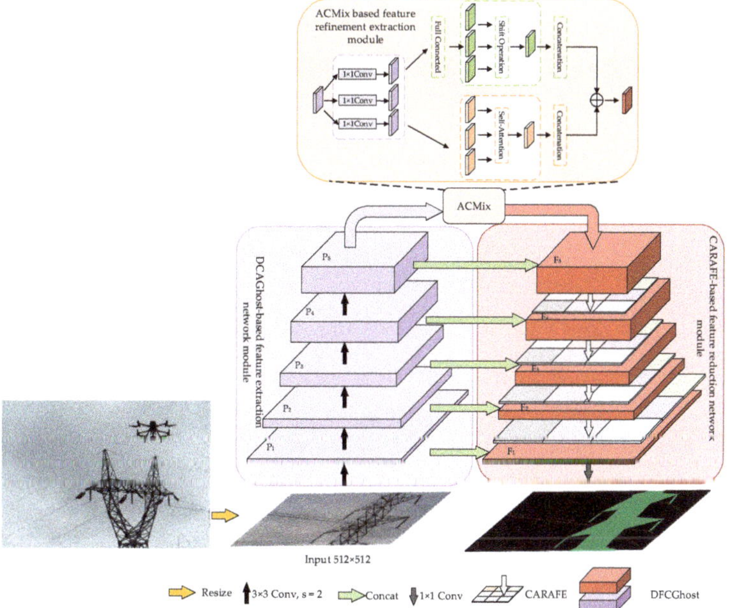

Figure 8. Transmission line segmentation model diagram.

Meanwhile, the overall computational flow of the method designed in this paper is shown in Figure 9. In the process, we mainly train the collected data for model input, and in the process of training, we perform model loss, optimization, updating of weight parameters, and testing of the model's performance. Finally, the training is completed and the optimal result is saved. In the model generalization capability validation, data from the test set are selected for validation. On the one hand, the performance metrics of the model under the test set are obtained, and on the other hand, the analysis of the image detection speed and the visualization of the detection results are performed.

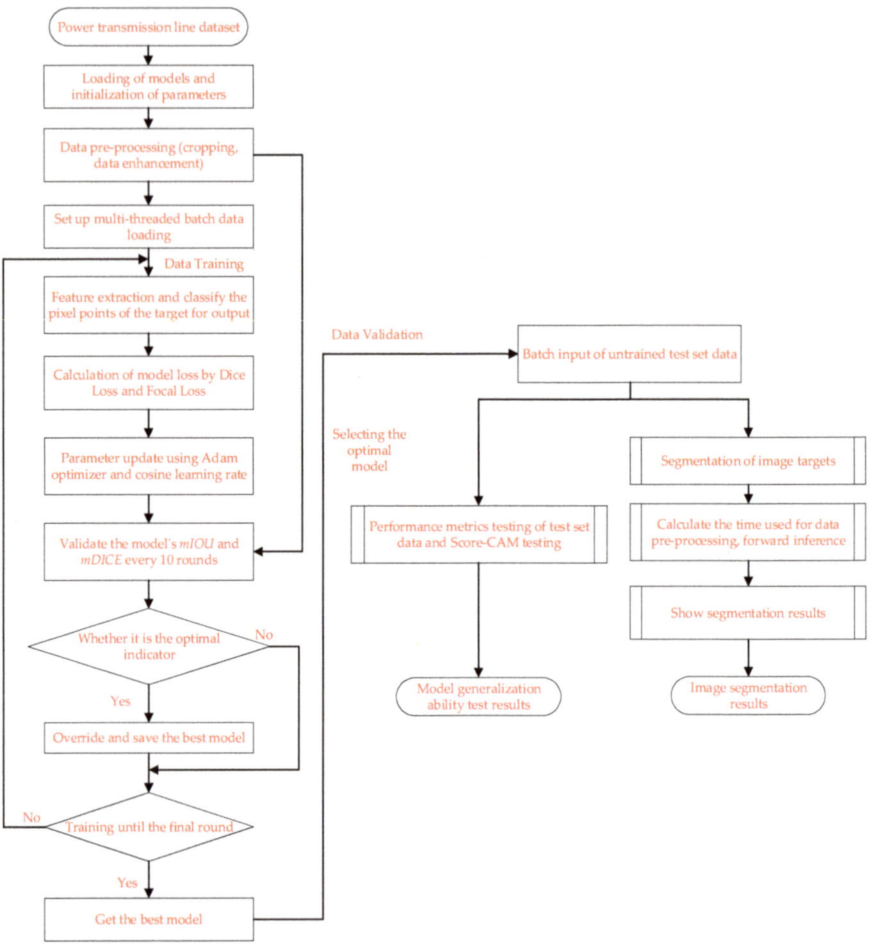

Figure 9. Transmission line segmentation flow chart.

3. Experimental Results and Analysis

3.1. Experimental Environment and Setting

The data are randomly divided into the training set, validation set, and test set in the ratio of 8:1:1. The training set is used to train the model parameters of the segmentation algorithm to obtain the training weights for this dataset. Validation sets are used to monitor the training process and prevent training overfitting. The test set is used to test the training effect and algorithm performance. No data augmentation was performed before training.

This experiment was conducted on an Ubuntu 18.08 system with Python version 3.8.0, CUDA version 11.2, and a deep learning framework based on PyTorch 1.8 environment for training and testing. The training was conducted with two NVIDIA GeForce RTX 3090-24G graphics cards and the graphics card used for the data tested in this article was the NVIDIA GeForce GTX 1050 Ti-4G. Table 4 contains the experimental parameters and settings.

Table 4. Parameter Settings.

Size of Image	Batch_Size	Momentum	Initial Lr	Min_Lr	Lr_Decay_Type	Optimizer	Iterations
512×512	4	0.9	e^{-4}	e^{-6}	cos	Adam	200

A comparison of the loss curves of the training and validation sets during the training process is shown in Figure 10. A total of 200 rounds were trained; the initial learning of the model was 0.0001, the momentum was 0.9, and the learning rate was optimized using cosine decay. The batch size was 4, and Iterative optimization of model parameters took place using Adam optimizer. Figure 10a shows the training loss convergence of the improved model and the original model. Figure 10b shows the validation loss convergence of the improved model and the original model. Among them, we will finally fuse all the improved models named TLSUNet (UNet + DFC-Ghost + ACmix + CARAFE), which is the red curve in Figures 10 and 11. From the figure, it can be seen that the improved model has faster loss convergence and lower loss value than the original model, The model is smooth in the region around 150 rounds. Figure 11a,b show the validation set change curves of the *mIOU* and *mDice* metrics of the model before and after the improvement. As can be seen from the curves, the *mIOU* and *mDice* indexes of the improved model are higher than those of the original model, verifying the effectiveness of the model improvement.

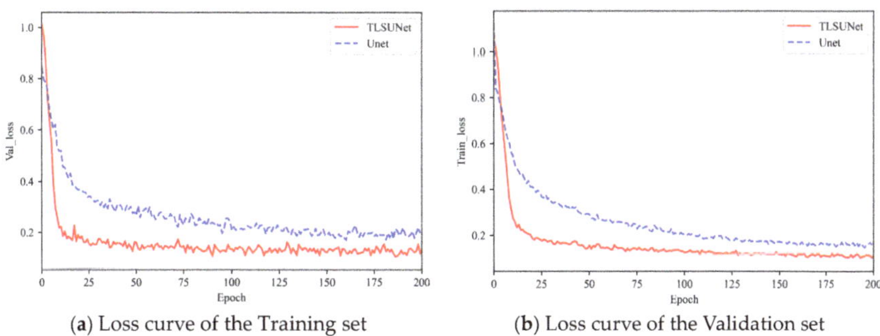

(a) Loss curve of the Training set   (b) Loss curve of the Validation set

Figure 10. Loss curve of the model.

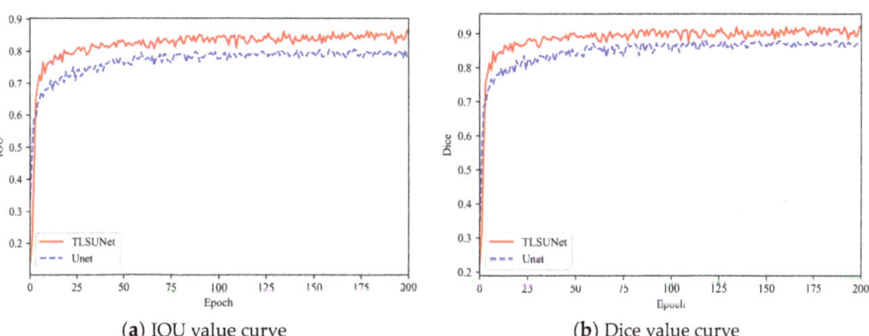

(a) IOU value curve   (b) Dice value curve

Figure 11. Evaluation metrics curve of the model.

3.2. Comparison of Splitting Accuracy Metrics

To verify the effectiveness of the improved algorithm, ablation experiments were conducted in the same environment, as shown in Table 5:

Table 5. Comparison of ablation experiments.

Model	mIOU	mDice	mPA	mRecall	mPrecision	GFLOPs (G)	Parameters (M)	Runtime (s)
UNet	79.75	87.83	86.81	86.81	88.89	457.706	24.89	0.0516
DFCGhost + UNet	81.37	88.99	87.47	87.47	90.61	29.866	4.773	0.0254
ACmix + UNet	83.21	90.11	88.60	88.60	91.77	453.431	25.721	0.0584
UNet + CARAFE	82.60	89.74	88.59	88.59	90.98	482.493	25.765	0.0929
UNet + DFCGhost + ACmix	85.14	91.50	90.07	90.07	93.00	30.992	4.722	0.0276
UNet + DFCGhost + CARAFE	85.22	91.56	90.49	90.49	92.67	41.235	5.952	0.0641
UNet + ACmix + CARAFE	86.33	92.25	90.92	90.92	93.70	480.344	26.478	0.1039
UNet + DFC-Ghost + ACmix + CARAFE	86.46	92.40	91.17	91.17	93.69	38.359	5.327	0.0396

(1) From the perspective of detection accuracy, the UNet model without any improvements as a baseline "+" indicates a mix of modules. *mIOU* has been improved by 6.71 compared to the original model, *mDice* by 4.57 compared to the original model, *mPA* by 4.36 compared to the original model, *mRecall* by 4.36 compared to the original model, and *mPrecision* is improved by 4.9 compared to the original model. At the same time, the results of different ablation experiments are better compared to the original model. The above enhancement verifies the feasibility of the improved scheme in terms of detection accuracy.

(2) From the perspective of detection speed, in this paper, by performing DFCGhost-based model lightweighting on both the encode and decode of the original model. The improved model on GFLOPs is only 8.3% of the computational effort of the original model, and the number of parameters is only 21.4% of the original model. The inference speed delay is reduced by 0.012 s. The above enhancement verifies the feasibility of the improved scheme in terms of detection speed.

3.3. Test Image Comparison

To verify the actual detection effect of the model, Table 6 exemplifies the detection results of the algorithm before and after the improvement of the test set. In Table 6, the first line is the real image under the aerial image, and the second line Ground Truth is the real label image after labeling the target through Labelme. It is mainly used for evaluating and comparing the predicted images during the training process. With the help of the second section, the mentioned evaluation index is used to judge the results between the real map and the predicted map. The following is the prediction result graph of the comparison model.

Table 6. Comparison of test set data.

Table 6. *Cont.*

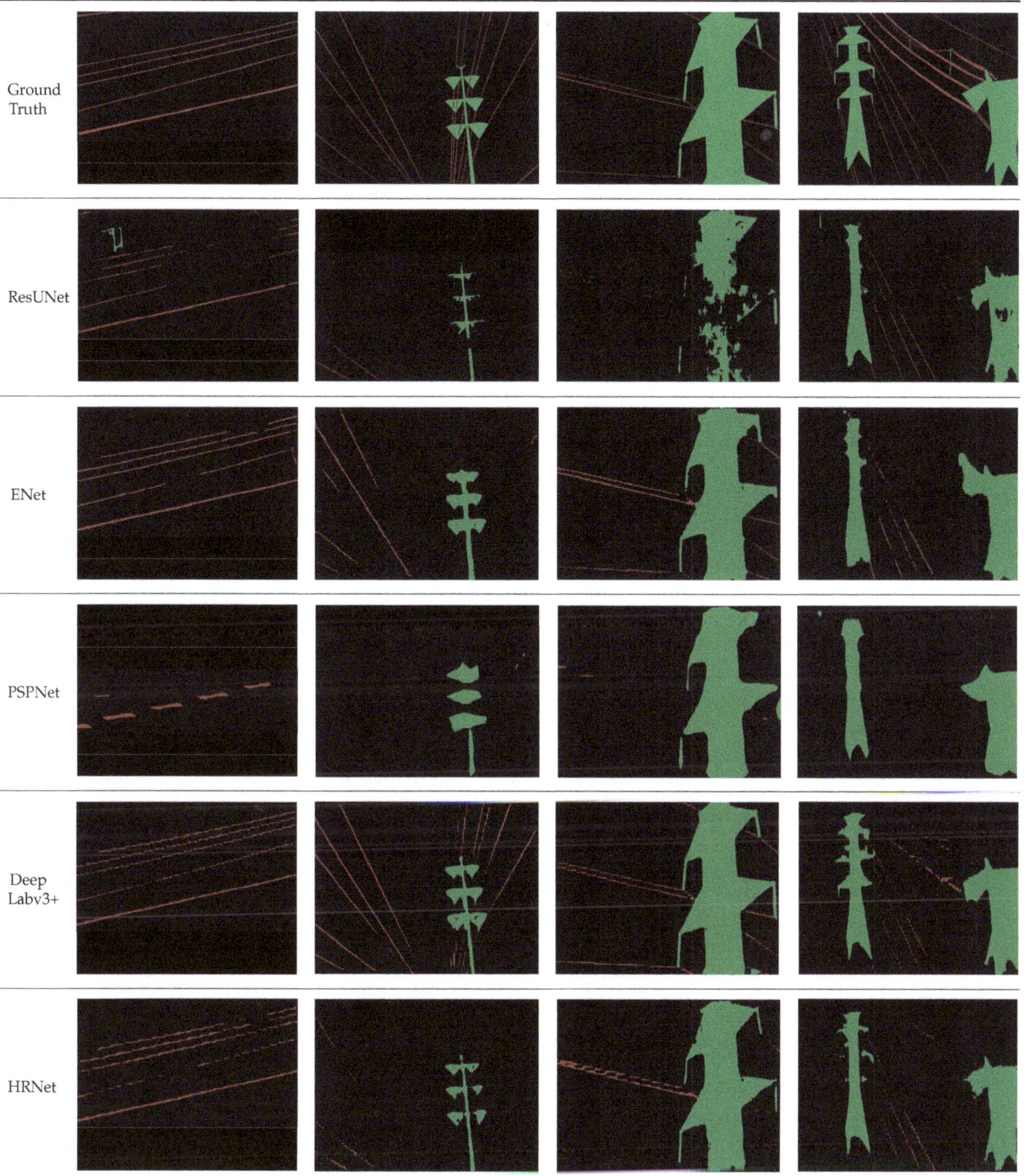

Table 6. *Cont.*

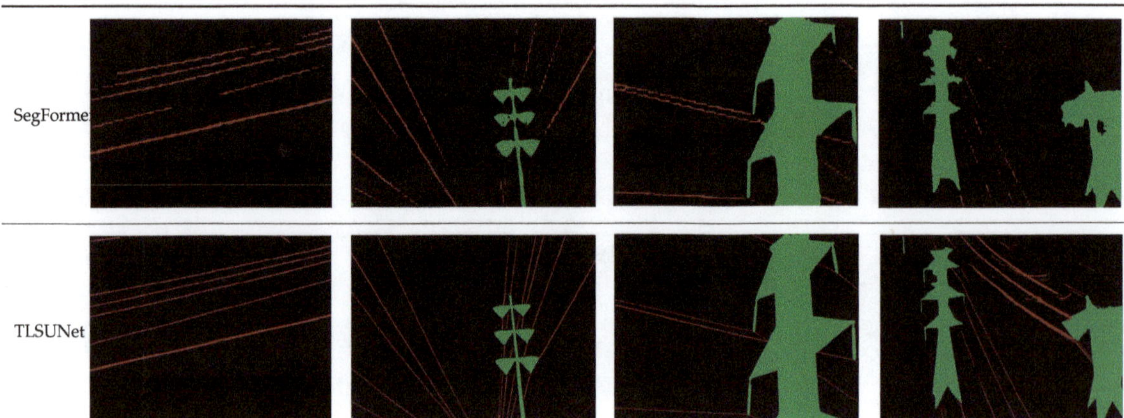

(1) The first of these images is a typical power line segmentation in a complex background. The fusion improvement model (TLSUNet) can better segment power lines in complex backgrounds, the rest of the models have intermittent segmentation or mis-segmentation problems.

(2) The second picture and the third picture show the division of power lines and towers with the sky in the background. The second picture is a wood-type tower and the other is a fence-type tower. The fusion improvement model (TLSUNet) model can better achieve the effective segmentation of power lines and different types of towers, while the rest of the network has the problem of missed segmentation.

(3) The fourth image shows the simultaneous segmentation of the pole tower and power line in a complex background, where the background is a more complex mountainous area and there is interference from the road color being similar to the pole tower. The fusion improvement model (TLSUNet) can better achieve the effective segmentation of power lines and towers. The above segmentation results also verify the feasibility of the improved scheme for image generalization ability detection.

3.4. Test Image with Score-CAM Comparison

Finally, this paper visualizes and compares the areas of interest of the images through the Score-cam heatmap. The principle of the Score-cam is mainly to weigh the feature map with the score of the target region to remove the dependence on the feature gradient (due to the complexity of the gradient information and the problem of gradient disappearance for activation functions such as Sigmoid and ReLU) [41]. The final result is obtained by taking a linear combination of weights and activation maps. The calculation principle is shown in Equation (15).

$$H_l^k = \sigma(UP(A_l^k))$$
$$C_l^k = f(X * H_l^k) - f(X) \tag{15}$$

In Equation (15), A_l^k denotes the size of the output feature map, l indicates the feature hierarchy of the output, and k denotes the number of channels corresponding to each feature layer. $\sigma(*)$ is the sigmoid activation function. The value interval is used to normalize the feature map. H_l^k denotes the size of the original output image size. $f(X)$ is the input feature map, and $f(X*H_l^k)$ is the weighted result of the input feature map. C_l^k is the region of interest of the obtained model for the input image.

Through the visualization of the heat map, we know that the heat map is concerned with the location of the target segmentation area, and the darker color represents the focus of attention on the region; as such, the proportion of attention to the outward diffusion is reduced. The first image in Table 7 shows the complex background power line segmentation detected earlier, and the heat map shows that the improved model (TLSUNet) has a clear focus on transmission lines, and the rest of the models all show breakpoints or a low level of color-based focus. The second image shows the segmentation of multiple transmission line towers. The improved model (TLSUNet) pays better attention to all the towers present in the image, especially the long-distance ones. The third and fourth sheets show two types of targets for the same diagram and the improved model (TLSUNet) has a clear focus on the pairs of transmission lines and towers. The heat map-based analysis results also verify the improved scheme's feasibility in enhancing the image segmentation capability.

Table 7. Comparison of heat map visualization.

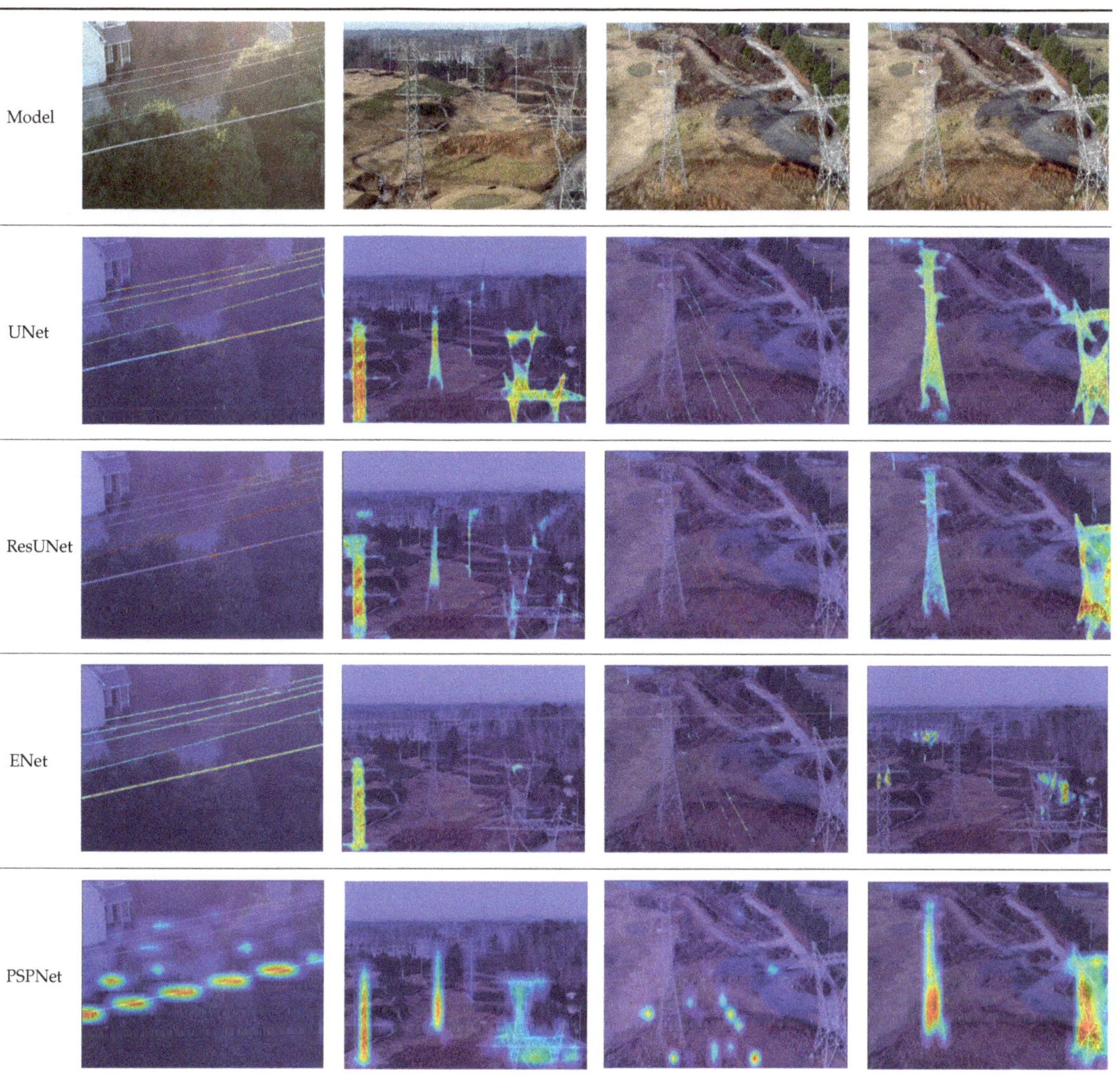

Table 7. *Cont.*

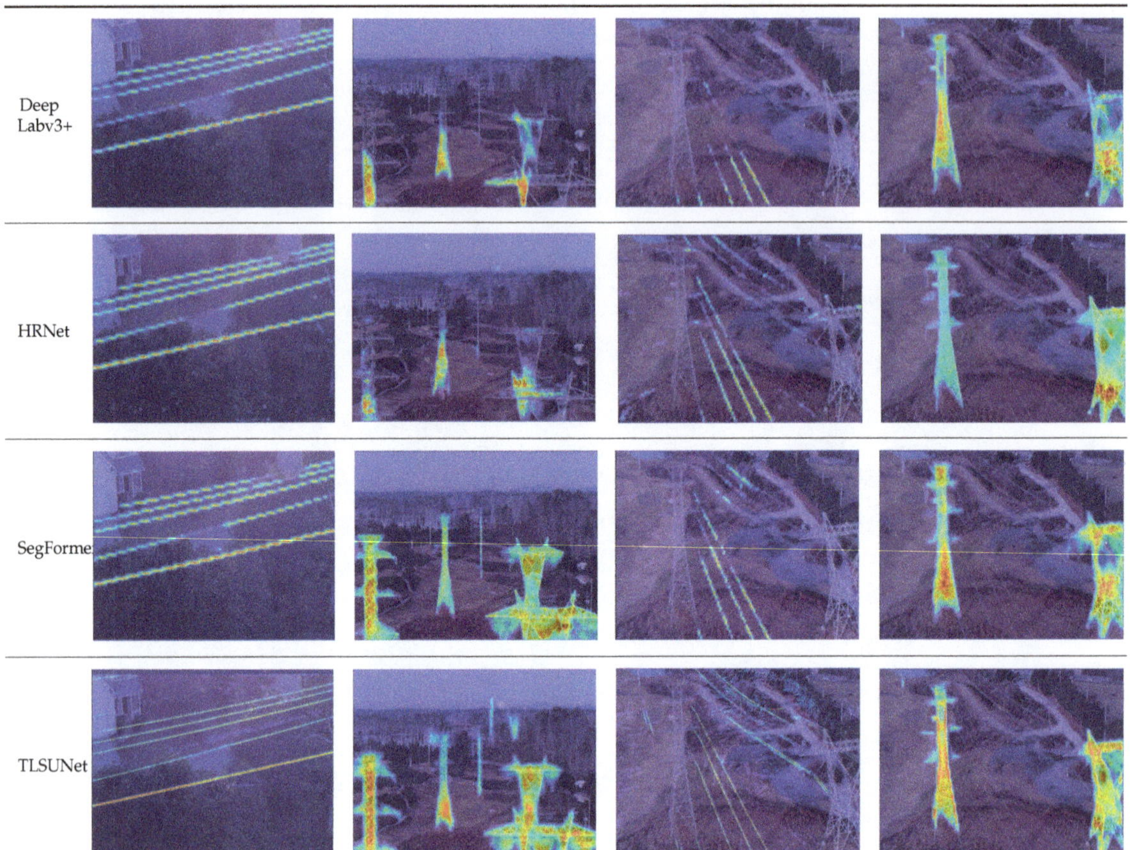

4. Discussion

To achieve efficient and accurate segmentation of power lines and power towers in transmission lines, in this paper we propose an improved UNet segmentation algorithm. In this study, we verified the effectiveness of the improved network in terms of segmentation accuracy and segmentation speed by comparing the basic segmentation network with the improved network, and have better solved the power lines and power towers in the transmission line under the complex background. In addition, on this basis, we will further conduct research on the following points:

(1) Use the researched lightweight segmentation network on the edge hardware to conduct test experiments to verify whether its inference speed can meet the requirements of normal inspection, and provide a reference for further performance optimization.

(2) It can be seen from Table 6 that at present, only the division of power lines and power towers can be realized, and more power transmission line equipment will be introduced in the future for detection, to further improve the demand for transmission line inspection.

(3) Further combine mobile edge terminals with UAVs to achieve fully autonomous line inspection requirements. At the same time, it is necessary to further consider the detection effect of the model in dense urban places.

5. Conclusions

For efficient segmentation of power lines and power towers in transmission lines in a complex context, this paper proposes a segmentation algorithm based on an improved UNet structure, and the following conclusions can be drawn by analyzing and comparing the effects of relevant factors on the segmentation effect through existing power transmission data sets.

(1) To address the lightweight problem of the model, this paper designs the DFCGhost convolutional feature extraction network, which is used for the compression of the number of parameters on the one hand, and enhances the feature extraction process in the horizontal and vertical directions at the same time, so that the model can be modeled with long-distance spatial information. The results indicate an improvement of 1.62 in $mIOU$ and 1.16 in $mDice$ of the model. Secondly, the complex background is the main factor affecting the model segmentation, so the deep semantic features are refined and weight extracted by combining the convolution module with local perceptual field enhancement and the transformer module with global perceptual field enhancement. The results showed that the model improved $mIOU$ by 3.46 and $mDice$ by 2.28. Finally, high-precision decoding of features is achieved by using CARAFE's feature parameter reconstruction to improve the usability of features. The results showed that the model had been improved by 2.85 for $mIOU$ and 1.91 for $mDice$.

(2) The results of the ablation experiments show that the model incorporating all improvements improves by 6.71 on $mIOU$, 4.57 on $mDice$, 4.8 on $mPrecision$, 4.36 on $mRecall$, and 4.36 on mPA for power lines and power towers.

(3) The fusion experimental model is tested on the computer side and the results show that the parameters of the lightweight model are only 8.3% of the computation of the original model, and the number of parameters is only 21.4% of the original model. The inference speed delay is reduced by 0.012 s. The test results can play a certain role in the intelligent inspection of power system automation.

At the same time, the power line segmentation task realized in this paper can provide a feasible technical solution and reference for UAV automatic line following inspection technology.

Author Contributions: Conceptualization, L.Q. and M.H.; methodology, L.Q.; software, M.H.; validation, M.H., X.D., S.Z. and H.L.; formal analysis, L.Q.; writing—original draft preparation, L.Q. and M.H.; writing—review and editing, M.H.; visualization, X.D., S.Z. and H.L.; supervision, K.L. and L.Q. All authors have read and agreed to the published version of the manuscript.

Funding: This work was supported by the National Key R & D Program of China (No. 2020YFB0905900).

Data Availability Statement: Dataset link: https://github.com/R3ab/ttpla_dataset (accessed on 14 December 2022).

Conflicts of Interest: The authors declare no conflict of interest.

References

1. Zhou, F.; Ren, G. Image classification and feature extraction of transmission line based on color space variable. *Power Syst. Prot. Control* **2018**, *46*, 89–98.
2. Otsu, N. A threshold selection method from gray-level histograms. *IEEE Trans. Syst. Man Cybern.* **1979**, *9*, 62–66. [CrossRef]
3. Zhao, L.; Wang, X.; Dai, D.; Long, J.; Tian, M.; Zhu, G. Automatic Extraction Algorithm of Power Line in Complex Background. *High Volt. Eng.* **2019**, *45*, 218–227.
4. Touzi, R.; Lopes, A.; Bousquet, P. A statistical and geometrical edge detector for SAR images. *IEEE Trans. Geosci. Remote Sens.* **1988**, *26*, 764–773. [CrossRef]
5. Zhao, L.; Wang, X.; Yao, H.; Tian, M.; Gong, L. Power Line Extraction Algorithm Based on Local Context Information. *High Volt. Eng.* **2021**, *47*, 2553–2563.
6. Duda, R.O.; Hart, P.E. Use of the Hough transformation to detect lines and curves in pictures. *Commun. ACM* **1972**, *15*, 11–15. [CrossRef]
7. Yuan, C.; Guan, Y.; Zhang, J.; Yuan, C. Power line extraction based on improved Hough transform. *Beijing Surv. Mapp.* **2018**, *32*, 730–733. [CrossRef]

8. Cao, H.; Zeng, W.; Shi, Y.; Xu, P. Power line detection based on Hough transform and overall least squares method. *Comput. Technol. Dev.* **2018**, *28*, 164–167.

9. Shan, H.; Zhang, J.; Cao, X.; Li, X.; Wu, D. Multiple auxiliaries assisted airborne power line detection. *IEEE Trans. Ind. Electron.* **2017**, *64*, 4810–4819. [CrossRef]

10. Zhao, L.; Wang, X.; Yao, H.; Tian, M. Survey of Power Line Extraction Methods Based on Visible Light Aerial Image. *Power Syst. Technol.* **2021**, *45*, 1536–1546.

11. Li, Z.; Zhang, Y.; Wu, H.; Suzuki, S.; Namiki, A.; Wang, W. Design and Application of a UAV Autonomous Inspection System for High-Voltage Power Transmission Lines. *Remote Sens.* **2023**, *15*, 865. [CrossRef]

12. Jenssen, R.; Roverso, D. Automatic autonomous vision-based power line inspection: A review of current status and the potential role of deep learning. *Int. J. Electr. Power Energy Syst.* **2018**, *99*, 107–120.

13. Zhang, Y.; Yuan, X.; Li, W.; Chen, S. Automatic power line inspection using UAV images. *Remote Sens.* **2017**, *9*, 824. [CrossRef]

14. Senthilnath, J.; Kumar, A.; Jain, A.; Harikumar, K.; Thapa, M.; Suresh, S. BS-McL: Bilevel segmentation framework with metacognitive learning for detection of the power lines in UAV imagery. *IEEE Trans. Geosci. Remote Sens.* **2021**, *60*, 1–12. [CrossRef]

15. Fan, Z.; Shi, L.; Xi, C.; Wang, H.; Wang, S.; Wu, G. Real-Time Power Equipment Meter Recognition Based on Deep Learning. *IEEE Trans. Instrum. Meas.* **2022**, *71*, 1–15. [CrossRef]

16. Dong, X.; Fu, R.; Gao, Y.; Qin, Y.; Ye, Y.; Li, B. Remote sensing object detection based on receptive field expansion block. *IEEE Geosci. Remote Sens. Lett.* **2021**, *19*, 1–5. [CrossRef]

17. Simonyan, K.; Zisserman, A. Very deep convolutional networks for large-scale image recognition. *arXiv* **2014**, arXiv:1409.1556.

18. He, K.; Zhang, X.; Ren, S.; Sun, J. Deep residual learning for image recognition. In Proceedings of the IEEE Conference on Computer Vision and Pattern Recognition, Las Vegas, NV, USA, 30 June 2016; pp. 770–778.

19. Yetgin, Ö.E.; Benligiray, B.; Gerek, Ö.N. Power line recognition from aerial images with deep learning. *IEEE Trans. Aerosp. Electron. Syst.* **2018**, *55*, 2241–2252. [CrossRef]

20. Yang, L.; Fan, J.; Xu, S.; Li, E.; Liu, Y. Vision-based power line segmentation with an attention fusion network. *IEEE Sens. J.* **2022**, *22*, 8196–8205. [CrossRef]

21. Han, G.; Zhang, M.; Li, Q.; Liu, X.; Li, T.; Zhao, L. A Lightweight Aerial Power Line Segmentation Algorithm Based on Attention Mechanism. *Machines* **2022**, *10*, 881. [CrossRef]

22. Xu, C.; Li, Q.; Zhou, Q.; Zhang, S.; Yu, D.; Ma, Y. Power line-guided automatic electric transmission line inspection system. *IEEE Trans. Instrum. Meas.* **2022**, *71*, 1–18. [CrossRef]

23. Gao, Z.; Yang, G.; Li, E.; Liang, Z.; Guo, R. Efficient parallel branch network with multi-scale feature fusion for real-time overhead power line segmentation. *IEEE Sens. J.* **2021**, *21*, 12220–12227. [CrossRef]

24. Choi, H.; Yun, J.P.; Kim, B.J.; Jang, H.; Kim, S.W. Attention-based multimodal image feature fusion module for transmission line detection. *IEEE Trans. Ind. Inform.* **2022**, *18*, 7686–7695. [CrossRef]

25. Abdelfattah, R.; Wang, X.; Wang, S. Ttpla: An aerial-image dataset for detection and segmentation of transmission towers and power lines. In Proceedings of the Asian Conference on Computer Vision, Kyoto, Japan, 4 December 2020.

26. Available online: https://github.com/r3ab/ttpla_dataset (accessed on 14 December 2022).

27. Ronneberger, O.; Fischer, P.; Brox, T. U-net: Convolutional networks for biomedical image segmentation. In Proceedings of the Medical Image Computing and Computer-Assisted Intervention—MICCAI 2015: 18th International Conference, Munich, Germany, 5–9 October 2015; Springer International Publishing: Cham, Switzerland, 2015; pp. 234–241.

28. Song, K.; Yang, G.; Wang, Q.; Xu, C.; Liu, J.; Liu, W.; Shi, C.; Wang, Y.; Zhang, G. Deep learning prediction of incoming rainfalls: An operational service for the city of Beijing China. In Proceedings of the 2019 International Conference on Data Mining Workshops (ICDMW), Beijing, China, 8–11 November 2019; pp. 180–185.

29. Paszke, A.; Chaurasia, A.; Kim, S.; Culurciello, E. Enet: A deep neural network architecture for real-time semantic segmentation. *arXiv* **2016**, arXiv:1606.02147.

30. Zhao, H.; Shi, J.; Qi, X.; Wang, X.; Jia, J. Pyramid scene parsing network. In Proceedings of the IEEE Conference on Computer Vision and Pattern Recognition, Honolulu, HI, USA, 21–26 July 2017; pp. 2881–2890.

31. Chen, L.-C.; Zhu, Y.; Papandreou, G.; Schroff, F.; Adam, H. Encoder-decoder with atrous separable convolution for semantic image segmentation. In Proceedings of the European Conference on Computer Vision (ECCV), Munich, Germany, 8–14 September 2018; pp. 801–818.

32. Sun, K.; Xiao, B.; Liu, D.; Wang, J. Deep high-resolution representation learning for human pose estimation. In Proceedings of the IEEE/CVF Conference on Computer Vision and Pattern Recognition, Long Beach, CA, USA, 15–20 June 2019; pp. 5693–5703.

33. Xie, E.; Wang, W.; Yu, Z.; Anandkumar, A.; Alvarez, J.M.; Luo, P. SegFormer: Simple and efficient design for semantic segmentation with transformers. *Adv. Neural Inf. Process. Syst.* **2021**, *34*, 12077–12090.

34. Howard, A.; Sandler, M.; Chu, G.; Chen, L.-C.; Chen, B.; Tan, M. Searching for mobilenetv3. In Proceedings of the IEEE/CVF International Conference on Computer Vision, Seoul, Republic of Korea, 27 October–2 November 2019; pp. 1314–1324.

35. Han, K.; Wang, Y.; Tian, Q.; Guo, J.; Xu, C.; Xu, C. Ghostnet: More features from cheap operations. In Proceedings of the IEEE/CVF Conference on Computer Vision and Pattern Recognition, Seattle, WA, USA, 14–19 June 2020; pp. 1580–1589.

36. Tan, M.; Le, Q. Efficientnet: Rethinking model scaling for convolutional neural networks. In Proceedings of the International Conference on Machine Learning, Long Beach, CA, USA, 10–15 June 2019; pp. 6105–6114.

37. Zhang, X.; Zhou, X.; Lin, M.; Sun, J. Shufflenet: An extremely efficient convolutional neural network for mobile devices. In Proceedings of the IEEE Conference on Computer Vision and Pattern Recognition, Salt Lake City, UT, USA, 18–22 June 2018; pp. 6848–6856.
38. Tang, Y.; Han, K.; Guo, J.; Xu, C.; Xu, C.; Wang, Y. GhostNetV2: Enhance Cheap Operation with Long-Range Attention. *arXiv* **2022**, arXiv:2211.12905.
39. Pan, X.; Ge, C.; Lu, R.; Song, S.; Chen, G.; Huang, Z.; Huang, G. On the integration of self-attention and convolution. In Proceedings of the IEEE/CVF Conference on Computer Vision and Pattern Recognition, New Orleans, LA, USA, 18–24 June 2022; pp. 815–825.
40. Wang, J.; Chen, K.; Xu, R.; Liu, Z.; Loy, C.C.; Liu, D. Carafe: Content-aware reassembly of features. In Proceedings of the IEEE/CVF International Conference on Computer Vision, Seoul, Republic of Korea, 27 October–2 November 2019; pp. 3007–3016.
41. Wang, H.; Wang, Z.; Du, M.; Yang, F.; Zhang, Z.; Ding, S.; Mardziel, P.; Hu, X. Score-CAM: Score-weighted visual explanations for convolutional neural networks. In Proceedings of the IEEE/CVF Conference on Computer Vision and Pattern Recognition Workshops, Seattle, WA, USA, 14–19 June 2020; pp. 24–25.

Article

An Automatic Visual Inspection of Oil Tanks Exterior Surface Using Unmanned Aerial Vehicle with Image Processing and Cascading Fuzzy Logic Algorithms

Mohammed A. H. Ali [1,*], Muhammad Baggash [2], Jaloliddin Rustamov [3], Rawad Abdulghafor [4], Najm Al-Deen N. Abdo [2], Mubarak H. G. Abdo [2], Talep S. Mohammed [2], Ameen A. Hasan [2], Ali N. Abdo [2], Sherzod Turaev [5] and Yusoff Nukman [1]

[1] Department of Mechanical Engineering, Faculty of Engineering, University of Malaya, Kuala Lumpur 50603, Malaysia
[2] Faculty of Engineering, Taiz University, Taiz 9674, Yemen
[3] Faculty of Computer Science & Information Technology, University of Malaya, Kuala Lumpur 50603, Malaysia
[4] Department of Computer Science, Faculty of Information and Communication Technology, International Islamic University Malaysia, Kuala Lumpur 53100, Malaysia
[5] Department of Computer Science and Software Engineering, College of Information Technology, United Arab Emirates University, Al Ain 15551, United Arab Emirates
[*] Correspondence: hashem@um.edu.my

Citation: Ali, M.A.H.; Baggash, M.; Rustamov, J.; Abdulghafor, R.; Abdo, N.A.-D.N.; Abdo, M.H.G.; Mohammed, T.S.; Hasan, A.A.; Abdo, A.N.; Turaev, S.; et al. An Automatic Visual Inspection of Oil Tanks Exterior Surface Using Unmanned Aerial Vehicle with Image Processing and Cascading Fuzzy Logic Algorithms. *Drones* **2023**, *7*, 133. https://doi.org/10.3390/drones7020133

Academic Editor: Seokwon Yeom

Received: 19 December 2022
Revised: 31 January 2023
Accepted: 9 February 2023
Published: 13 February 2023

Abstract: This paper presents an automatic visual inspection of exterior surface defects of oil tanks using unmanned aerial vehicles (UAVs) and image processing with two cascading fuzzy logic algorithms. Corrosion is one of the defects that has a serious effect on the safety of the surface of oil and gas tanks. At present, human inspection, and climbing robots inspection are the dominant approach for rust detection in oil and gas tanks. However, there are many shortcomings to this approach, such as taking longer, high cost, and covering less surface area inspection of the tank. The purpose of this research is to detect the rust in oil tanks by localizing visual inspection technology using UAVs, as well as to develop algorithms to distinguish between defects and noise. The study focuses on two basic aspects of oil tank inspection through the images captured by the UAV, namely, the detection of defects and the distinction between defects and noise. For the former, an image processing algorithm was developed to improve or remove noise, adjust the brightness of the captured image, and extract features to identify defects in oil tanks. Meanwhile, for the latter aspect, a cascading fuzzy logic algorithm and threshold algorithm were developed to distinguish between defects and noise levels and reduce their impact through three stages of processing: The first stage of fuzzy logic aims to distinguish between defects and low noise generated by the appearance of objects on the surface of the tank, such as trees or stairs, and reduce their impact. The second stage aims to distinguish between defects and medium noise generated by shadows or the presence of small objects on the surface of the tank and reduce their impact. The third stage of the thresholding algorithm aims to distinguish between defects and high noise generated by sedimentation on the surface of the tank and reduce its impact. The samples were classified based on the output of the third stage of the threshold process into defective or non-defective samples. The proposed algorithms were tested on 180 samples and the results show its superiority in the inspection and detection of defects with an accuracy of 83%.

Keywords: oil tank; automatic visual inspection; unmanned aerial vehicle; camera; fuzzy logic

1. Introduction

Oil products are one of the most important sources of energy and have a vital impact on countries' economic sectors as demand for it increases gradually with continuous developing of other industrial and commercial sectors. This results in a need for storage tanks to store as much oil as possible.

The exterior surface of oil tanks is commonly affected by corrosion and dust due to chemical reactions between the surface and both oil and air substances, necessitating continuous inspection, and monitoring.

Corrosion is one of the biggest problems for companies in the oil sector due to the cost of repairing or replacing the damaged parts with non-corroded ones. Corrosion can be defined as the destructive attack of a substance by interaction with its environment [1] which promotes the tendency of these unstable metals to return to their more stable natural form. Regular inspection and monitoring of these tanks are the most important ways to reduce risk and corrosion because they help to early detect the damage, prolong the life of these tanks, and prevent the closure of the oil facility or the suspension of production processes.

In addition, such inspections provide protection against legal and financial accountability for leakage caused by the corrosion process and its destructive impact on the environment. Analysis of major refinery accidents over the past 35 years has shown that loss of containment due to corrosion has contributed up to 25% of these accidents [2]. It has been noticed that corrosion causes 42% of the failure mechanisms in all engineering structures [2]. The damage corrosion causes maintenance costs to be increase in the range of 3–5% of the total products' costs in developed countries [3]. In the oil and gas industry alone, the cost of repairing damage caused by corrosion is $1.372 billion which includes surface pipelines ($589 million), expenses of pipelining ($463 million), and another $320 million in corrosion-related capital expenditure [4]. In addition, corrosion can significantly reduce the annual income by up to $10 billion during maintenance time [5], e.g., in US, the annual cost of corrosion damage is $170 billion [6].

Despite improvements in the design process and the selection of metals for better construction of these tanks, this is not enough to ensure their safety. The introduction of modern inspection techniques, such as climbing robots and manned UAVs, has been rapidly evolving in an effort to solve the problem with more sophisticated techniques.

UAVs are used in several fields, including inspection, and monitoring, searching for missing persons, and monitoring illegal immigrants, monitoring vital infrastructure, and detecting hidden corrosion in aluminum structures and checking railway surface defects, etc. Several researchers are developing UAV inspection systems for oil and gas tank inspection, as they can play a significant role in reducing inspection time, cost, and risks to overall required maintenance.

The traditional inspection and maintenance process is very expensive and, due to its complex nature and dangerous environment, time consuming, but it is necessary to avoid the catastrophic risks that may be inflicted on the environment and humans (as a result of the effects of corrosion and the consequences of neglecting the inspection process and monitoring). The industry's oil and gas plants need to be maintained regularly to keep their components running with high safety and efficiency. Regular testing and equipment inspection has a great effect on the costs of maintenance and daily operational processes [7].

The structure of the oil and gas and oil industry is so complex with high hazards, thus maintenance, inspection, and repairs in such places include high risks to employers. The maintenance operators must regularly climb up high-rise oil facilities, such as storage tanks, flare stacks, boilers, chimneys, and cooling towers, to inspect their surfaces. Based on the performance of the inspection, two common methods are used to inspect the oil and gas industry, namely, manual, and climbing robot inspections.

The climbing robot inspection system is nowadays used widely to inspect oil and gas facilities [8]. This system has resulted in high cost savings in daily operational costs. There are many kinds of climbing robot systems that can access high rise buildings; they imitate mammals, reptiles, and insects when climbing, using several movement methods to climb such as jumping sliding, extension, and swinging. Many problems can occur during climbing robots' operations in high-rise buildings, such as flexibility, motors overheating, power-consumption stability, slippage on the surface being climbed, and climbing between neighboring surfaces.

The current climbing robots used to inspect outdoor storage tanks have limited movement and commonly work based on remote control which has resulted in a decrease in flexibility. As an example, the MATS climbing robot with 5-DOF has excellent maneuverability, but needs a place to dock. A prototype model called Walloid, that is able to choose an adhesion method which increases the robustness and flexibility needed for industrial applications, has been designed for offshore oil and gas facility inspection [9].

A robot based on bio-inspired principle, called Sticky Bot, has adhesive material on the bottom of its feet to enable it to hold onto all surface types [10]. The non-destructive test is usually performed to inspect the metallic plates for corrosion and determine the presence of defects without damaging the surface. A climbing robot for corrosion observation on cooling towers used by the oil and gas industry has been developed by a fusion of wheel electrodes and adhesion operation [11]. Such robots have increased performance efficiency in comparison with humans, but need a special mechanical design for materials used in climbing and good analysis of system dynamics. Climbing robot applications are confined to some types of structures such as those with cylinder shapes.

The alternative solution is to use UAVs to inspect surfaces with a simple and straight-forward mechanism. UAV technology has been widely utilized in the gas and oil industry to inspect high-rise facilities with a better efficiency and sustainability, in comparison with climbing robots or manual methods. UAV inspection depends on data analysis of a range of sensors data that need to be acquired, processed, stored, and well analyzed. Unlike climbing robots that need a suspension system and scaffolds, UAVs can move freely and perform inspections with high efficiency and reliability.

UAV inspection is accomplished using high-definition (HD) cameras and infra-red (IR) sensors that can carry out the risk-based inspection (RBI) for gas and oil equipment with the standards of API RP-580, API 579-1, and ASME FFS-1. UAVs can also test piping with API 570 (pipe inspection code) and tanks of standard API 653 (tank inspection repair reconstruction) [12].

One of the first drone inspection systems was introduced in 2010 to inspect an onshore oil refinery in the UK. This helped operators to gain an understanding of the condition of the equipment without any need for shutdown or exposing the operators to a risky situation [13]. Such systems allow engineers to inspect high-rise critical equipment in the oil and gas industry (such as vents, ducting, pipes, and chimneystacks), reduce the time for maintenance, and prioritize components' maintenance without a need to shut down the facilities. The UAV flies manually under the control of a certified pilot who enables the drone to fly along the facilities which require inspection using normal/thermal cameras and sensor of hydrocarbon leakage determination, etc. The acquired data is then analyzed to find defects on the surfaces such as corrosion, hairline cracks, and leakages. Since the drones can carry cameras with 4K video recording and optical zoom, along with various other sensors, there is no need to fly the drone too close to the inspected structures or other risk areas.

Drone-based inspection in the oil and gas industry is getting more attention due to four reasons, namely, (i) its ability to inspect areas that are potentially hazardous; (ii) it is a cost effective as well as efficient inspection method; (iii) its ability to inspect a large area in short time; and (iv) the operation of the drone does not require a highly skilled inspector. Many third-party companies are now offering UAV inspection solutions for the oil and gas industry.

This paper is aimed at developing a UAV-based visual inspection system for high-rise oil tanks. Such facilities must be continuously inspected to avoid hazardous surface leakage once it has appeared. The contribution of this paper is related to a combination of theoretical and experimental techniques. The theoretical aspect relies on developing a classification algorithm based on the fusing of image processing and two cascading fuzzy logic (FL) and threshold processes. The experimental works present defect detection on high-rise oil tanks as a challenging subject which needs further improvement through research.

2. Automatic Visual Inspection System

The inspection of the oil tanks is performed through four main stages as shown in Figure 1, namely, UAV scanning, image processing, AI, and thresholding process.

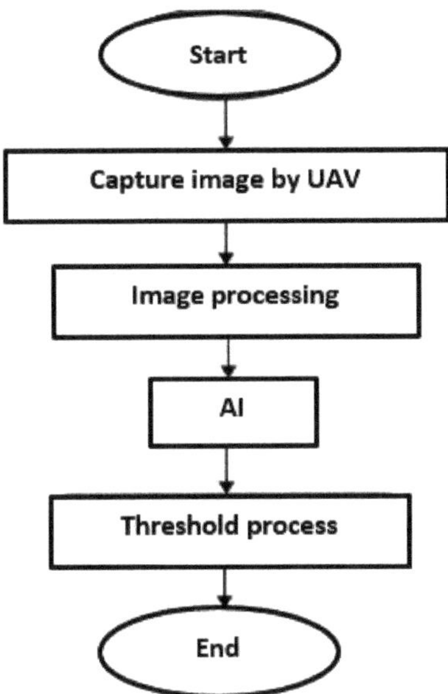

Figure 1. Visual inspections stages for oil tanks.

2.1. UAV Scanning

The Pro Mavic UAV, shown in Figure 2, was used for inspection of oil tanks in this work. It can fly for a period of up to 27 min at a speed of 40 mph. It also uses data that is recorded with GPS information to ensure that the UAV lands in an accurate location.

Figure 2. Pro Mavic UAV.

The camera built into the Mavic Pro UAV is the smallest 3-axis camera, which has the feature of recording both images and video. With a 90-degree tilt of the camera, it produces video with blurry side scenes, or with black bars.

The camera supports 4K shooting at 30 frames per second, as well as 1080p full HD video shooting at 96 frames per second, so it is expected to support slow motion video shooting. The controller is connected to the UAV within a range of 4.3 miles, with live scenes from the plane at a 1080p display quality.

2.2. Image Processing-based Defects Detection

Images taken by UAVs are often inconsistent or lack specific behaviors and trends. The image is likely to contain many errors and distortions making it very complicated to handle. Therefore, image processing is required to remove the noise to accurately assess the defects on the oil tanks.

Image processing consists of several steps to prepare clear images that are contrast-adjusted and free from the blurring which results from the heterogeneity of lighting, the appearance of objects on the sides of the images, etc.

This project is focused on inspection all sides of the tank except for the stairs, and the top and bottom.

2.2.1. Pre-Processing of Captured Image

Four main operations are applied on the captured image in pre-processing stages, namely, cropping, resizing, RGB (red green blue) conversion and brightness adjusting as shown in Figure 3. Cropping involves the removal of unwanted parts in an image. The areas located under the tank have no significance and thus are most likely to be removed.

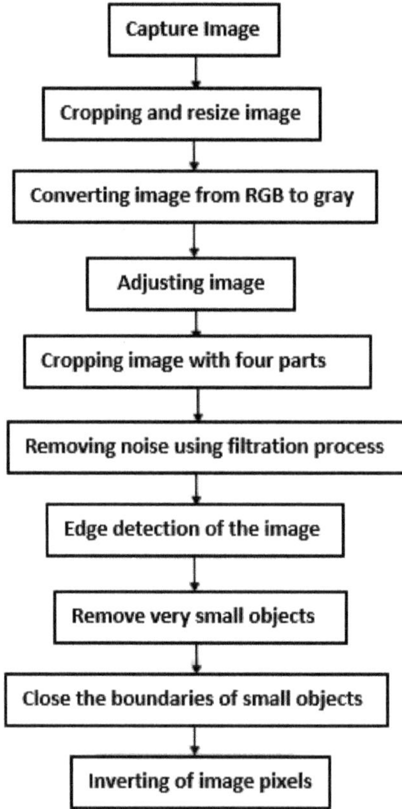

Figure 3. Image Pre-processing operations.

Resizing controls image dimensions, length and width, and allocates a fixed size value to all input images, reducing execution time and data processing speed. This process allows for easy image handling and the option of obtaining a square image that can be divided into four equal parts. The images after implementation of cropping and resizing operations are shown in Figure 4b.

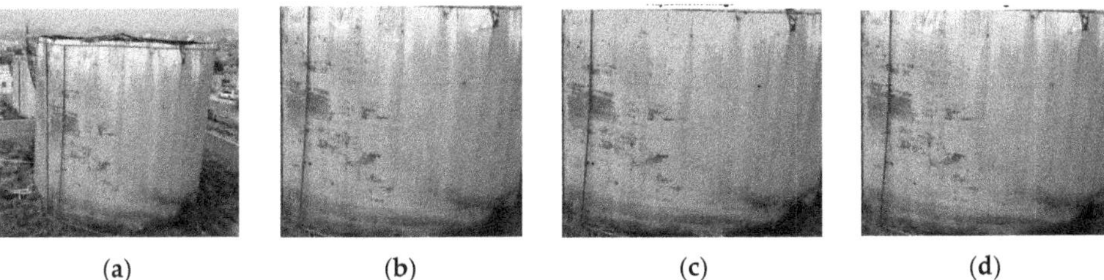

(a)   (b)   (c)   (d)

Figure 4. Image cropping and resizing: (**a**) Original image of the whole tank captured by the UAV (**b**) the image after implementation of cropping and resizing operation (**c**) the image after conversion process from RGB to gray (**d**) the image after adjustment.

The RGB color image is converted to gray to reduce the image size in order to increase the processing speed and facilitate its handling with some image processing instructions. The image after conversion process from RGB to gray as shown in Figure 4c,d.

Brightness allows adjustment of the high and low pixel values that affect the homogeneity of the image. Figure 4d shows an image after adjustment operation.

2.2.2. Image Processing of the Prepared Image

The image in the pre-processing stage needs a further process using image processing tools to prepare it for defects detection. The image has been divided into four parts of equal dimension, as shown in Figure 5, to increase classification accuracy, reduce the effect of noise, and choose an appropriate threshold if the image is inhomogeneous.

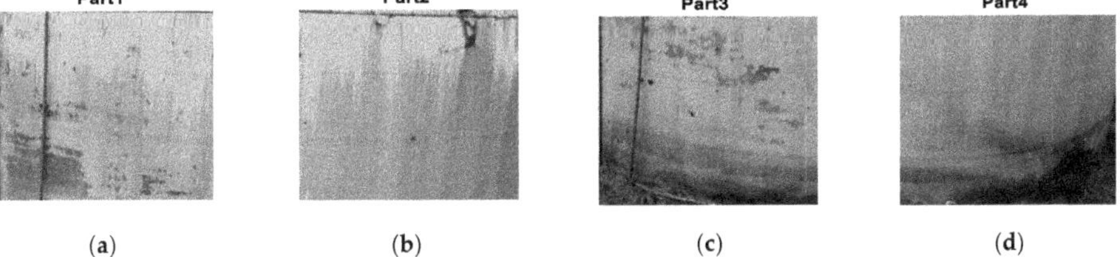

(a)   (b)   (c)   (d)

Figure 5. The image after division into four equal parts. (**a**) pre-processed image part 1. (**b**) pre-processed image part 2. (**c**) pre-processed image part 3. (**d**) pre-processed image part 4.

The importance of the division is outlined in the following points:

- When the threshold value is specified it will be determined according to the grayscale values of the whole image, while if it is divided into four parts it will be determined for a specific area. Thus, picture noise effects will occur on one part rather than the whole image.
- When filters are used, one can obtain an enhanced and higher quality image with four parts rather than if they are used on the entire image.

The performance of the edge filter is improved with the four image parts as it depends on the threshold value in the classification process. The smaller the image size, the higher the filtration efficiency. Accuracy of the fuzzy logic algorithm in the classification process will be higher with four image parts.

- **Filtration of Image:** filtering is a technique used to eliminate noise and unwanted things from images. Two main filters are used in this work to eliminate the effects of noise from i402mages, namely Gaussian and Prewitt filters as follows:

Gaussian Filter

The Gaussian filter uses a 2D convolution operator which is suitable for blurring images and eliminating noise as show in Figure 6:

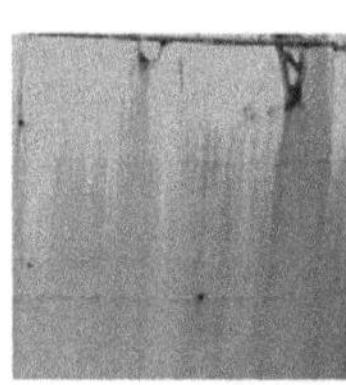

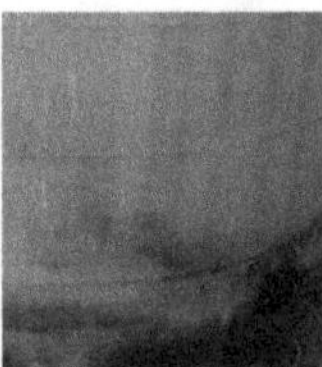

Figure 6. Image after applying Gaussian filtration.

Prewitt Filter

The Prewitt filter is utilized to detect two kinds of object edges, namely, vertical and horizontal as shown in Figure 7. The detection of edge is performed by calculation of the pixels gradient in the images.

Figure 7. Image after applying prewitt edge detection.

- **Morphological image processing**: Morphology operations help to extract useful features of the detected object such as shape, convex-hull skeletons, and boundaries. It depends on the division of images into small pieces, called a structuring element. The structuring element is a kind of array that defines the current processed pixels and their neighbors. It is a typically preferred method for choosing the element that has the same shape as the required, e.g., for finding lines, a function called "strel" can be used to extract it as shown in Figure 8.

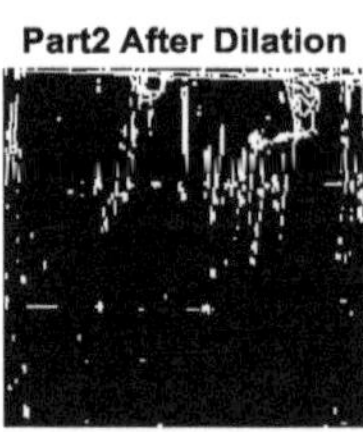
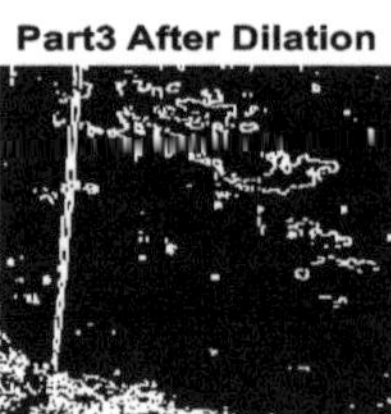
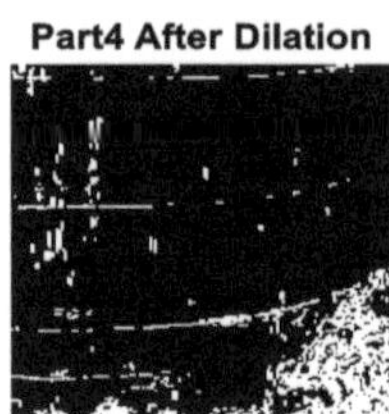

Figure 8. Image after applied dilation.

Dilation: The dilation operation enlarges the boundaries of the foreground pixels in grayscale images, by increasing their pixels size and reducing the holes size within such boundaries as shown in Figure 8.

- **Bwareaopen:** The instruction bwareaopen is a morphological operation that works based on removing all connected components with fewer pixels than a specified value of pixels from a binary or grayscale image, which results in another binary or grayscale image as shown in Figure 9.

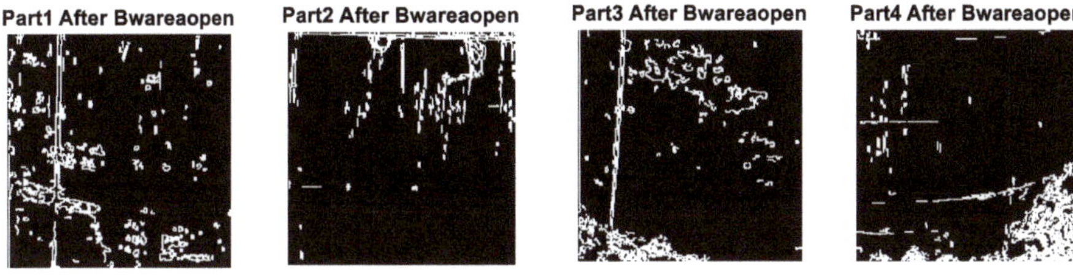

Figure 9. Image after applied bwareaopen.

- **Filling**: The imfill function performs the filling of an object to make it similar to the foreground in the binary images. As shown in Figure 10, imfill modifies those pixels in the connected background that have a value of zeros making them similar to the foreground pixels which have values of ones. The process will stop when the boundary of the object is reached.

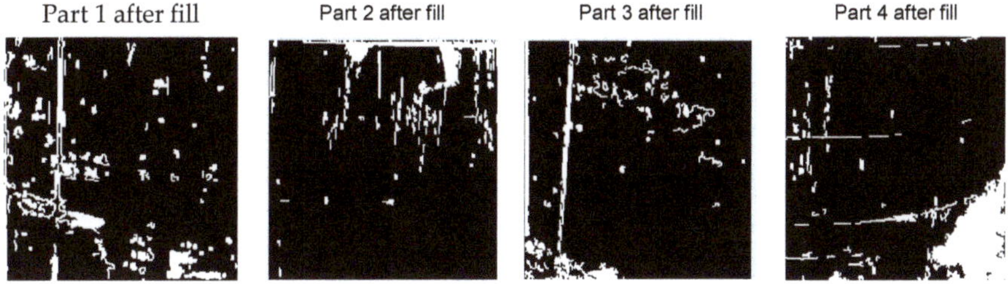

Figure 10. Image after applied filling process.

- **Inverting Image:** Inverting the image is to make the black pixel similar to the defect instead of the white pixel in order be more visible to the viewer as shown in Figure 11.

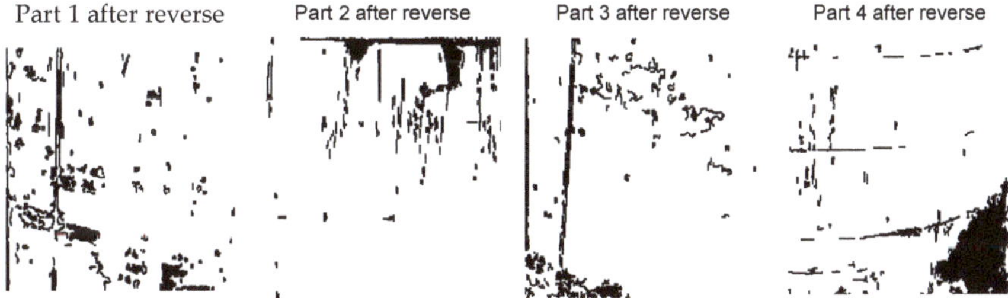

Figure 11. Image after applied reverse process.

2.3. Fuzzy Logic Based Classification

The fuzzy logic algorithm is one of the best AI algorithms that resembles human behavior in thinking and decision-making [14,15]. During the inspection of oil tanks there are three main sources of noise that appear on the surface of the tank: heterogeneity of illumination; the presence of objects, and the presence of sediments or dirt. These factors are resulted in three different levels of noise as follows:

The first level is low noise which is noise caused by the presence of objects on the surface of the tank, such as ladders, trees, or valves at the bottom of the tank, or other objects on the tank's surface. The second level is medium noise, which is that caused by small objects or shade from asymmetric lighting or small sediments. The third level is high noise, caused by dirt or large sediments.

In the presence of low noise, one can distinguish easily between the noise and defects, however, distinguishing between defects and high-level noise is more complicated. To overcome this problem, the fuzzy logic algorithm is implemented in two cascading stages to distinguish defects from low and medium noise to minimize its impact on the inspection process and help detect defects as shown in Figures 12–18.

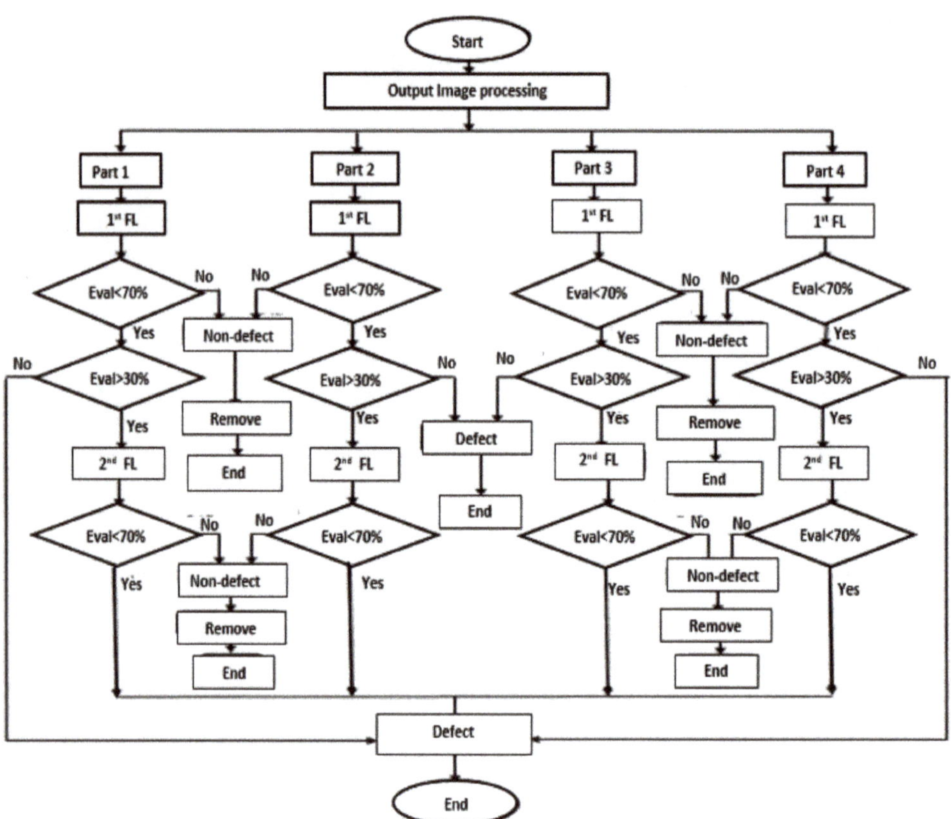

Figure 12. Flow chart diagrams for the cascading fuzzy logic process.

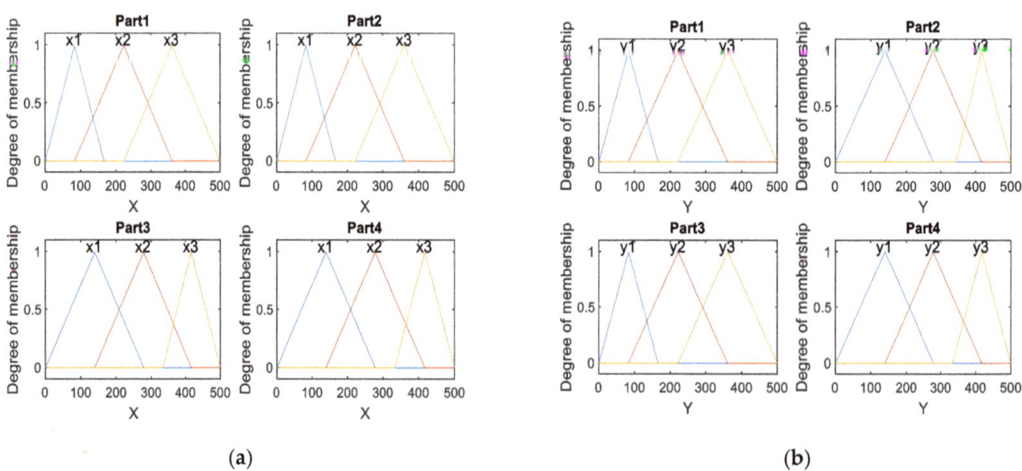

Figure 13. The input sets of the first stage fuzzy logic: (**a**) input sets(x); (**b**) input sets(y).

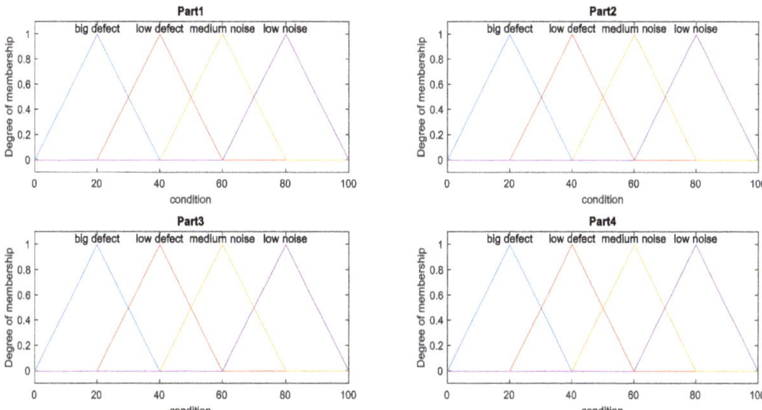

Figure 14. Output set conditions.

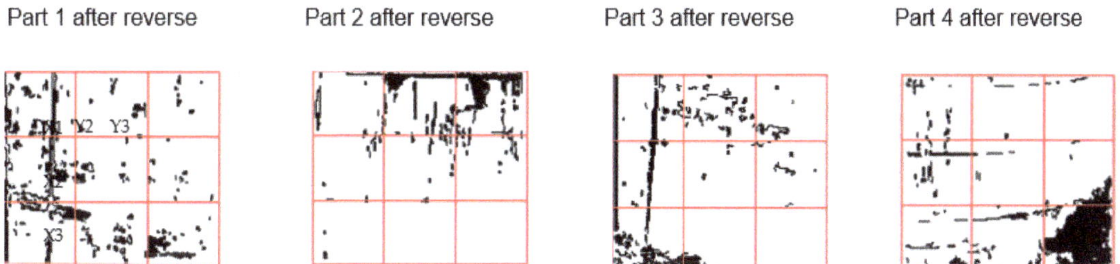

Figure 15. Regions of input sets in first stage fuzzy logic.

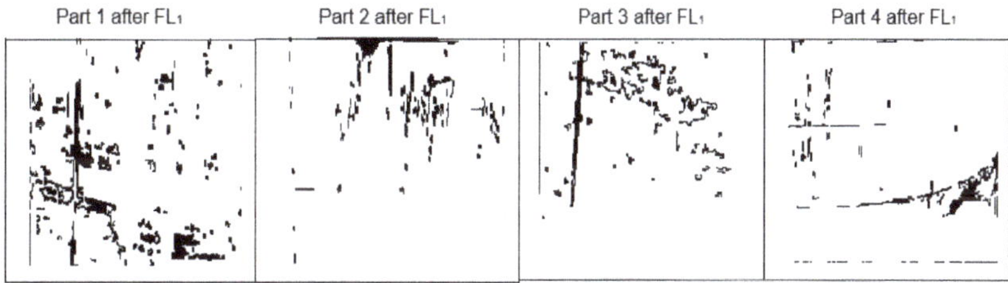

Figure 16. Image after applied FL1.

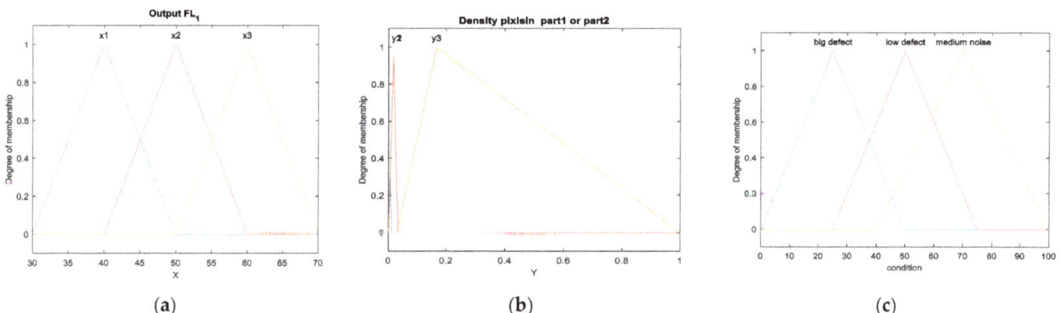

Figure 17. Input and output sets of second FL stage: (**a**) first input set (**b**) second input set (**c**) output set (condition).

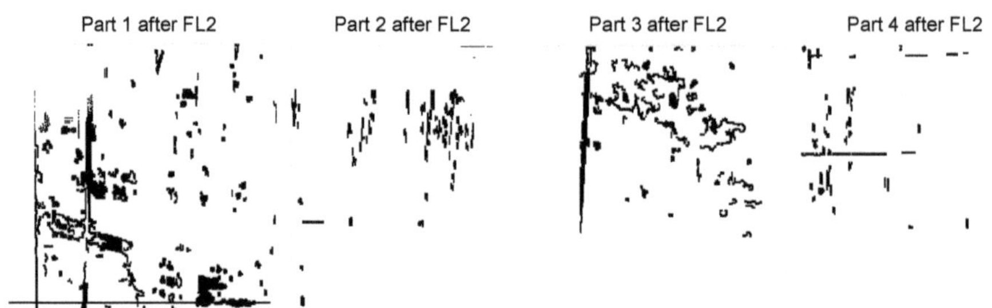

Figure 18. Image after applied FL2.

2.3.1. First Stage Fuzzy Logic Algorithm

The first stage fuzzy logic aims to distinguish between defects and low noise and remove such noise from the image. Low noise is located mostly at the bottom or the corners of images. This stage of fuzzy logic is applied to distinguish between defects and low noise according to the image's pixel location. After identifying the position of low noise that is affecting the inspection, the process of noise elimination will be applied.

The x and y coordinate systems are used as two input sets of the first stage in fuzzy logic inference system which are measured in pixels as shown in Figure 13. The image has been divided into four parts and each part divided into nine regions based on x and y locations as shown in Figure 15. Each region has a length and width that are described in x and y directions, respectively, as shown in Figure 13. The degree to which these regions are affected by noise varies from place to place. The side regions of the image are considered to be those most affected by noise, while regions in the middle of the image are less affected.

The first input set x utilizes three linguistic variables: x = {x1, x2, x3}, where x1, x2, and x3 have a total range of 500 pixels in the x coordinate system, as illustrated in Figure 13a. In a similar way, the y input set utilizes three linguistic variables: y = {y1, y2, y3}, where y1, y2, and y3 x3 have a total range of 500 pixels in the y coordinate system, as illustrated in Figure 13b. The membership functions of the input sets are shown in Figure 13. The shape of membership is chosen to be in triangular form after conducting some trials.

The classification of the detected defects into noise or defects is utilized as the output set of the first stage in the fuzzy logic inference system as shown in Figure 14. It is called "condition" and has four linguistic variables: Condition = {big defect, low defect, medium noise, low noise} with a range equal to 100%, starting from the big defect as the maximum range and ending with low noise as the minimum as shown in Figure 14. The four parts of the images have the same output membership function as shown in Figure 14.

In the output set of the first stage of fuzzy logic, the low noise can be considered if the condition set has a value above 70%, which will be eliminated. However, if the condition has a value less than 30%, the decision is to consider it as a defect.

We were unable to decide the condition values located in the range (30–70)% at this first stage, thus these values will be classified at the second fuzzy logic stage. The first stage fuzzy logic rules are formed for each part of the image separately based on x and y input sets' locations which are mapped to output set as shown as follows: Numbers in tables (1,2,3,4) indicate the output sets linguistic variables: (4) for low noise, (3) for medium noise, (2) for low defect and (1) for big defect.

1- Part first rule

The rules are illustrated in Table 1.

Table 1. Fuzzy rules in image_part 1 in first FL stage.

	Y3	Y2	Y1
X3	3	2	1
X2	3	2	2
X1	4	4	4

2- Part second rule
The rules are illustrated in Table 2.

Table 2. Fuzzy rules in image_part 2 in first FL stage.

	Y3	Y2	Y1
X3	4	3	3
X2	3	2	2
X1	4	2	1

3- Part third rule
The rules are illustrated in Table 3.

Table 3. Fuzzy rules in image_part 3 in first FL stage.

	Y3	Y2	Y1
X3	3	3	4
X2	2	2	3
X1	1	2	3

4- Part fourth rule
The rules are illustrated in Table 4.

Table 4. Fuzzy rules in image_part 4 in first FL stage.

	Y3	Y2	Y1
X3	1	2	3
X2	2	2	3
X1	4	4	4

The image after the implementation of the first stage fuzzy logic is shown in Figure 16.

2.3.2. Second Stage Fuzzy Logic Algorithm

In the second step of fuzzy logic inference system, the output crisp values of the first stage fuzzy logic which is located in the range (70–30)% are used as the input set of the second fuzzy logic. The regions surrounding the central region of the tank are the most difficult to distinguish between low defects and medium noise. The fuzzy logic second stage is intended to detect the intermediate noise produced by the shade, which can be found at the bottom, top, left, or right of the tank. This stage is applied to the four parts separately.

The input set of second stage fuzzy logic is the outputs of first stage fuzzy logic which have values between 30% and 70%. The four parts of the images have the same input membership function as shown in Figure 17. The second stage fuzzy logic in the classification process also depends on the pixel density in regions that the first stage fuzzy logic could not classify.

Table 5 shows a range of linguistic variables to the input sets X and Y.

Table 5. Range variables.

Range x1, y1	[1 166]
Range x2, y2	[166 334]
Range x3, y3	[334 500]

According to the concentration of black pixels in these regions, the decision was made to classify as follows:

The number of black pixels between 0 and 2000 is very small and can be considered as a defect. If the number of black pixels is between 1600 and 8000, we can consider it as medium noise caused by a shadow, because the area on which the shadow is located looks more homogeneous. Thus, when the edge detection filter is used with such shadow area, it will only show the edges of the shadow, which is larger and has higher number of black pixels in comparison with defect. If the number of black pixels is higher than 6000, this can be considered as a defect but impure. Through experience, the largest number of black pixels that can be considered as a defect is half of the inspected image area, e.g., if the image pixels area is 2MP, then the number of black pixels (the defect) will not exceed 1MP, half of the image size.

To calculate the percentage of pixel density, Equation (1) is used: The percentage of pixel density is

$$Pd\% = \frac{N_{tbp}}{H_{image}} \tag{1}$$

where N_{tbp} is the number of true black pixels in the inspected image and H_{image} is half the area of the inspected image.

The fuzzy logic second stage has two input sets; the first is the output of the first stage of fuzzy logic and is expressed by the variable X2, while the second input is the black pixel density, expressed by the variable Y2.

X2 is the first input set with three linguistic variables: X2 = {x12, x22, x32}, where x12, x22, and x32 are ranges of the output first stage fuzzy logic which are located between 30% and 70% as shown in Figure 17a with equal range in all parts image.

Y2 is the second input set with three linguistic variables: Y2 = {y12, y22, y32}, where y12, y22, and y32 are ranges of the black pixel density ranged between 0 -1 as shown in Figure 17b.

The output set of the fuzzy logic second stage classifies the defects on the object into medium noise-2 (non-defects), big defects-2, and low defects-2. It is called condition2 with three linguistic variables: Condition2 = {big defect_2, low defect_2, medium noise_2} with a range of 100%, which starts from the big defect as a small value in the range and ends with medium noise as a maximum of the range as shown in Figure 17c.

As the four parts of the images have the same input and output membership function in the second FL stage, only one part has been represented in Figure 17. Table 6 shows the relationship between the first input set and output that are used to build the rules of the classification process.

Table 6. Rules of second stage FL.

Input Sets of Second Stage (X)	Output	Less Than 2000 Pixels	(1600–8000) Pixel	More Than 6000
x1 (30_50)%		Big defect 2	Low defect_2	Big defect_2
x2 (40_60)%		Big defect 2	Medium noise_2	Big defect_2
x3 (50_70)%		Low defect_2	Medium noise_2	Big defect_2

The rules of the fuzzy logic second stage for first, second, third, and fourth parts are formed as shown in Table 7:

Table 7. Fuzzy rules in image_parts 1,2,3, and 4 in second FL stage.

	Y12	**Y22**	**Y32**
X12	1	2	1
X22	1	3	1
X32	2	3	3

Where 1 is "big defect" in output sets; 2 is "low defects" in output sets; and 3 is "medium noise" in output sets.

The second stage of the fuzzy logic in the classification process depends on two important factors: the first is the location of the pixels that have met the 30–70% condition of the output set in the first stage; the second factor is the density of the pixels in the output of the first fuzzy logic stage. The second stage of fuzzy logic is implemented to distinguish the medium noise from defects, and then will be eliminated from the output set images once they have a value greater than 70%.

Output values which are less than 70% will be classified as big defect and low defect as shown in Figure 17. Thus, there is a need for thresholding or anther fuzzy logic stage. Figure 18 shows the output of fuzzy logic second stage.

2.4. Thresholding Process

After the second stage of the fuzzy logic is implemented, the four parts of the image are collated back into one image to prepare them for a new stage of processing. The collated image is input to the third stage of processing (threshold process) as shown in Figure 19.

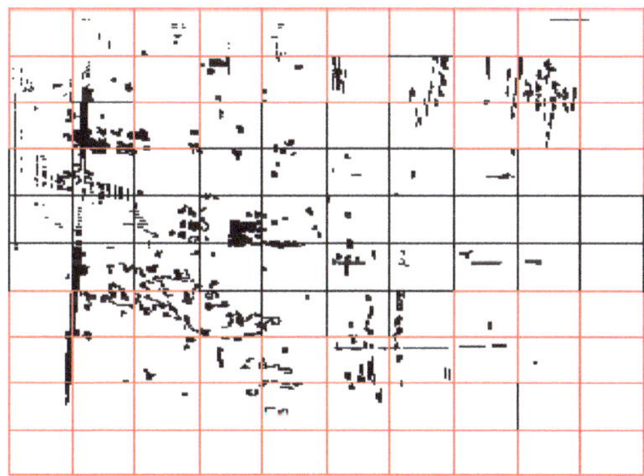

Figure 19. Input image to the third stage of processing (threshold process).

The image is then divided into 100 equidimensional cells as shown in Table 8. The cells where the high noise is concentrated are located above and below the image as shown in Figure 19. The group of cells located above and below the image is shown in Table 8.

Through the experimental values, the threshold value in the cells located was estimated at 70% of total black pixels that resulted from the four parts of the image from the second stage FL (3000–3500 pixels in this case). The flow chart of threshold processing is illustrated in Figure 20.

Table 8. The group of cells being processed in the threshold process.

1	2	3	4	5	6	7	8	9	10
11	12	13	14	15	16	17	18	19	20
21	22	23					28	29	30
61	62	63					68	69	70
71	72	73	74	75	76	77	78	79	80
81	82	83	84	85	86	87	88	89	90
91	92	93	94	95	96	97	98	99	100

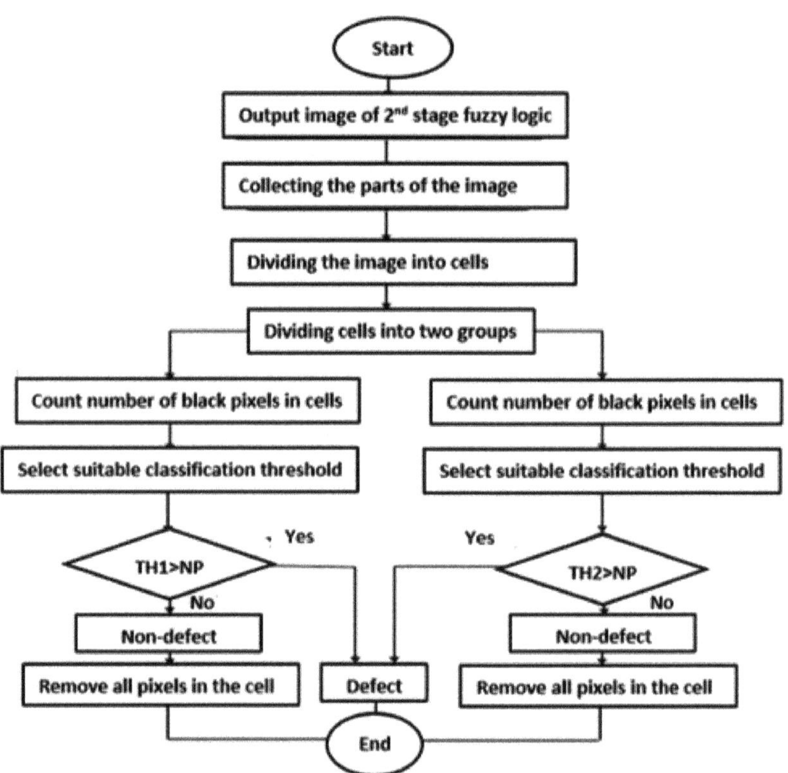

Figure 20. Flow chart diagrams for the third stage of processing (threshold process).

This stage in the classification process depends on the pixel density within the cells. All cells where the number of black pixels is greater than the threshold value will be deleted. Figure 21 shows the final image after the thresholding process. Figure 22 shows the original image and the stages it went through during processing.

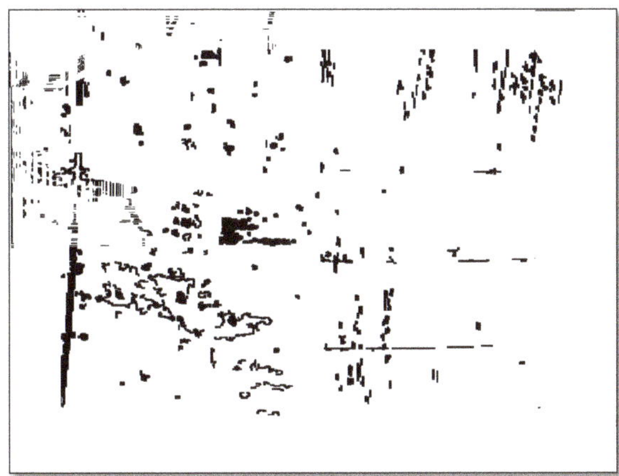

Figure 21. Final image after thresholding process.

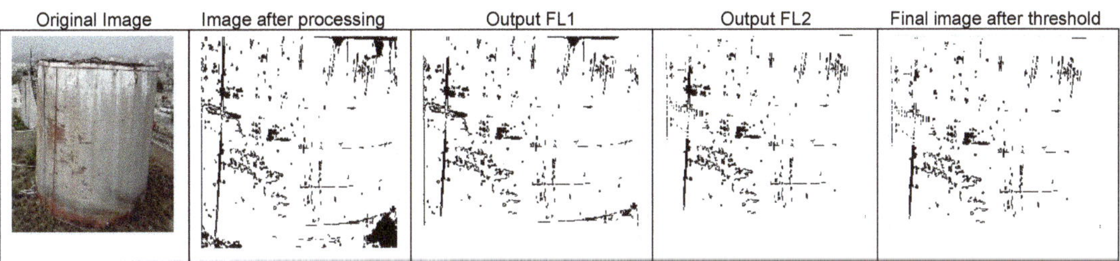

Figure 22. Results of all processing stages.

3. Experimental Results and Discussion

To show evidence of the proposed inspection system, many experiments (as depicted in Figure 23) were conducted on real oil tanks with various parameters and conditions to evaluate the proposed algorithm.

Figure 23. Samples of oil tanks used for experiments.

3.1. Experimental Results

To check the performance of the inspection system, it must be tested on a wide range of samples with different parameters with several tests run under different conditions.

Hence, several tests for the inspection system were conducted during different hours of the day as well as testing the system on tanks of different colors, shapes, and sizes. The worst inspection cases dealt with those parameters that show the limitations of the proposed algorithms.

Although several experiments were performed before making a decision, only the significant and important cases will be discussed. The samples presented in this sub-section have the following features: One contains pure defects and low noise, which is one of the easiest inspection cases that the system can detect. Another contains all low, medium, and high noise levels in different regions with different levels of defects, which is one of the worst cases of inspection due to it being difficult for the system to detect, which made us focus on it more during the inspection.

As shown in Figure 24, the results indicate that there is no noise in the middle region of the tank in all images, while some defects may exist without any noise. As shown in Figure 24b,d,e,f, the lower regions are the lowest noise regions, and in Figure 24a,c the lower areas and the tank corners are the lowest noise areas. The three noise levels may not combine into one sample in real systems.

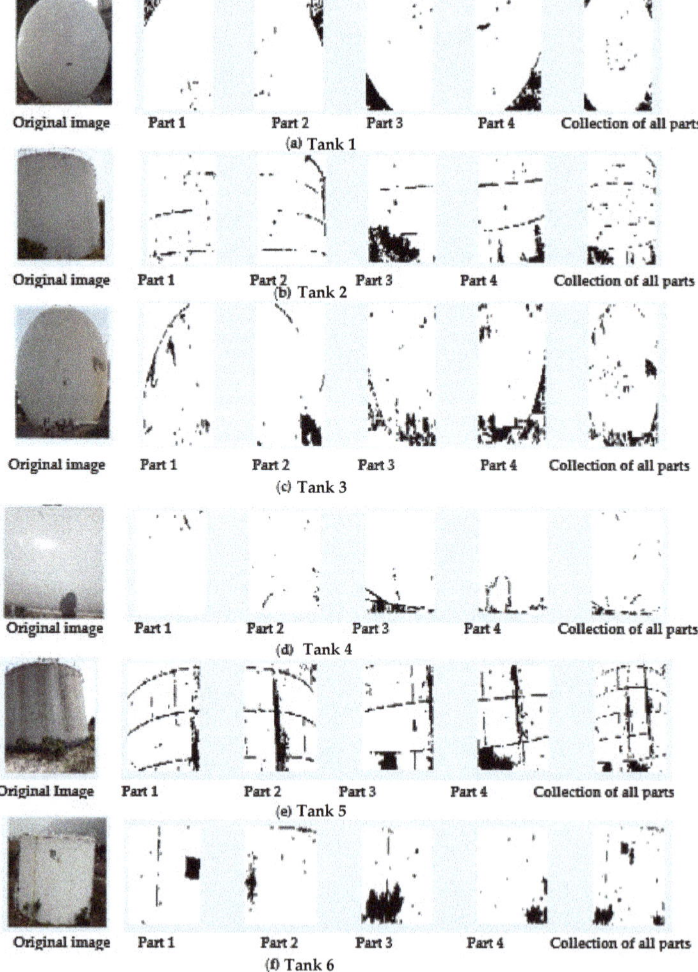

Figure 24. Samples after image processing process.

The algorithm of image processing as described in Section 2.2 was implemented to detect the surface defects of the six types of oil tanks in Figure 23. The image processing results for all samples are shown in Figure 24.

As mentioned above, the image processing algorithm cannot eliminate all noise from an image and thus is unable to make a definitive decision if the tank has defects or noise. Noises affecting the classification process have been divided into three different levels, namely, low, medium, and high noise. These noises cannot be eliminated completely by the image processing algorithm due to several factors affecting the processing, such as heterogeneity of illumination on the surface of the tank, the presence of objects appearing on the surface of the tank, and the presence of sediments or dirt on the surface of the tank.

It is clear that the peripheral, lower regions, and corners of the tanks are strongly affected by noise as shown in Figure 23. In Figure 24, the results indicate that there is no noise in the middle region of the tank in all images, while some defects may exist without any noise affecting it. As shown in Figure 24, the lower regions are the noisiest regions (low noise), while the tank corners are the lowest noise regions. One cannot distinguish between defects and the three levels of noise until the algorithms of cascading fuzzy logic and the thresholding have been implemented on the image.

The fuzzy logic inference system as designed in Section 2.3 is applied to remove the noise effects caused by the appearance of the above-mentioned factors on the surface of the tank. The fuzzy logic first stage aims to remove the low noise as shown in Figure 25. All the outputs of the first stage of fuzzy logic have values between 30 and 70% which means that values below 30% can be classified as pure defects, while those above 70% can be classified as low noise. All the samples shown in Figure 23 will have the same output values that are between 30 and 70% during the implementation of the first stage of the fuzzy logic as shown in Figure 25. In the first sample as shown in Figure 25a, the first part has output values confined between 40 and 80%, which indicates the presence of low noise and the absence of pure defects, whereas the second, third, and fourth parts have the same output values confined between 20 and 80%, indicating the presence of pure defects and low noise.

Similarly, in the second sample as shown in Figure 25b, the first part has output values confined between 40 and 65% which indicates that there is no low noise and pure defects, while the second, third, and fourth parts have the same output values confined between 20 and 80%, and this indicates the presence of pure defects and low noise.

In the third sample as shown in Figure 25c, the first part has output values confined between 40 and 78%, (indicates the presence of low noise with no pure defects); the second part has output values confined between 20 and 60% (indicates the presence of pure defects with no low noise), while the third and fourth parts have the same output values, ranging from 20 to 80% (indicates the presence of pure defects and low noise).

In the fourth sample as shown in Figure 25d, the first part has output values confined between 40 and 67% (indicates the absence of any low noise and pure defects), the second part has output values confined between 22 and 60% (indicates the presence of pure defects with no low noise), the third and fourth parts have the same output values, which range between 20 and 80% (indicates the presence of pure defects and low noise).

In the fifth sample as shown in Figure 25e, the first part has output values confined between 40 and 80% (indicates the presence of low noise with no pure defects), however, the second, third, and fourth parts have confined output values between 20 and 80% (indicates the presence of pure defects and low noise).

In the sixth sample shown in Figure 25f, the first part has output values confined between 40 and 78% (indicates the presence of low noise with no pure defects), and the second, third, and fourth parts have confined output values between 20 and 80% (indicates the presence of pure defects and low noise).

As shown in Figure 26, the results indicate that there is some medium noise which exists between the output values of the first stage of the fuzzy logic located between 30 and 70%. The regions around the central region of the tank are the most difficult to differentiate between defects and medium noise.

The second stage of the fuzzy logic inference system as designed in Section 2.3.2, was applied to remove the medium noise effects caused by small dirt, heterogeneity in lighting, and the appearance of small objects on the surface of the tank. The fuzzy logic second stage aims to eliminate medium-scale noise as illustrated in Figure 27. All the outputs of the fuzzy logic second stage have values between 0 and 70% and this means that values less than 70% can be classified as defects, while values above 70% can be classified as medium noise.

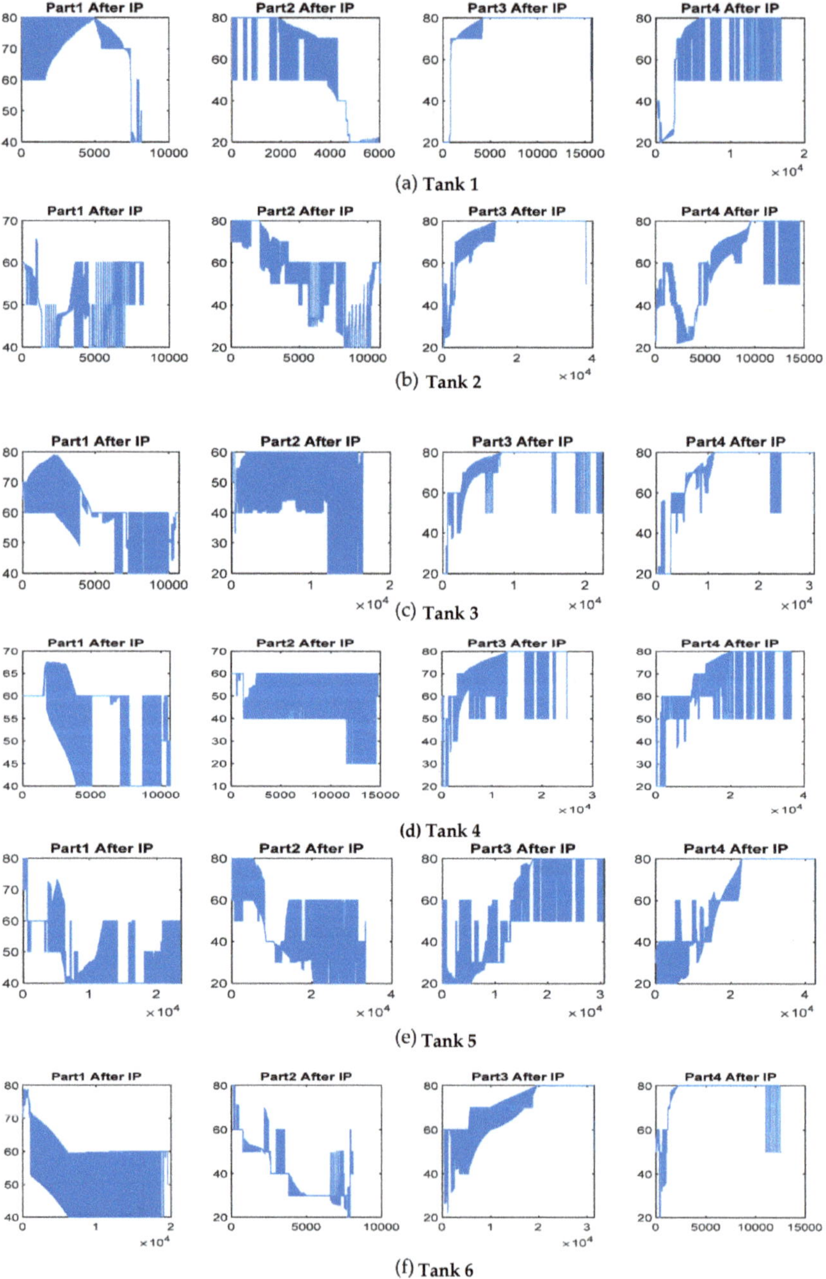

Figure 25. Evaluation of the first stage of fuzzy logic before eliminating low noise.

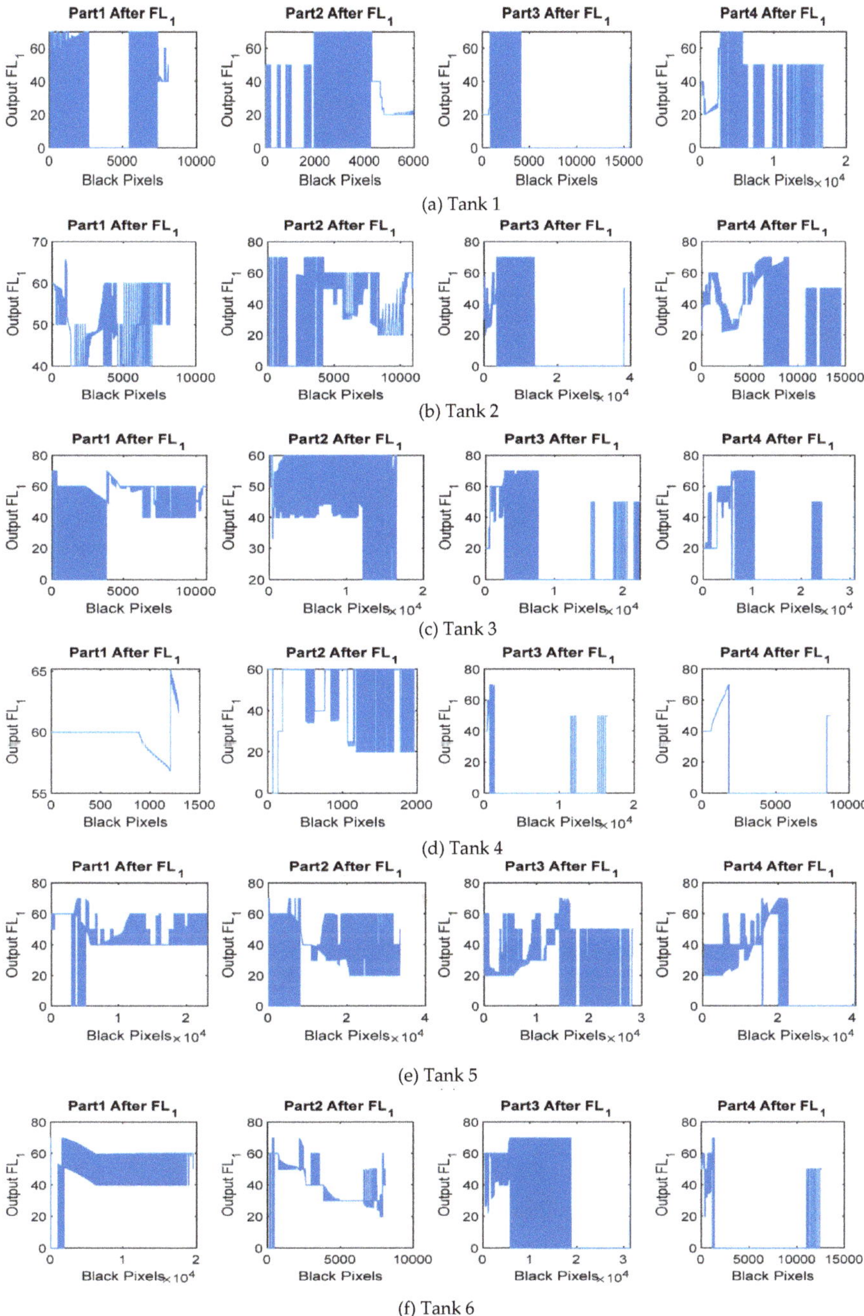

Figure 26. Evaluation of the first stage of fuzzy logic after eliminating low noise.

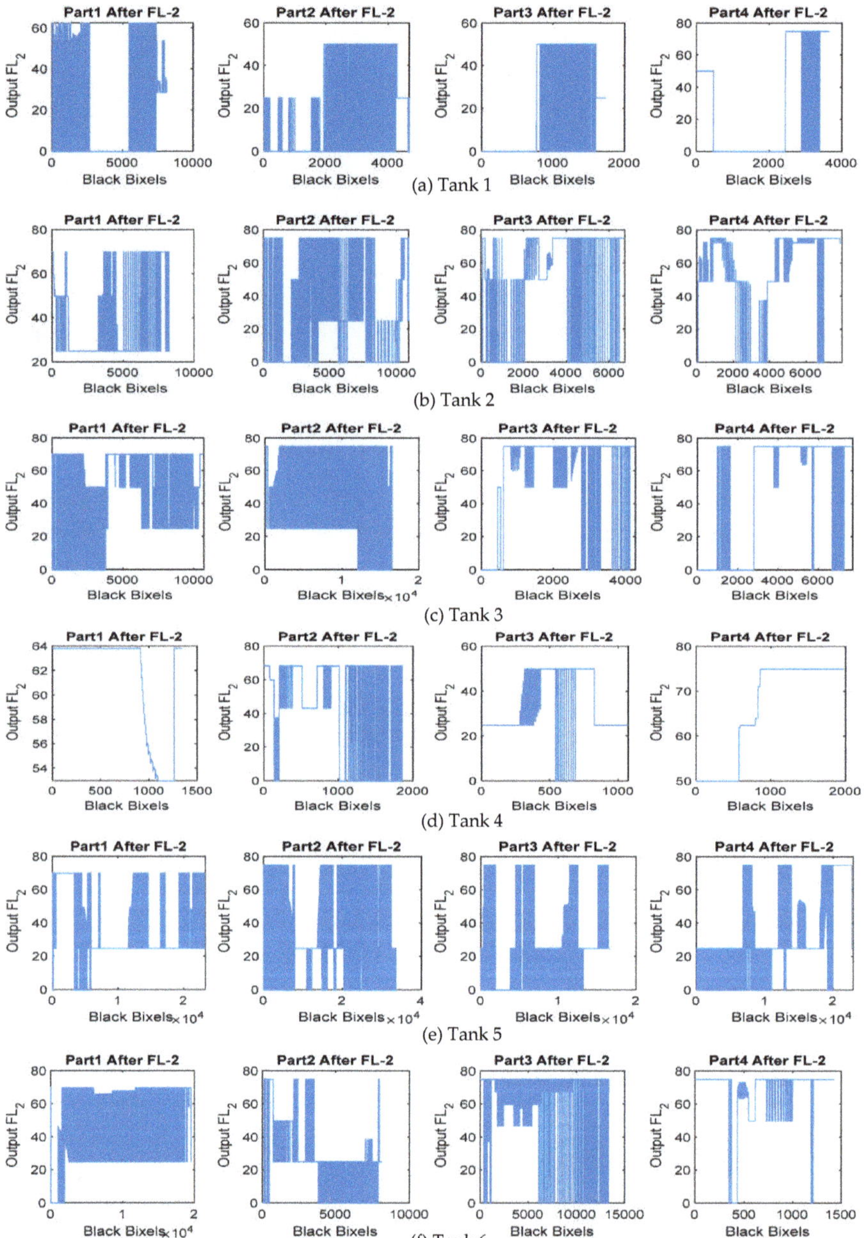

Figure 27. Evaluation of the second stage of fuzzy logic before eliminating low noise.

All samples, as shown in Figure 23, have the same output values between 0 and 70% after the second stage of fuzzy logic was executed. In the first sample as shown in Figure 27a, only the fourth part contains output values greater than 70%, which indicates the presence of medium noise. In the second and third samples, as shown in Figure 27b,c, the second, third, and fourth parts have output values greater than 70%, which indicates the presence of medium noise. In the fourth sample, as shown in Figure 27d, only the fourth part contains output values greater than 70%, indicating the presence of medium noise. In the fifth and sixth sample, as shown in Figure 27e,f, the second, third, and fourth parts have output values greater than 70% (indicates the presence of medium noise).

Figure 23 shows the presence of high noise in some images. The results indicate the presence of some high noise caused by large sediment or dirt on the surface of the tank as shown in Figure 28. The third stage of the threshold algorithm aims to remove the high noise that can be located above and below the tank as shown in Section 2.4. As shown in Figure 28, there are numbers of cells that show black pixel density.

(a) Tank 1

(b) Tank 2

(c) Tank 3

(d) Tank 4

Figure 28. *Cont.*

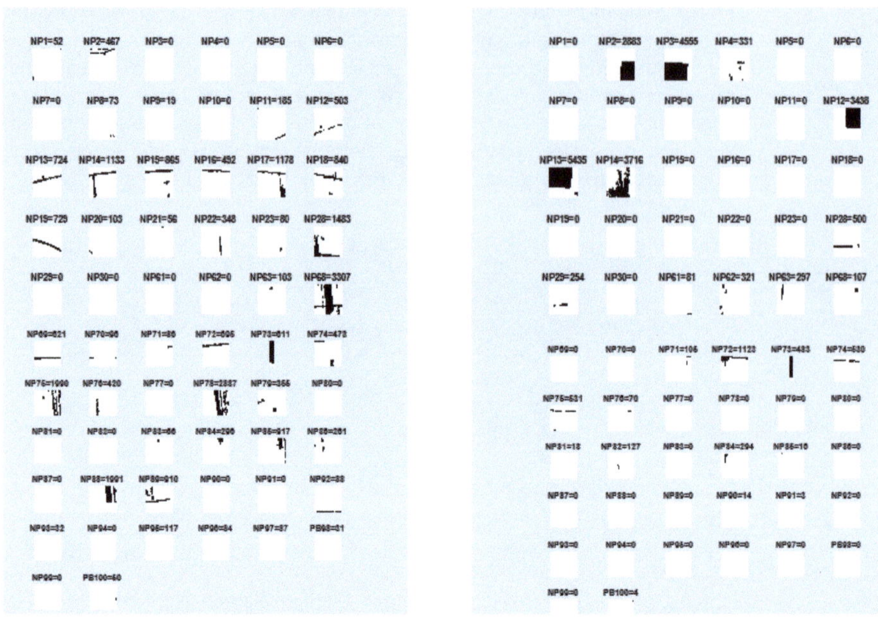

(e) Tank 4   (f) Tank 6

Figure 28. Pixel density in cells processed using the thresholding process.

The threshold value in the cells located above the tank is estimated at 3500 black pixels, while those located below the tank are estimated at 3000 black pixels. Cells in which the black pixel exceeds the threshold value are classified as high noise, while others are classified as defects.

In the first, second, third, and fourth samples, as shown in Figure 28a–d, there are no cells where the number of black pixels exceeds the threshold value, indicating the absence of high noise. In the fifth sample, as shown in Figure 28e, cell 68 is the only one where the number of black pixels exceeds the threshold value, and this indicates the presence of high noise in this cell. In the sixth sample, as shown in Figure 28f, there are several cells (2,3,12,13,14) where the number of black pixels exceeds the threshold value, which indicates the presence of high noise.

3.2. Results Discussion and Evaluation

To measure the reliability and accuracy of the proposed UAV-based inspection process, the experiments for all the above cases were repeated 30 times for each tank with a total of 180 trials. This test shows the limits of the visual inspection system capability to detect defects and distinguish them from noise. The final classification process was performed by visually viewing the samples and classifying them into defective pure or defective impure samples.

Pure defects are those that are free of any noise while impure defects still contain some noise. The current study of the samples shown in Figure 23 proves that the first, second, third, fourth, and sixth samples contain pure defects and only the fifth sample contains impure defects. Figure 29 shows the classification results after executing the three stages of processing for each tank with 30 trials, in which the proposed algorithms gave an average of the right decisions equal to 83.33%, 86.66%, 80%, 86.66%, 76.66%, and 86.66% for trials in tanks 1, 2, 3, 4, 5, and 6, respectively, as shown in Figure 29. Thus, a successful classification accuracy among all trials in the six tanks is around 83.33%.

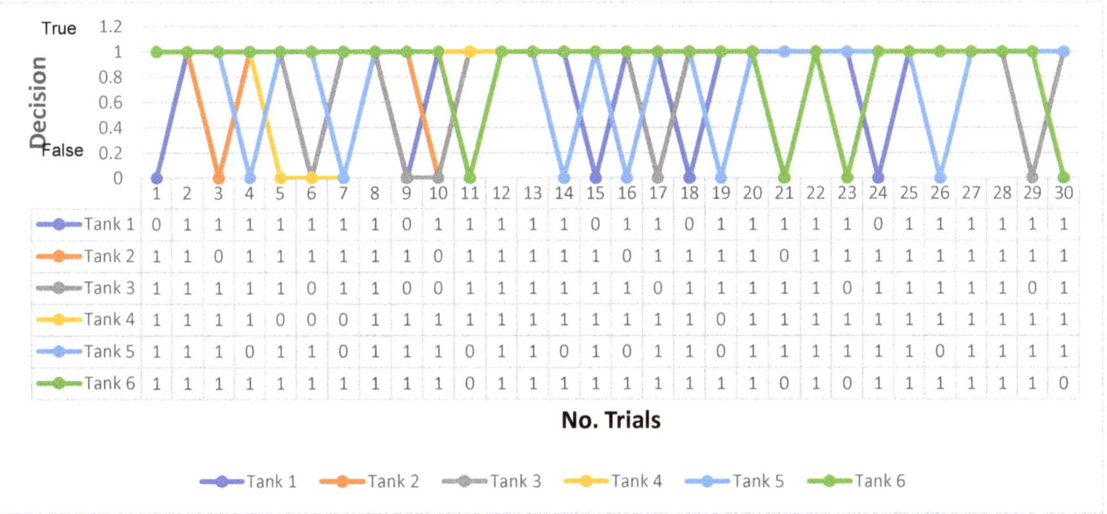

	1	2	3	4	5	6	7	8	9	10	11	12	13	14	15	16	17	18	19	20	21	22	23	24	25	26	27	28	29	30
Tank 1	0	1	1	1	1	1	1	1	0	1	1	1	1	1	0	1	1	0	1	1	1	1	1	0	1	1	1	1	1	1
Tank 2	1	1	0	1	1	1	1	1	1	0	1	1	1	1	1	0	1	1	1	1	0	1	1	1	1	1	1	1	1	1
Tank 3	1	1	1	1	1	0	1	1	0	0	1	1	1	1	1	1	0	1	1	1	1	1	0	1	1	1	1	1	0	1
Tank 4	1	1	1	1	0	0	0	1	1	1	1	1	1	1	1	1	1	1	0	1	1	1	1	1	1	1	1	1	1	1
Tank 5	1	1	1	0	1	1	0	1	1	1	0	1	1	0	1	0	1	1	0	1	1	1	1	1	1	0	1	1	1	1
Tank 6	1	1	1	1	1	1	1	1	1	1	0	1	1	1	1	1	1	1	1	0	1	0	1	1	1	1	1	1	1	0

Figure 29. The results of classifying samples of tanks 1–6 (1 indicates right decision on defect detection and 0 indicates false decision on defect detection).

An average of 26.77% of the trials were associated with big noise that was wrongly classified, even though most of the noise was removed. The reasons behind the incorrect classification results are as follows: Some of the noise from the first stage of processing still remained in the second stage and was wrongly classified as less than 30% or higher than 70%, so such noise was not entered into second stage.

There are some failures that appear as a result of the uncontrolled and random cropping process. In order to overcome this dilemma, the input image must be cropped with a strong focus, making it more proportional and compatible with the specified restrictions, so it is necessary to crop an exact image during the execution of this process.

The fuzzy logic inference system as designed in Section 2.2 was applied to remove the noise effects. The fuzzy logic first stage was aimed at eliminating the effects of the low-scale noise as depicted in Figure 25.

4. Conclusions

This study has contributed to research on automatic inspection and defects detection of oil and gas tanks. This method includes the use of a drone capable of moving in all directions to ensure safe movement between tanks. It also includes a high-resolution imaging camera equipped with Wi-Fi technology, which is fixed in the front of the UAV and can be rotated by means of a control device to make it perpendicular to the tank so that the captured image is accurate. An image processing algorithm was developed with appropriate filters to extract the features of the inspected objects such as cracks, defects, and edges of objects on the samples, but it was still affected by several levels of noise. Three levels of noise were eliminated by using three stages of processing, two stages using the cascading fuzzy logic algorithm, and the third stage using the thresholding algorithm. The cascade fuzzy logic algorithm was implemented in two stages to distinguish between the low and medium noise from the defects. The first and second stages were able to eliminate the low and medium noise levels, respectively, while the third stage eliminates the high noise level. Low noise was calculated from output of the fuzzy logic first stage that had high condition values (first stage output set) of 70%, while output values with low condition values less than 30% were classified as pure defects. Medium noise was calculated from the output of the second stage fuzzy logic that had a condition (second stage output set) value greater than 70%. High noise was determined from the threshold stage output, where cells in the upper part of the image were classified as high noise if their density value was greater than 70% of the total black pixels, while cells in the lower part of the image were classified as high noise if their density value are more than 75% black pixels. Then the samples were categorized based on the third stage output of the thresholding process into defective or non-defective samples.

The results illustrate that the proposed inspection system is able to detect the defects with several types of oil tanks. The system was tested on 20 samples and the results showed its superiority in the inspection and detection of defects with an accuracy of 80%.

Author Contributions: Conceptualization, M.A.H.A., N.A.-D.N.A., A.A.H., M.H.G.A., T.S.M. and A.N.A.; Data curation, M.A.H.A.; Formal analysis, M.B. and M.A.H.A.; Funding acquisition, M.A.H.A., R.A. and Y.N.; Investigation, M.A.H.A.; Methodology, M.A.H.A., N.A.-D.N.A., A.A.H., M.H.G.A., T.S.M. and A.N.A.; Project administration, M.A.H.A.; Resources, M.A.H.A.; Software, M.A.H.A., N.A.-D.N.A., A.A.H., M.H.G.A., T.S.M. and A.N.A.; Supervision, M.B. and M.A.H.A.; Validation, M.A.H.A., N.A.-D.N.A., A.A.H., M.H.G.A., T.S.M. and A.N.A.; Visualization, M.A.H.A.; Writing—original draft, M.B., M.A.H.A., N.A.-D.N.A., A.A.H., M.H.G.A., T.S.M., A.N.A., R.A. and Y.N.; Writing—review and editing, M.A.H.A., J.R. and S.T. All authors have read and agreed to the published version of the manuscript.

Funding: This works is supported by Universiti Malaya (UM) and Ministry of Transportation (MOT) through Private-Research University grants, PV045-22 and GPF020A-2023.

Data Availability Statement: Data sharing not applicable.

Conflicts of Interest: The authors declare no conflict of interest.

References

1. Roberge, P.R. *Handbook of Corrosion Engineering*; McGraw-Hill: New York, NY, USA, 2000.
2. Alum, M.; Eze, T. The New Faces of Corrosion and Damage Detection in Oil and Gas Facilities: A Brief of What has Worked So Far and How it Can Work for You. In Proceedings of the SPE Nigeria Annual International Conference and Exhibition, Virtual, 11–13 August 2020.
3. Uhlig, H.H. The cost of corrosion in the United States. *Chem. Engng. News* **1946**, *27*, 2764. [CrossRef]
4. Benjamin, A.; Cunha, D.; Campello, G.C.; Roveri; Silva, R.; Guerreiro, J.N.C. Fatigue Life Assessment of a Drilling Riser Containing Corrosion Pits. Proceedings of Offshore Technology Conference (OTC), NACE International Oil and Gas Production, Houston, TX, USA, 5–8 May 2008.
5. Champion Technologies. *Corrosion Mitigation for Complex Environments*; Champion Technologies: Houston, TX, USA, 2012.
6. Tuttle, R.N. Corrosion in oil and gas production. *J. Pet. Technol.* **1987**, *39*, 756–762. [CrossRef]
7. Felsch, T.; Strauss, G.; Perez, C.; Rego, J.M.; Maurtua, I.; Susperregi, L.; Rodríguez, J.R. Robotized Inspection of Vertical Structures of a Solar Power Plant Using NDT Techniques. *Robotics* **2015**, *4*, 103–119. [CrossRef]
8. Berns, K.; Hillenbrand, C.; Luksch, T. Climbing robots for commercial applications–a survey. In Proceedings of the 6th International Conference on Climbing and Walking Robots CLAWAR, London, UK, 17 September 2003; pp. 17–19.
9. Moghaddam, A.F.; Lange, M.; Mirmotahari, O.; Hovin, M. Novel mobile climbing robot agent for offshore platforms, World Academy of Science. *Eng. Technol.* **2012**, *68*, 1353–1359.
10. Kim, S.; Spenko, M.; Trujillo, S.; Heyneman, B.; Santos, D.; Cutkosky, M.R. Smooth vertical surface climbing with directional adhesion. *IEEE Trans. Robot.* **2008**, *24*, 65–74.
11. Cleavinger, K.W.J. Flare system inspections for OLEFINS facilities. In Proceedings of the AIChE 2012 Spring National Meeting, Houston, TX, USA, 2–5 April 2012.
12. Cohen, M. 7 Facts That Make the Oil and Gas Asset Inspections Risky and Costly. 17 May 2017. Available online: https://info.qii.ai/blog/7-facts-that-make-the-oil-and-gas-inspections-risky-and-costly (accessed on 11 February 2023).
13. Shukla, A.; Karki, H. Application of robotics in onshore oil and gas industry—A review Part I. *Robot. Auton. Syst.* **2016**, *75*, 490–507. [CrossRef]
14. Ali, M.A.H.; Lun, A.K. A cascading fuzzy logic with image processing algorithm–based defect detection for automatic visual inspection of industrial cylindrical object's surface. *Int. J. Adv. Manuf. Technol.* **2019**, *102*, 81–94. [CrossRef]
15. Ali, M.A.H.; Alshameri, M.A. An intelligent adjustable spanner for automated engagement with multi-diameter bolts/nuts during tightening/loosening process using vision system and fuzzy logic. *Int. J. Adv. Manuf. Technol.* **2019**, *101*, 2795–2813. [CrossRef]

drones

Review

An Overview of Drone Applications in the Construction Industry

Hee-Wook Choi, Hyung-Jin Kim, Sung-Keun Kim and Wongi S. Na *

Department of Civil Engineering, Seoul National University of Science and Technology,
Seoul 01811, Republic of Korea; cheesew01@seoultech.ac.kr (H.-W.C.); hyungjin4843@naver.com (H.-J.K.);
cem@seoultech.ac.kr (S.-K.K.)
* Correspondence: wongi@seoultech.ac.kr

Abstract: The integration of drones in the construction industry has ushered in a new era of efficiency, accuracy, and safety throughout the various phases of construction projects. This paper presents a comprehensive overview of the applications of drones in the construction industry, focusing on their utilization in the design, construction, and maintenance phases. The differences between the three different types of drones are discussed at the beginning of the paper where the overview of the drone applications in construction industry is then described. Overall, the integration of drones in the construction industry has yielded transformative advancements across all phases of construction projects. As technology continues to advance, drones are expected to play an increasingly critical role in shaping the future of the construction industry.

Keywords: drone application; unmanned aerial vehicle; smart construction; aerial inspections; structure maintenance

Citation: Choi, H.-W.; Kim, H.-J.; Kim, S.-K.; Na, W.S. An Overview of Drone Applications in the Construction Industry. *Drones* **2023**, *7*, 515. https://doi.org/10.3390/drones7080515

Academic Editor: Diego González-Aguilera

Received: 4 July 2023
Revised: 1 August 2023
Accepted: 2 August 2023
Published: 3 August 2023

1. Introduction

In recent years, the construction industry has witnessed a remarkable transformation fueled by technological advancements. Among these innovations, drones have emerged as game-changers, redefining the way construction projects are planned, executed, and maintained. Equipped with sophisticated sensors, cameras, and GPS (global positioning system) technology, drones offer unparalleled capabilities to capture real-time data, generate accurate 3D models, and conduct remote inspections. This review paper aims to provide a comprehensive overview of the applications of drones in construction, shedding light on their impact across different project phases and highlighting the potential benefits they bring to the table.

Drones have rapidly evolved from being mere novelties to indispensable tools in the construction sector. By utilizing different types of drones, construction professionals can optimize their workflow, improve project coordination, and mitigate risks [1,2]. Surveying drones, equipped with high-resolution cameras and LiDAR (light detection and ranging) sensors facilitate precise mapping, topographical analysis, and site planning [3–5]. These drones capture detailed aerial imagery and generate comprehensive 3D models, enabling architects and engineers to make informed decisions about building placement, design optimization, and resource utilization [6,7].

Inspection drones, on the other hand, provide an unprecedented advantage in assessing hard-to-reach or hazardous areas of construction sites. Equipped with thermal cameras, high-resolution imaging systems, and even artificial intelligence, these drones enable efficient and accurate inspections of infrastructure, buildings, and equipment [8–10]. By swiftly identifying structural defects, monitoring construction quality, and ensuring compliance with safety regulations, inspection drones contribute to enhanced project transparency, reduced manual labor requirements, and improved overall project outcomes.

Beyond the core phases of design and construction, drones continue to revolutionize the maintenance stage. Regular inspections using drones enable proactive maintenance planning, identifying potential issues early on, and preventing costly repairs. By conducting detailed assessments of buildings, bridges, and infrastructure, drones contribute to the longevity and resilience of constructed assets. Additionally, the integration of drones in the maintenance phase allows for the swift identification and resolution of defects, leading to improved safety, and reduced downtime [8,11].

The utilization of drones in the construction industry represents a transformative leap towards achieving higher levels of efficiency, safety, and sustainability. By harnessing their data acquisition, monitoring, and inspection capabilities, construction professionals can make informed decisions, improve project outcomes, and optimize resource utilization. As drone technology continues to evolve, it is expected to play an increasingly pivotal role in reshaping the construction industry, fostering innovation, and driving the adoption of smart, resilient construction practices.

In this study, drone technology is reviewed by dividing the actual construction process into three different phases. The architecture of the paper is organized as follows. Section 2 describes the three different types of drones used in the construction industry with the advantages and disadvantages of using each of the drone types. Section 3 discusses the drones used at the designing phase, Section 4 reviews how drones are used at the construction phase and Section 5 overviews the maintenance phase, summarizing how drones are applied to ensure effective maintenance. Then, the study concludes with sections including challenges and opportunities and future directions.

2. Drone Types for Application in Construction Industry

There are various type of drones used in the construction industry for various purposes. In reference [6], the study conducted surveys from many construction companies to find out how drones were used for their construction projects and it showed that the most popular use of drones was in capturing progress photos, followed by taking promotional videos, conducting inspections, and enhancing site management (Figure 1).

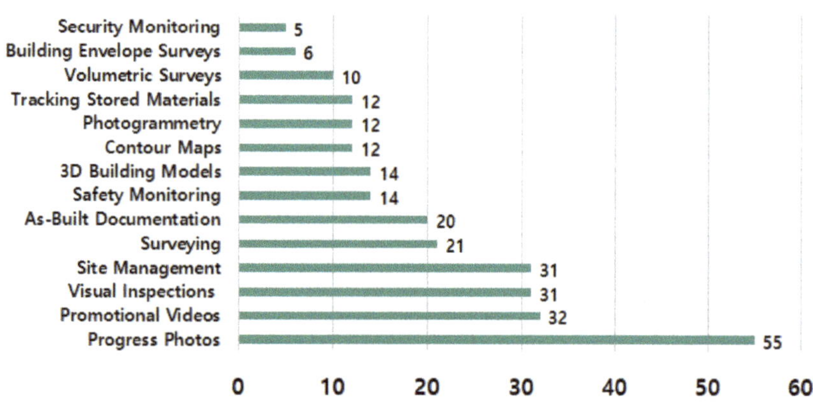

Figure 1. Drone applications result from survey companies [6].

As shown in Figure 2, fixed-wing drones, rotary-wing drones, and hybrid drones are three types of drones or unmanned aerial vehicles (UAVs) that are commonly used for various applications, including in the construction and maintenance industries. Each type of drone has its unique advantages and disadvantages, and choosing the right drone for a specific application depends on several factors, such as the size of the area to be covered, the required payload capacity, and the environmental conditions in which the drone will be operated (Table 1). Fixed-wing drones are ideal for covering large areas quickly [12], while rotary-wing drones are more suitable for close-range inspections and operations in confined spaces [13–15]. Hybrid drones offer a more versatile and adaptable solution, but may be more complex and expensive than fixed-wing or rotary-wing drones [16]. Ultimately, the right drone for a specific application should be chosen based on a careful analysis of the requirements of the task at hand.

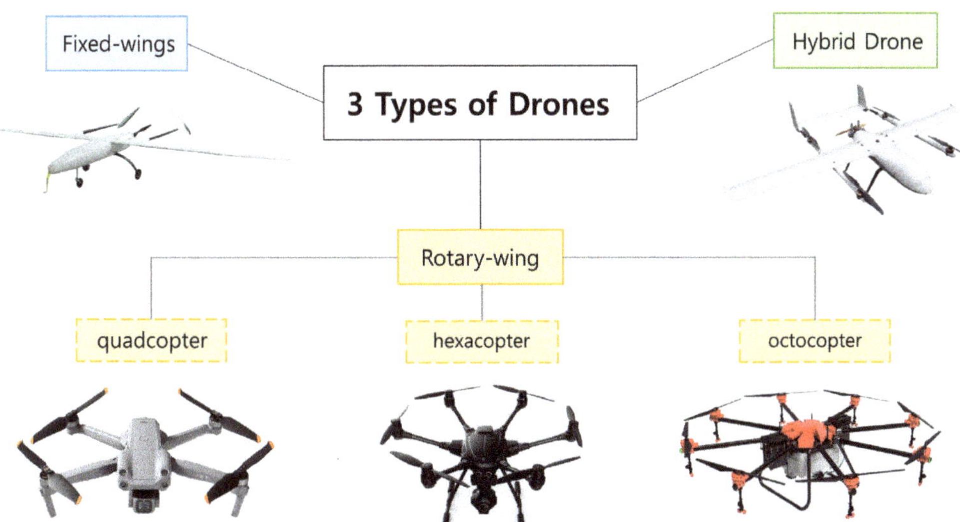

Figure 2. Three different types of drones used in construction industry.

Table 1. Three types of drones.

Types of Drones	References	Brief Summary
Rotary-wing Drones	Calantropio, A et al. [17] Villanueva, J.R.E et al. [18] Templin, T et al. [19] Anders, N et al. [20]	Large-scale topographic surveys
	Yi, W et al. [21] El Tin, F et al. [22]	Aerial inspections and monitoring of construction sites
	Sonkar, S et al. [23]	Capturing images in difficult weather
	Khan, S et al. [24]	UAV platform research
	Chae, M. H et al. [25] Sujit, P.B et al. [26]	Pilot's expertise needs
	Jin, J. W et al. [27]	High initial cost of fixed-wing drones
Fixed-wing Drones	Yang, H et al. [28]	Detailed inspections available
	Altınuç, K. O et al. [29]	Safe take-off and landing scenarios in case of failure
	Freimuth, H et al. [30] Kim, S.S. [31]	Accessible for small-scale civil engineering projects or businesses with limited resources
	Deng, C et al. [32]	Limited flight time
	Boon, M.A et al. [33] Thibbotuwawa, A et al. [34] Eck, C. [35] Li, X et al. [36]	Structural issues impact quality and stability
	Al-Rawabdeh et al. [37] Jacob-Loyola, N et al. [38] Motawa, I. et al. [39] Khaloo, A et al. [40] Lindner, G et al. [41]	High-resolution mapping, limiting advanced data collection

Table 1. *Cont.*

Types of Drones	References	Brief Summary
Hybrid Drones	Panigrahi, S et al. [42] Gunarathna, J.K et al. [43]	Benefits of long flights
	Saeed, A.S et al. [44] Yuksek, B et al. [45]	Increase detailed data collection
	Nguyen, K.D et al. [46]	Designed with numerical simulations

2.1. Fixed-Wing Drones

Fixed-wing drones offer several advantages for civil engineering applications but also come with a few disadvantages. They have longer flight times compared to rotary-wing drones. Their efficient forward flight allows them to cover larger areas and remain airborne for an extended period, which is beneficial for large-scale surveying and mapping projects. These drones can cover larger distances in a single flight due to their higher speed and endurance. This increased coverage area makes them ideal for large-scale topographic surveys [17–20], aerial inspections, and monitoring of construction sites [21,22]. Fixed-wing drones are generally more stable in windy conditions than multi-rotor drones. Their aerodynamic design and ability to withstand gusts allows them to maintain stability and capture high-quality imagery even in challenging weather [23]. They have a higher payload capacity, enabling them to carry heavier equipment such as high-resolution cameras and LiDAR sensors. This capability allows them to capture detailed aerial data for the precise mapping, 3D modeling, and volumetric analysis of construction sites. However, fixed-wing drones have some disadvantages to consider. They have limited maneuverability and cannot hover or fly in tight spaces like rotary-wing drones. This restricts their ability to inspect vertical structures or perform close-range inspections in congested areas. Furthermore, it requires a relatively longer runway or open area for take-off and landing compared to vertical take-off and landing (VTOL) drones. This can be a constraint in sites with limited space or challenging terrain [24].

Operating fixed-wing drones often requires skilled pilots due to their advanced flight characteristics and longer flight distances. Pilots need expertise in planning flight paths, conducting pre-flight checks, and coordinating with air traffic authorities, if necessary [25,26]. In general, fixed-wing drones are more expensive than rotary-wing drones due to their sophisticated design and advanced flight capabilities. The initial investment required for a fixed-wing drone system can be a barrier for small-scale civil engineering projects or businesses with limited budgets [27]. While fixed-wing drones offer significant advantages in terms of flight time, coverage area, and stability, their limited maneuverability, longer take-off/landing requirements, complex operation, and higher initial cost should be considered when selecting the appropriate drone for civil engineering applications.

2.2. Rotary-Wing Drones

Rotary-wing drones, also known as quadcopters and multi-rotor drones, have their own set of advantages and disadvantages when used in the civil engineering field. Rotary-wing drones provide excellent maneuverability and the ability to hover, making them well-suited for close-range inspections of vertical structures and operating in tight spaces. Their agility allows for detailed inspections of construction sites and infrastructure [28], providing valuable data for engineers and project managers. These drones have shorter take-off and landing requirements compared to fixed-wing drones [29]. They can perform vertical take offs and landings, eliminating the need for a runway or open area. This makes them more suitable for operating in confined construction sites or areas with limited space. Rotary-wing drones are relatively easier to operate compared to fixed-wing drones. They can be flown by pilots with less training and experience, making them accessible for small-scale civil engineering projects or businesses with limited resources [30,31].

The main disadvantage of rotary-wing drones is their limited flight time. They have shorter endurance due to the energy-intensive nature of hovering and maneuvering. This restricts their coverage area and makes them less suitable for large-scale surveys or monitoring projects that require long flight times [32]. Furthermore, it can be affected by wind and gusts more than fixed-wing drones. Their small size and lightweight construction make them more susceptible to wind disturbances, which can impact flight stability and the quality of captured data [33–36]. Compared to fixed-wing drones, rotary wing drones have lower payload capacities. They can carry lighter equipment such as small cameras or sensors, which may limit their capabilities for high-resolution mapping or advanced data collection tasks [37–41]. In summary, rotary-wing drones offer advantages in terms of maneuverability, close-range inspections, and ease of operation. They are suitable for small-scale projects and operations in confined spaces. However, their limitations include shorter flight times, susceptibility to wind, and lower payload capacities.

2.3. Hybrid Drones

Hybrid drones combine the features and capabilities of both fixed-wing and rotary-wing drones. They can take off and land vertically like rotary-wing drones, allowing them to operate in tight spaces and perform close-range inspections. At the same time, they can transition to fixed-wing flight for efficient forward flight, enabling them to cover larger areas and achieve longer flight times. This flexibility makes them suitable for a wide range of civil engineering applications. Hybrid drones offer extended flight times compared to traditional rotary-wing drones. By transitioning to fixed-wing flight, they can conserve energy and cover larger distances in a single flight. This is advantageous for conducting large-scale surveys, mapping, and monitoring projects that require longer flight durations [42,43]. The combination of vertical take-off and landing capability and fixed-wing flight allows them to carry larger cameras, LiDAR sensors, or other equipment. This enhances their capacity for detailed data collection, such as in high-resolution mapping or 3D modeling of construction sites [44,45].

However, hybrid drones have some disadvantages to consider. They are generally more complex to operate compared to single-mode drones. Pilots require specific training and expertise to handle the transition between vertical and fixed-wing flight modes, as well as understanding the nuances of operating a hybrid system. Furthermore, hybrid drones may have higher initial costs compared to single-mode drones as the integration of both fixed-wing and rotary-wing capabilities requires additional engineering and design, leading to a potentially higher purchase price [46]. This cost factor may limit their accessibility for smaller civil engineering projects or businesses with limited budgets.

3. Drone Application during the Designing Phase of Construction

3.1. Suitable Site Selection

Choosing the right site for any construction project is one of the first steps before constructing a structure. It involves assessing various potential sites to determine the most suitable location for the project. Drones play a significant role in this process by providing valuable data and insights through aerial imagery and data collection. One of the primary advantages of using drones for site selection and evaluation is the ability to capture high-resolution aerial imagery [47,48]. Drones equipped with cameras can capture detailed photographs and videos of the prospective sites from different angles and altitudes. This imagery provides a comprehensive overview of the site, allowing project managers, architects, and engineers to assess its characteristics and potential. The aerial imagery obtained from drones enables the evaluation of factors such as accessibility, proximity to transportation networks, and neighboring infrastructure. By analyzing this information, stakeholders can determine if the site meets the project's logistical requirements. They can identify any limitations or challenges related to site access, which can impact construction activities and the transportation of materials and equipment [49].

3.2. Land Surveying and Mapping

Surveying and mapping construction sites using drones offers significant advantages over traditional methods, revolutionizing the field of land surveying and providing valuable data for design and construction processes. Surveying with drones involves capturing high-resolution aerial imagery that provides a comprehensive aerial perspective of the construction site. The data collected by drones, including images and measurements from sensors such as LiDAR [50–53] or thermal sensors [51], enables accurate assessments of the site's topography, existing structures, and boundaries. Precise measurements of distances, elevations, and contours can be obtained, contributing to the creation of detailed 3D models and accurate calculations. Additionally, drones assist in establishing survey control points for precise georeferencing, ensuring the accuracy and reliability of subsequent mapping activities.

Mapping with drones encompasses the generation of accurate maps and models of the construction site using aerial imagery and photogrammetry techniques [49,54]. High-resolution aerial imagery captured by drones covers large areas efficiently, providing a visual representation of the site's features. Photogrammetry algorithms process the overlapping images to create 2D and 3D maps, including topographic maps that depict elevations, contours, slopes, and other topographic features [50,55–57]. Orthomosaic maps, created by stitching together multiple images, offer geometrically accurate and orthorectified representations of the site, facilitating precise measurements, distance calculations, and visual analysis. Moreover, drones assist in asset inventory by mapping existing structures, utilities, and vegetation, allowing designers and planners to incorporate them into their design processes [58,59].

The general process for using drones in surveying and mapping a construction site involves several key steps. It begins with pre-flight planning, where the survey area is defined, flight paths are determined, and necessary permits and safety measures are ensured [25,60,61]. During the flight, the drone captures aerial imagery and data using onboard sensors. Once the data are collected, it is processed using specialized software to generate accurate maps, models, and orthomosaics [48,59,62,63], where Figure 3 shows a general process of creating a digital terrain model regarding UAV photogrammetric process and field survey parameters [64].

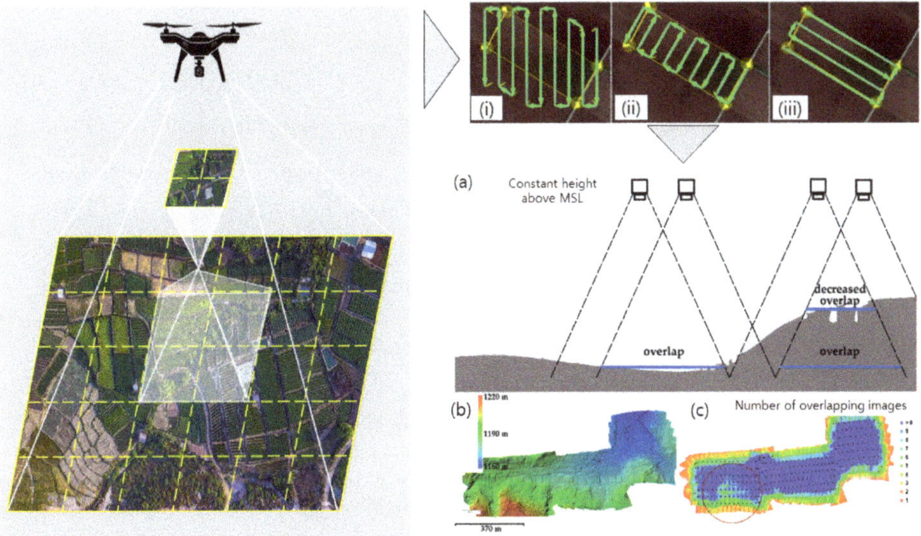

Figure 3. A general process for generating a digital terrain model with a drone (cf. [64]).

Overall, the utilization of drones in surveying and mapping offers significant benefits, including efficient data collection, high accuracy, comprehensive visual information, and enhanced decision-making capabilities. The combination of surveying and mapping with drones provides valuable insights for design, engineering, and asset management processes, ultimately improving the efficiency and quality of construction projects.

4. Drone Application during the Construction Phase

The use of drones in the construction industry has been growing rapidly in recent years. Drones offer numerous benefits during the construction phase, such as improving safety, enhancing efficiency, and reducing costs. Figure 4 shows the percentage of content for the references used in this section at construction phase where we can see that "support for rescue operations", at 31.6%, is most commonly used drone application.

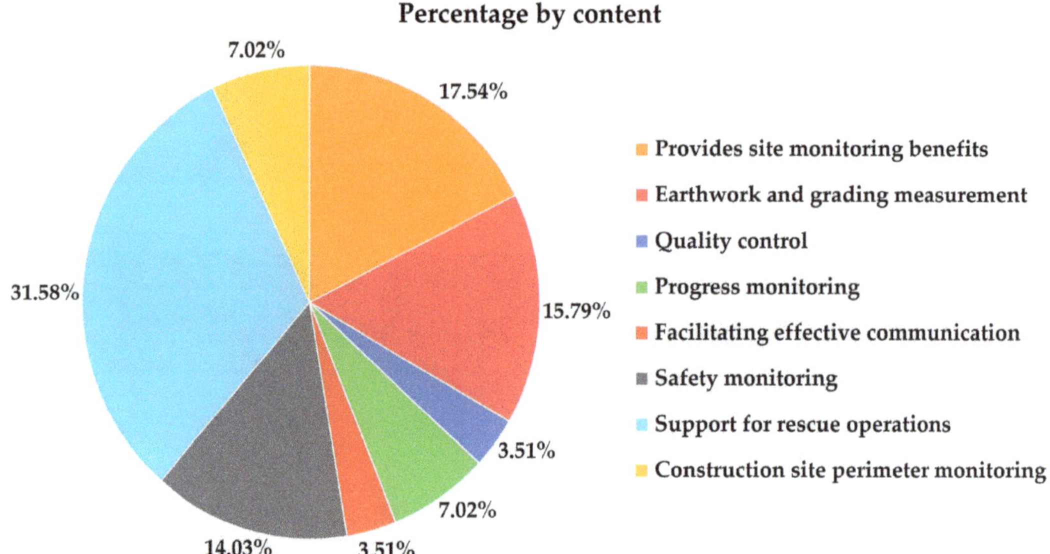

Percentage by content

Figure 4. Percentage of content for drone applications at construction phase.

4.1. Earthwork and Grading Monitoring

Earthwork and grading monitoring using drones in the construction phase has revolutionized the way construction projects are executed and managed. Drones equipped with high-resolution cameras and advanced sensors offer a range of benefits, including increased efficiency [65–67], improved accuracy [68–71], and enhanced safety [72–74]. Figure 5 shows a case study of a construction project of an apartment building complex in Seoul, Republic of Korea for a 771 household capacity where a drone was used to 3D model the area [70]. Using the UAV platform in this study, four primary analysis and visualization types were performed. These were automatic volume calculation with cut-and-fill volume data, height difference review by comparing two terrain models from different time stamps, site monitoring through 2D/3D visualization, and documentation of the project from start to completion. Traditional methods of earthwork and grading monitoring often rely on manual measurements, which are prone to human error. Drones, on the other hand, offer exceptional accuracy and precision. They capture precise measurements and detailed images of the site, allowing for accurate volume calculations [75–78], cut and fill analysis [77,79,80], and slope monitoring [81–83]. This level of accuracy helps minimize rework, optimizes resource allocation, and ensures compliance with design specifications.

Figure 5. Drone data processing results of 3D point cloud (**left**) and orthomosaic (**right**) [70].

4.2. Quality Control and Progress Monitoring

Drones play a crucial role in quality control by capturing high-resolution imagery and data that allow for thorough inspections and defect detection. The detailed imagery enables inspectors to identify even minor defects, such as cracks, corrosion, or surface imperfections, that might be missed during ground inspections [84,85]. By comparing the captured data with the construction plans or 3D models, inspectors can quickly identify any deviations or errors in the construction process. This early detection of defects enables timely rectification, ensuring that the project meets the required quality standards. Drones also facilitate the systematic documentation and tracking of identified defects, providing a clear record of issues that need to be addressed.

Drones provide an efficient and accurate method for monitoring construction progress throughout the project's lifecycle. By regularly capturing aerial imagery or conducting photogrammetry surveys, drones enable project managers to assess the status of different construction activities [86–89]. The captured data can be compared against the project timeline, enabling progress tracking and the identification of any delays or bottlenecks. Real-time progress monitoring allows for proactive decision-making and resource allocation adjustments to keep the project on schedule. Additionally, drones facilitate effective communication among the construction team, enabling stakeholders to visualize and understand the progress of the project [90,91]. This visual documentation of the construction site's evolution aids in coordination, reducing the likelihood of misunderstandings, and fosters a shared understanding of the project's status among all stakeholders.

By leveraging drones for quality control and progress monitoring, construction projects can significantly improve efficiency, minimize rework, and ensure that the project is delivered on time and within the specified quality standards. The use of drones enhances the accuracy and thoroughness of inspections, enabling the identification of defects and deviations from design plans. Real-time progress monitoring enables project managers to proactively address any issues or delays, optimizing resource allocation and ensuring the project stays on track. Ultimately, drones contribute to better construction outcomes, improved project coordination, and enhanced communication among all project stakeholders.

4.3. Safety Monitoring

Safety monitoring using drones during the construction phase has emerged as a valuable tool for enhancing safety practices and mitigating potential hazards. Drones equipped with advanced cameras, sensors, and data processing capabilities offer several benefits for safety monitoring in construction. By capturing high-resolution imagery and videos, drones can identify unsafe conditions such as unstable structures, debris, equipment malfunctions, or improper use of personal protective equipment [92–96]. The aerial perspective allows inspectors to assess the overall safety of the site, identify potential risks, and take necessary preventive measures [97–99]. Figure 6 shows a case study for using a drone for safety monitoring in a high-rise building construction project in Santiago, Chile. From the study, it was possible to identify safety issues using drone images as shown in the figure, such as a lack of guardrails and worker without safety rope, to ensure a safer environment for the workers.

Figure 6. UAV images for safety monitoring at a construction site in Chile: (**a**) lack of guardrails; (**b**) worker without safety rope; and (**c**) lack of guardrails (Images by Jhonattan G. Martinez) [92].

In the event of an emergency or incident on a construction site, drones can quickly provide real-time situational awareness. By capturing live video feeds and aerial imagery, drones assist emergency response teams in assessing the situation, identifying access points, and planning rescue operations [100–108]. Drones equipped with thermal cameras can aid in locating missing persons or hotspots in fire incidents [109–117]. This real-time information helps expedite emergency response efforts, ensuring the safety of personnel on-site. Beyond safety monitoring, drones can enhance site security by providing surveillance capabilities. Drones equipped with cameras and sensors can monitor the construction site perimeter and detect unauthorized access [47,49,118,119]. The live video feeds and recorded footage can be used for investigations, enhancing site security and protecting valuable construction assets.

4.4. Material Tracking and Delivery

Drones are also being used in various industries for delivering materials including the construction industry. [120–128]. They can deliver materials quickly and efficiently, reducing the time and cost associated with traditional delivery methods. This is particularly useful in areas with limited access or where heavy machinery cannot be used.

5. Drone Application during the Maintenance Phase

The use of drones in the maintenance of structures has been increasing in recent years. Drones offer numerous benefits in structure maintenance, such as improving safety, enhancing efficiency, and reducing costs. It can provide real-time data on the condition of structures, allowing maintenance teams to make informed decisions and adjust maintenance schedules accordingly. Drones equipped with high-resolution cameras, LiDAR technology, and thermal cameras can detect defects and damage in structures that might not be visible to the naked eye (Figure 7). This can help maintenance teams detect problems early, before they become major issues. Furthermore, it is possible for drones to be used for repair and restoration of structures. They can be used to apply coatings, sealants, and other materials to structures in a fraction of the time it would take using traditional methods. The following subsections discuss the research of the aforementioned areas.

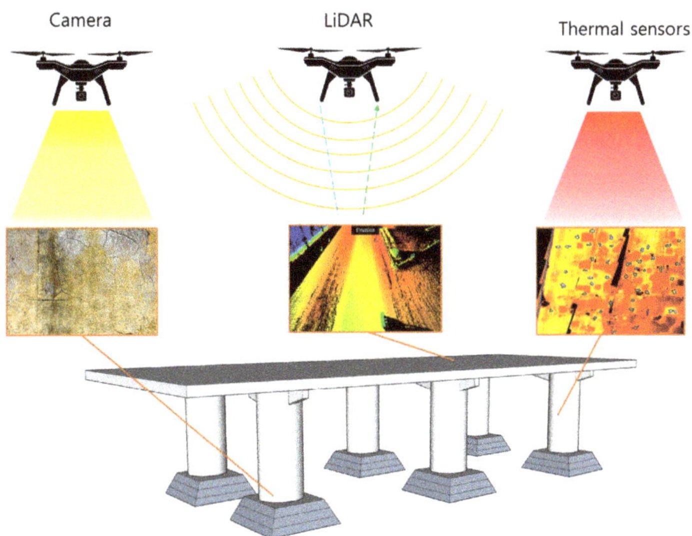

Figure 7. Difference of drone equipped with camera, LiDAR, and thermal sensor [129,130].

5.1. High Resolution Camera-Based Inspection with Drone

Drones have revolutionized the field of bridge maintenance. The ability to inspect bridges from the air provides engineers and maintenance crews with valuable data that can be used to ensure the safety and structural integrity of bridges. Drones equipped with cameras are particularly useful for bridge inspections as they can capture high-resolution images of the structure, allowing for a more detailed analysis of damage. One of the main advantages of using drones for bridge inspections is the increased safety they provide [130–137]. Drones can access areas that are difficult or dangerous for humans to reach, such as the underside of bridges or high above the ground. This reduces the need for maintenance workers to use scaffolding or other equipment, which can be expensive and time-consuming to set up. Additionally, drones can be operated remotely, reducing the risk of injury to maintenance workers who would otherwise have to climb the bridge structure to perform inspections. Another benefit of using drones for bridge inspections is the speed at which they can complete inspections. Drones can quickly fly over the bridge, capturing images and videos that can be analyzed in real-time. This allows for a faster turnaround time for inspection reports, reducing downtime for the bridge and minimizing disruption to traffic.

The high-resolution images and videos captured by drones can reveal details that may be missed during a visual inspection by human inspectors. This can include cracks or other signs of wear that are not immediately visible to the naked eye [138–140]. By providing a more detailed analysis of the bridge, maintenance crews can identify potential problems before they become serious issues, reducing the need for costly repairs and extending the lifespan of the bridge. In addition to inspections, drones can also be used for ongoing monitoring of bridge conditions. They can be programmed to fly over the bridge at regular intervals and capture data on changes in the structure over time. This can help maintenance workers identify potential problems before they become a serious issue. For example, drones can be used to monitor changes in the condition of the bridge after extreme weather events such as floods, typhoons, and earthquakes [141–143].

One of the challenges of using drones for bridge maintenance is the need for skilled operators. Drones require a trained operator who can maneuver the drone safely and capture high-quality images and videos. Additionally, the analysis of the data captured by the drone requires specialized knowledge and expertise. Therefore, it is essential to have a team of skilled professionals to operate the drones and analyze the data.

The concept of using a drone equipped with a camera for damage detection of structures can be seen in Figure 8. Once the images are captured from the camera, it is transferred to a computer for image processing to enhance the quality of the images, improve contrast, and reduce noise or distortion. Then, various image analysis algorithms can be employed to automatically detect and locate cracks in the processed images. These algorithms typically involve edge detection, texture analysis, or pattern recognition techniques to identify regions that indicate crack presence. Here, detected cracks can be classified based on their characteristics, such as length, width, orientation, or severity. It is worth noting that the effectiveness of high-resolution cameras for crack detection depends on various factors, such as the image quality, lighting conditions, surface texture, and the expertise of the image analysts. It is important to establish appropriate standards and guidelines for image capture, processing, and analysis to ensure accurate and reliable crack detection results. Additionally, the integration of advanced technologies such as artificial intelligence (AI) and machine learning can enhance crack detection capabilities by training algorithms to recognize and classify cracks more accurately, thereby improving the efficiency and effectiveness of the process.

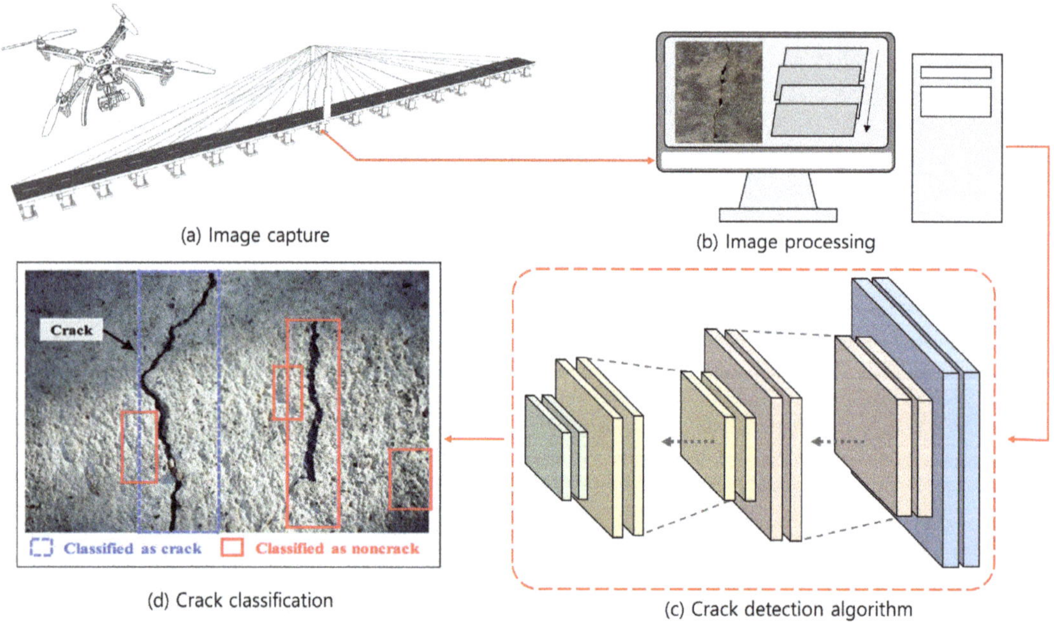

Figure 8. General concept of crack damage detection using a drone [144].

5.2. Drone Equipped with LiDAR for Structure Maintenance

LiDAR is a remote sensing technology that uses laser light to measure distances and create detailed 3D maps of the surroundings. It is often referred to as the optical equivalent of radar, as it uses light instead of radio waves. In a typical LiDAR system, a laser emitter emits short pulses of laser light, usually in the infrared range. These pulses of light are directed towards the target object. When the laser light hits an object, a fraction of the light is reflected back towards the LiDAR system where the time taken for the laser light to return is calculated to measure the distance between the system and the object. By combining these distance measurements with the known position and orientation of the LiDAR device, a point cloud of the surrounding environment can be generated, which represents the location and shape of objects.

Drones equipped with LiDAR technology can be a powerful tool for structure maintenance to ensure safety as they offer a wide range of advantages and opportunities for efficient and effective inspection and monitoring of various structures, including buildings, bridges, and industrial facilities. The ability to generate accurate 3D point cloud data of structures can be utilized for the precise mapping, modeling, and visualization of the structure, providing valuable insights for maintenance and assessment purposes [145–148]. These data can be processed and analyzed to detect and identify various structural issues, including cracks, deformations, and corrosion [131,149]. By comparing the captured data with reference models or previous scans, changes in the structure's condition can be identified, allowing for timely maintenance interventions and the prevention of further deterioration or failure.

While the utilization of LiDAR-equipped drones for structure maintenance holds immense potential, it is important to consider the challenges associated with this technology. Data processing complexity is one such challenge, as the captured LiDAR data require specialized software and expertise to convert it into useful information for analysis [150,151]. Skilled operators are required to operate the drones and process the collected data effectively. Another challenge lies in adverse weather conditions. Rain, fog, or other inclement weather can affect the performance of LiDAR systems, potentially reducing data quality or hindering data collection altogether. This limitation necessitates careful planning and scheduling of drone operations to ensure optimal weather conditions for data collection. Sensor accuracy is another aspect that requires attention. While LiDAR sensors offer high precision, variations in sensor quality or calibration can impact the accuracy of the captured data. The regular calibration and maintenance of the LiDAR system are crucial to ensure reliable and accurate measurements. Signal interference is also a consideration when using LiDAR-equipped drones. Obstacles such as trees, power lines, or other structures in the vicinity can obstruct the LiDAR signals, leading to incomplete or distorted data. Proper flight planning and obstacle avoidance algorithms are essential to mitigate these interference issues and ensure data integrity.

To address these challenges, researchers are exploring advanced data processing techniques to streamline the analysis of LiDAR data and extract meaningful information more efficiently. Improved sensor technologies and calibration methods are being developed to enhance the accuracy and reliability of LiDAR measurements [152–155]. Moreover, advancements in drone navigation and obstacle avoidance systems are being pursued to ensure safer and more precise operations, even in complex environments [156–160]. Integration with other sensing technologies, such as thermal imaging or multispectral cameras, is also being explored to enhance the inspection capabilities of LiDAR-equipped drones and enable a more comprehensive assessment of structural conditions.

5.3. Drones Equipped with Thermal Camera for Structure Maintenance

Thermal cameras mounted on drones enable the capture of thermal images, providing valuable insights into temperature variations and potential issues within the structures being inspected [118,161–163]. Thermal cameras capture infrared radiation emitted by objects, allowing the visualization of temperature variations. By detecting these temperature differences, thermal camera drones can identify various structural issues, including insulation problems, moisture infiltration, electrical faults, and thermal bridges. The ability to identify these anomalies at an early stage enables proactive maintenance and prevents further damage or deterioration.

Once thermal images are acquired, image processing and analysis techniques are employed to extract valuable information and identify potential anomalies. Image enhancement, noise reduction, and temperature calibration are crucial steps in preparing the thermal data for analysis. Advanced algorithms and machine learning approaches can be applied to automate the anomaly detection process, improving efficiency and reducing the burden on inspectors.

Several case studies and real-world applications have demonstrated the effectiveness of thermal camera drones in structure maintenance. These applications include monitoring the energy efficiency of buildings, assessing the integrity of infrastructure, inspecting solar panels, and identifying insulation or HVAC (heating, ventilation and air conditioning) system failures. The use of thermal camera drones for structure maintenance offers significant advantages in terms of accessibility, early anomaly detection, and cost-effectiveness. With continued advancements in drone technology, thermal imaging capabilities, and data analysis techniques, the integration of thermal camera drones into standard maintenance practices holds immense potential for enhancing the safety and longevity of structures.

6. Challenges and Opportunities

With the opportunities provided by the wide application of drones using in civil engineering projects, there are still challenges that need to be addressed to fully leverage the potential of drone technology in this field. For all three phases of design, construction, and maintenance, one of the crucial challenges is its limited flight time and range as battery life remains a significant constraint for drones. Most commercial drones have relatively short flight durations, limiting their ability to cover large construction sites or inspect extensive infrastructure. This limitation hampers their overall operational efficiency, requiring frequent battery replacements and recharging. Currently, research and development efforts are continuously improving battery energy density and recharge rates, allowing drones to operate for longer periods and cover greater distances. The adoption of alternative power sources, such as fuel cells, solar panels, or wireless charging, could potentially eliminate the need for frequent battery replacements, enhancing operational efficiency. Environmental conditions such as strong winds, rain or fog can impede drone operations affecting project time which is another challenge for drones. Drones are susceptible to turbulence caused by high winds, and precipitation can damage sensitive electronic components, leading to potential downtime and increased maintenance costs. Designing drones with robust structures, waterproofing, and advanced navigation systems can enable them to withstand harsh weather elements and continue operations in challenging environments safely. With the opportunities that lie ahead with drone technologies to improve safety at construction sites, monitoring processes, surveys, 3D modeling and more, civil engineers can harness the capabilities of drones to drive innovation, optimize project management, and promote safer and more sustainable infrastructure development.

7. Future Directions

Over the years, drone technology has evolved and become more sophisticated, offering a wide range of applications in areas such as those shown in this study. From this, we can predict that future drones will be equipped with advanced automation and AI capabilities to conduct missions with minimal human intervention such as autonomous flight planning and obstacle avoidance. AI-powered drones can autonomously navigate complex terrains, identify and assess potential hazards, and conduct advanced data analysis. This automation streamlines data collection, data processing, and reporting, enabling civil engineers to make informed decisions faster and more accurately. Furthermore, improvements in battery technology and drone design will lead to extended flight times to allow drones to cover large areas in a single flight, making them more effective for tasks where an extensive reviewer works on energy storage systems, as can be found in reference [164].

An extremely encouraging prospect for the future of drone applications in the civil industry involves incorporating 5G connectivity. The advanced capabilities of 5G networks, including substantially higher data transfer rates, remarkably low latency, and expanded capacity, enable seamless real-time communication between drones and ground stations. Through the implementation of 5G-enabled drones, civil engineers gain the ability to remotely oversee construction sites and infrastructure with enhanced accuracy and efficiency. This seamless exchange of high-resolution data and live video feeds empowers them to make agile decisions, enabling swift responses to dynamic project conditions. Collaborative swarm technology is another area which needs to be further researched in the future for drones as their ability to operate in swarms will revolutionize various civil industries. Drones working together will enhance efficiency and data collection capabilities which can be applied at construction sites. Such technology could possibly increase the demand for drones and with the increase in numbers, one cannot ignore that there will be a strong emphasis on making drones more environmentally friendly and sustainable. This could involve using bio-inspired designs, energy-efficient propulsion systems, Micro Air Vehicles [165], and materials with reduced environmental impact.

8. Conclusions

In conclusion, the review of drone applications in the construction industry underscores their significant contributions across various phases of the construction process, including design, construction, and maintenance. The utilization of different types of drones has proven to be immensely beneficial in enhancing efficiency, accuracy, and safety within the industry.

During the design phase, drones equipped with high-resolution cameras and advanced mapping capabilities have revolutionized site surveys and aerial mapping. These drones enable construction professionals to gather precise data, generate accurate 3D models, and assess topography. This, in turn, facilitates informed decision-making and enhances the overall design process. In the construction phase, drones have played a vital role in monitoring construction progress, conducting inspections, and ensuring safety. Equipped with real-time video transmission and thermal imaging cameras, drones provide a comprehensive and timely overview of the construction site, identifying potential issues for increasing productivity. Furthermore, drones have also demonstrated their utility in the maintenance phase of construction projects. By conducting routine inspections of structures, buildings, and infrastructure, drones efficiently detect and identify any damages, enabling proactive maintenance, cost reduction, and the prolongation of asset lifespan.

Overall, the integration of drones in the construction industry has brought about transformative advancements in efficiency, accuracy, and safety across all phases of the construction process. As technology continues to advance, it is expected that drones will increasingly play a critical role in shaping the future of the construction industry, empowering professionals to achieve higher productivity, minimize risks, and deliver projects of exceptional quality.

Author Contributions: H.-W.C. was responsible for the conceptualization and methodology of the work; H.-J.K. carried out the investigation and participated in writing the manuscript. S.-K.K. supervised the manuscript and W.S.N. supervised and carried out the reviewing and editing of the manuscript. All authors have read and agreed to the published version of the manuscript.

Funding: This research was conducted with the support of the "National R&D Project for Smart Construction Technology (RS-2020-KA157089, 4th year)" funded by the Korea Agency for Infrastructure Technology Advancement under the Ministry of Land, Infrastructure and Transport, and managed by the Korea Expressway Corporation.

Institutional Review Board Statement: Not applicable.

Informed Consent Statement: Not applicable.

Data Availability Statement: Not applicable.

References

1.   Shanti, M.Z.; Cho, C.S.; de Soto, B.G.; Byon, Y.J.; Yeun, C.Y.; Kim, T.Y. Real-time monitoring of work-at-height safety hazards in construction sites using drones and deep learning. *J. Saf. Res.* **2022**, *83*, 364–370. [CrossRef]

2. Zhang, Y.; Zhang, K. Design of Construction Site Dust Detection System Based on UAV Flying Platform. In Proceedings of the IEEE International Conference on Control Science and Electric Power Systems (CSEPS), Shanghai, China, 28–30 May 2021; pp. 148–151.

3. Olivatto, T.F.; Inguaggiato, F.F.; Stanganini, F.N. Urban mapping and impacts assessment in a Brazilian irregular settlement using UAV-based imaging. *Remote Sens. Appl. Soc. Environ.* **2023**, *29*, 100911.

4. Fan, J.; Saadeghvaziri, M.A. Applications of drones in infrastructures: Challenges and opportunities. *Int. J. Mech. Mechatron. Eng.* **2019**, *13*, 649–655.

5. Tao, C.; Zhu, K.; Chen, B.; Zhao, Y. UAV-assisted ground signal map construction based on 3-D spatial correlation. In Proceedings of the GLOBECOM 2020—2020 IEEE Global Communications Conference, Taipei, Taiwan, 7–11 December 2020; pp. 1–5.

6. Tatum, M.C.; Liu, J. Unmanned aircraft system applications in construction. *Procedia Eng.* **2017**, *196*, 167–175. [CrossRef]

7. Liu, Q.; Chen, B.; Yan, Y.; Zhang, C. A measurement method for construction planning and completion based on UAV photography. In Proceedings of the 2022 4th International Conference on Intelligent Control, Measurement and Signal Processing (ICMSP), Hangzhou, China, 8–10 July 2022; pp. 148–151.

8. Zhang, C.; Wang, F.; Zou, Y.; Dimyadi, J.; Guo, B.H.; Hou, L. Automated UAV image-to-BIM registration for building façade inspection using improved generalised Hough transform. *Autom. Constr.* **2023**, *153*, 104957. [CrossRef]

9. Hamledari, H.; Sajedi, S.; McCabe, B.; Fischer, M. Automation of Inspection Mission Planning Using 4D BIMs and in Support of Unmanned Aerial Vehicle–Based Data Collection. *J. Constr. Eng. Manag.* **2021**, *147*, 04020179. [CrossRef]

10. Ruiz, R.D.B.; Lordsleem, A.C., Jr.; Rocha, J.H.A.; Irizarry, J. Unmanned Aerial Vehicles (UAV) as a Tool for Visual Inspection of Building Facades in AEC+FM Industry. *Constr. Innov.* **2021**, *22*, 1155–1170. [CrossRef]

11. Petritoli, E.; Leccese, F.; Ciani, L. Reliability Degradation, Preventive and Corrective Maintenance of UAV Systems. In Proceedings of the 2018 5th IEEE International Workshop on Metrology for AeroSpace (MetroAeroSpace), Rome, Italy, 20–22 June 2018; pp. 430–434.

12. Sowing UAV of Fixed Wings. Available online: https://www.uavos.com/products/fixed-wing-uavs/sitaria-e/ (accessed on 15 June 2023).

13. Sowing UAV of Quadcopter. Available online: https://www.ubuy.dk/en/product/1E71NHFK-dji-air-2s-drone-quadcopter-uav-with-3-axis-gimbal-camera-5-4k-video-1-inch-cmos-sensor-4-directions (accessed on 8 June 2023).

14. Sowing UAV of Hexacopter. Available online: https://www.best-quadcopter.com/reviews/2017/03/yuneec-typhoon-h-pro-realsense-obstacle-avoiding-drone/ (accessed on 28 June 2023).

15. Sowing UAV of Octocoptor. Available online: https://kr.made-in-china.com/co_agriculture-drone/image_30L-Payload-RC-8-Propellers-Professional-Plant-Protection-Drone_eeeessrsy_2f1j00AkmWNvsHpOrU.html (accessed on 1 June 2023).

16. Sowing UAV of Hybrid. Available online: https://www.viewprouav.com/product/tiger-shark-f380-heavy-lift-hybrid-vtol-fixed-wing-with-3-8m-wingspan-10kg-load-capacity.html (accessed on 3 June 2023).

17. Calantropio, A.; Chiabrando, F.; Sammartano, G.; Spanò, A.; Teppati Losè, L. UAV Strategies validation and remote sensing data for damage assessment in post-disaster scenarios. *Int. Arch. Photogramm. Remote Sens. Spat. Inf. Sci.* **2018**, *42*, 121–128. [CrossRef]

18. Villanueva, J.R.E.; Martínez, L.I.; Montiel, J.I.P. DEM Generation from Fixed-Wing UAV Imaging and LiDAR-Derived Ground Control Points for Flood Estimations. *Sensors* **2019**, *19*, 3205.

19. Templin, T.; Popielarczyk, D.; Kosecki, R. Application of Low-Cost Fixed-Wing UAV for Inland Lakes Shoreline Investigation. *Pure Appl. Geophys.* **2018**, *175*, 3263–3283. [CrossRef]

20. Anders, N.; Masselink, R.; Keesstra, S.; Suomalainen, J. High-Res Digital Surface Modeling using Fixed-Wing UAV-based Photogrammetry. In Proceedings of the Geomorphometry 2013, Nanjing, China, 16–20 October 2013; pp. 2–5.

21. Yi, W.; Liming, C.; LingYu, K.; Jie, Z.; Miao, W. Research on application mode of large fixed-wing UAV system on over-head transmission line. In Proceedings of the IEEE International Conference on Unmanned Systems (ICUS), Beijing, China, 27–29 October 2017; pp. 88–91.

22. El Tin, F.; Sharf, I.; Nahon, M. Fire Monitoring with a Fixed-wing Unmanned Aerial Vehicle. In Proceedings of the 2022 International Conference on Unmanned Aircraft Systems (ICUAS), Dubrovnik, Croatia, 21–24 June 2022; pp. 1350–1358.

23. Sonkar, S.; Kumar, P.; George, R.C.; Yuvaraj, T.P.; Philip, D.; Ghosh, A.K. Real-time object detection and recognition using fixed-wing Lale VTOL UAV. *IEEE Sens. J.* **2022**, *22*, 20738–20747. [CrossRef]

24. Khan, S.; Bendoukha, S.; Naeem, W.; Iqbal, J. Experimental validation of an integral sliding mode-based LQG for the pitch control of a UAV-mimicking platform. *Adv. Electr. Electron. Eng.* **2019**, *17*, 275–284.

25. Chae, M.H.; Park, S.O.; Choi, S.H.; Choi, C.T. Commercial Fixed-Wing Drone Redirection System using GNSS Deception. *IEEE Trans. Aerosp. Electron. Syst.* **2023**, 1–15. [CrossRef]

26. Sujit, P.B.; Saripalli, S.; Sousa, J.B. Unmanned Aerial Vehicle Path Following: A Survey and Analysis of Algorithms for Fixed-Wing Unmanned Aerial Vehicless. *IEEE Control Syst. Mag.* **2014**, *34*, 42–59.

27. Jin, J.W.; Miwa, M.; Shim, J.H. Design and construction of a Quad tilt-Rotor UAV using servo motor. *J. Eng. Educ. Res.* **2014**, *17*, 17–22.

28. Yang, H.; Lee, Y.; Jeon, S.-Y.; Lee, D. Multi-rotor drone tutorial: Systems, mechanics, control and state estimation. *Intell. Serv. Robot.* **2017**, *10*, 79–93. [CrossRef]

29. Altınuç, K.O.; Khan, M.U.; Iqbal, J. Avoiding contingent incidents by quadrotors due to one or two propellers failure. *PLoS ONE* **2023**, *18*, e0282055. [CrossRef]

30. Freimuth, H.; König, M. Planning and Executing Construction Inspections with Unmanned Aerial Vehicles. *Autom. Constr.* **2018**, *96*, 540–553. [CrossRef]
31. Kim, S.S. Opportunities for construction site monitoring by adopting first personal view (FPV) of a drone. *Smart Struct. Syst.* **2018**, *21*, 139–149.
32. Deng, C.; Wang, S.; Huang, Z.; Tan, Z.; Liu, J. Unmanned Aerial Vehicles for power line inspection: A cooperative way in platforms and communications. *J. Commun.* **2014**, *9*, 687–692. [CrossRef]
33. Boon, M.A.; Drijfhout, A.P.; Tesfamichael, S. Comparison of a fixed-wing and multi-rotor UAV for environmental mapping applications: A case study. *Int. Arch. Photogramm. Remote Sens. Spat. Inf. Sci.* **2017**, *42*, 47.
34. Thibbotuwawa, A.; Bocewicz, G.; Radzki, G.; Nielsen, P.; Banaszak, Z. UAV Mission Planning Resistant to Weather Uncertainty. *Sensors* **2020**, *20*, 515. [CrossRef] [PubMed]
35. Eck, C.; Imbach, B. Aerial magnetic sensing with an UAV helicopter. *Int. Arch. Photogramm. Remote Sens. Spat. Inf. Sci.* **2011**, *38*, 81–85. [CrossRef]
36. Li, X.; Wu, J. Extracting High-Precision Vehicle Motion Data from Unmanned Aerial Vehicle Video Captured under Various Weather Conditions. *Remote Sens.* **2022**, *14*, 5513. [CrossRef]
37. Al-Rawabdeh, A.; He, F.; Moussa, A.; El-Sheimy, N.; Habib, A. Using an Unmanned Aerial Vehicle-Based Digital Imaging System to Derive a 3D Point Cloud for Landslide Scarp Recognition. *Remote Sens.* **2016**, *8*, 95. [CrossRef]
38. Jacob-Loyola, N.; Munoz-La Rivera, F.; Herrera, R.F.; Atencio, E. Unmanned aerial vehicles (uavs) for physical progress monitoring of construction. *Sensors* **2021**, *21*, 4227. [CrossRef] [PubMed]
39. Motawa, I.; Kardakou, A. Unmanned aerial vehicles (UAVs) for inspection in construction and building industry. In Proceedings of the the 16th International Operation and Maintenance Conference, Cairo, Egypt, 18–20 November 2018.
40. Khaloo, A.; Lattanzi, D.; Cunningham, K.; Dell'Andrea, R.; Riley, M. Unmanned aerial vehicle inspection of the Placer River Trail Bridge through image-based 3D modelling. *Struct. Infrastruct. Eng.* **2018**, *14*, 124–136. [CrossRef]
41. Lindner, G.; Schraml, K.; Mansberger, R.; Hübl, J. Uav monitoring and documentation of a large landslide. *Appl. Geomat.* **2016**, *8*, 1–11. [CrossRef]
42. Panigrahi, S.; Krishna, Y.S.S.; Thondiyath, A. Design, Analysis, and Testing of a Hybrid VTOL Tilt-Rotor UAV for Increased Endurance. *Sensors* **2021**, *21*, 5987. [CrossRef]
43. Gunarathna, J.K.; Munasinghe, R. Development of a quad-rotor fixed-wing hybrid unmanned aerial vehicle. In Proceedings of the 2018 Moratuwa Engineering Research Conference (MERCon), Moratuwa, Sri Lanka, 30 May–1 June 2018; pp. 72–77.
44. Saeed, A.S.; Younes, A.B.; Cai, C.; Cai, G. A Survey of Hybrid Unmanned Aerial Vehicles. *Prog. Aerosp. Sci.* **2018**, *98*, 91–105. [CrossRef]
45. Yuksek, B.; Vuruskan, A.; Ozdemir, U.; Yukselen, M.A.; Inalhan, G. Transition flight modeling of a fixed-wing VTOL UAV. *J. Intell. Robot. Syst.* **2016**, *84*, 83–105. [CrossRef]
46. Nguyen, K.D.; Ha, C.; Jang, J.T. Development of a new hybrid drone and software-in-the-loop simulation using px4 code. In *Intelligent Computing Theories and Application, Proceedings of the International Conference on Intelligent Computing, Wuhan, China, 15–18 August 2018*; Springer: Cham, Switzerland, 2018; pp. 84–93.
47. Asadi, K.; Kalkunte Suresh, A.; Ender, A.; Gotad, S.; Maniyar, S.; Anand, S.; Noghabaei, M.; Han, K.; Lobaton, E.; Wu, T. An integrated UGV-UAV system for construction site data collection. *Autom. Constr.* **2020**, *112*, 103068. [CrossRef]
48. Congress, S.S.C.; Puppala, A.J. A Road Map for Geotechnical Monitoring of Transportation Infrastructure Assets Using Three-Dimensional Models Developed from Unmanned Aerial Data. *Indian Geotech. J.* **2021**, *51*, 84–96. [CrossRef]
49. Bang, S.; Kim, H.; Kim, H. UAV-Based Automatic Generation of High-Resolution Panorama at a Construction Site with a Focus on Preprocessing for Image Stitching. *Autom. Constr.* **2017**, *84*, 70–80. [CrossRef]
50. Kim, P.; Park, J.; Cho, Y. As-Is Geometric Data Collection and 3D Visualization through the Collaboration between UAV and UGV. In Proceedings of the 36th International Symposium on Automation and Robotics in Construction (ISARC), Banff, AB, Canada, 21–24 May 2019; Al-Hussein, M., Ed.; International Association for Automation and Robotics in Construction (IAARC): Banff, AB, Canada, 2019; pp. 544–551.
51. Giordan, D.; Adams, M.S.; Aicardi, I.; Alicandro, M.; Allasia, P.; Baldo, M.; De Berardinis, P.; Dominici, D.; Godone, D.; Hobbs, P.; et al. The use of unmanned aerial vehicles (UAVs) for engineering geology applications. *Bull. Int. Assoc. Eng. Geol.* **2020**, *79*, 3437–3481. [CrossRef]
52. Bemis, S.P.; Micklethwaite, S.; Turner, D.; James, M.R.; Akciz, S.; Thiele, S.T.; Bangash, H.A. Ground-based and UAV-Based photogrammetry: A multi-scale, high-resolution mapping tool for structural geology and paleoseismology. *J. Struct. Geol.* **2014**, *69*, 163–178. [CrossRef]
53. Westoby, M.J.; Brasington, J.; Glasser, N.F.; Hambrey, M.J.; Reynolds, J. Structure-from-Motion photogrammetry: A low-cost, effective tool for geoscience applications. *Geomorphology* **2012**, *179*, 300–314. [CrossRef]
54. Ajayi, O.G.; Palmer, M.; Salubi, A.A. Modelling farmland topography for suitable site selection of dam construction using unmanned aerial vehicle (UAV) photogrammetry. *Remote Sens. Appl. Soc. Environ.* **2018**, *11*, 220–230.
55. Dominici, D.; Alicandro, M.; Massimi, V. Uav photogrammetry in the post-earthquake scenario: Case studies in l'aquila. *Geomat. Nat. Hazards Risk* **2017**, *8*, 87–103. [CrossRef]
56. Park, J.; Im, S.; Lee, K.H.; Lee, J.O. Vision-based SLAM system for small UAVs in GPS-denied environments. *J. Aerosp. Eng.* **2012**, *25*, 519–529. [CrossRef]

57. Chesley, J.T.T.; Leier, A.L.L.; White, S.; Torres, R. Using unmanned aerial vehicles and structure-from-motion photogrammetry to characterize sedimentary outcrops: An example from the Morrison Formation, Utah, USA. *Sediment. Geol.* **2017**, *354*, 1–8.
58. Hammad, A.W.A.; da Costa, B.B.F.; Soares, C.A.P.; Haddad, A.N. The Use of Unmanned Aerial Vehicles for Dynamic Site Layout Planning in Large-Scale Construction Projects. *Buildings* **2021**, *11*, 602. [CrossRef]
59. Han, K.K.; Golparvar-Fard, M. Potential of Big Visual Data and Building Information Modeling for Construction Performance Analytics: An Exploratory Study. *Autom. Constr.* **2017**, *73*, 184–198. [CrossRef]
60. Siebert, S.; Teizer, J. Mobile 3D mapping for surveying earthwork projects using an unmanned aerial vehicle(UAV) system. *Autom. Constr.* **2014**, *41*, 1–14. [CrossRef]
61. Razzaq, S.; Xydeas, C.; Mahmood, A.; Ahmed, S.; Ratyal, N.I.; Iqbal, J. Efficient optimization techniques for resource allocation in UAVs mission framework. *PLoS ONE* **2023**, *18*, e0283923. [CrossRef]
62. Han, Y.; Feng, D.; Wu, W.; Yu, X.; Wu, G.; Liu, J. Geometric shape measurement and its application in bridge construction based on UAV and terrestrial laser scanner. *Autom. Constr.* **2023**, *151*, 104880. [CrossRef]
63. Keyvanfar, A.; Shafaghat, A.; Rosley, M.S. Performance comparison analysis of 3D reconstruction modeling software in construction site visualization and mapping. *Int. J. Archit. Computing.* **2022**, *20*, 453–475.
64. Jiménez-Jiménez, S.I.; Ojeda-Bustamante, W.; Marcial-Pablo, M.; Enciso, J. Digital Terrain Models Generated with Low-Cost UAV Photogrammetry: Methodology and Accuracy. *ISPRS Int. J. Geo-Inf.* **2021**, *10*, 285.
65. Baker, W.J., III; Meehan, C.L. Spatial Interpolation of UAV Survey Data for Lift Thickness Determination during Earthwork Construction. In Proceedings of the Geo-Congress 2023, Los Angeles, CA, USA, 26–29 March 2023; pp. 336–346.
66. Kim, J.; Lee, S.; Seo, J.; Lee, D.-E.; Choi, H. The Integration of Earthwork Design Review and Planning Using UAV-Based Point Cloud and BIM. *Appl. Sci.* **2021**, *11*, 3435. [CrossRef]
67. Kavaliauskas, P.; Židanavičius, D.; Jurelionis, A. Geometric Accuracy of 3D Reality Mesh Utilization for BIM-Based EarthworkQuantity Estimation Workflows. *ISPRS Int. J. Geo-Inf.* **2021**, *10*, 399. [CrossRef]
68. Hasegawa, H.; Sujaswara, A.A.; Kanemoto, T.; Tsubota, K. Possibilities of Using UAV for Estimating Earthwork Volumes during Process of Repairing a Small-Scale Forest Road, Case Study from Kyoto Prefecture, Japan. *Forests* **2023**, *14*, 677. [CrossRef]
69. Rachmawati, T.S.N.; Park, H.C.; Kim, S. A Scenario-Based Simulation Model for Earthwork Cost Management Using Unmanned Aerial Vehicle Technology. *Sustainability* **2023**, *15*, 503.
70. Park, H.C.; Rachmawati, T.S.N.; Kim, S. UAV-Based High-Rise Buildings Earthwork Monitoring—A Case Study. *Sustainability* **2022**, *14*, 10179. [CrossRef]
71. Ostrovskyi, A.; Kolb, I.; Vivat, A.; Lozynskyi, V.; Zhyvchuk, V. Simplified method of obtaining data for calculating the volume of earthworks based on aerial survey materials from UAVs. In Proceedings of the International Conference of Young Professionals «GeoTerrace-2021», Lviv, Ukraine, 4–6 October 2021; Volume 1, pp. 1–5.
72. Kim, D.; Liu, M.; Lee, S.; Kamat, V.R. Remote proximity monitoring between mobile construction resources using camera-mounted UAVs. *Autom. Constr.* **2019**, *99*, 168–182. [CrossRef]
73. Kim, D.; Yin, K.; Liu, M.; Lee, S.; Kamat, V.R. Feasibility of a Drone-Based On-Site Proximity Detection in an Outdoor Construction Site. In Proceedings of the ASCE International Workshop on Computing in Civil Engineering 2017, Seattle, WA, USA, 25–27 June 2017; pp. 392–400.
74. Wu, J.; Peng, L.; Li, J.; Zhou, X.; Zhong, J.; Wang, C.; Sun, J. Rapid safety monitoring and analysis of foundation pit construction using unmanned aerial vehicle images. *Autom. Constr.* **2021**, *128*, 103706.
75. Kim, Y.H.; Shin, S.S.; Lee, H.K.; Park, E.S. Field Applicability of Earthwork Volume Calculations Using Unmanned Aerial Vehicle. *Sustainability* **2022**, *14*, 9331.
76. Lee, K.; Lee, W.H. Earthwork Volume Calculation, 3D Model Generation, and Comparative Evaluation Using Vertical and High-Oblique Images Acquired by Unmanned Aerial Vehicles. *Aerospace* **2022**, *9*, 606. [CrossRef]
77. Lee, S.B.; Han, D.; Song, M. Calculation and Comparison of Earthwork Volume Using Unmanned Aerial Vehicle Photogrammetry and Traditional Surveying Method. *Sens. Mater.* **2022**, *34*, 4737–4753.
78. Ajayi, O.G.; Ajulo, J. Investigating the Applicability of Unmanned Aerial Vehicles (UAV) Photogrammetry for the Estimation ofthe Volume of Stockpiles. *Quaest. Geogr.* **2021**, *40*, 25–38. [CrossRef]
79. Astor, Y.; Utami, R.; Gustaman, F.A.; Ramdani, M.A.; Pangestu, Y.I. Implementation of Unmanned Aerial Vehicle (UAV) in the Sand Mine Project. In *Conference on Broad Exposure to Science and Technology 2021 (BEST 2021)*; Atlantis Press: Amsterdam, The Netherlands, 2022; pp. 66–71.
80. Villalobos, N.; Alzraiee, H. Earthwork Surface Creation and Volume Computation Using UAV 3D Mapping. In Proceedings of the Construction Research Congress 2022, Arlington, VA, USA, 9–12 March 2022; pp. 1243–1252.
81. Subramaniam, T.R.; Suhaimi, N.N.A.; Paizol, A.; Nor, A.H.M. Determination of Slope Stability using (UAV) Unmanned Aerial Vehicle. *Multidiscip. Appl. Res. Innov.* **2022**, *3*, 292–301.
82. Pasternak, G.; Zaczek-Peplinska, J.; Pasternak, K.; Jóźwiak, J.; Pasik, M.; Koda, E.; Vaverková, M.D. Surface Monitoring of an MSW Landfill Based on Linear and Angular Measurements, TLS, and LIDAR UAV. *Sensors* **2023**, *23*, 1847.
83. Cho, J.; Lee, J.; Lee, B. Application of UAV Photogrammetry to Slope-Displacement Measurement. *KSCE J. Civ. Eng.* **2022**, *26*, 1904–1913.
84. Kielhauser, C.; Manzano, R.R.; Hoffman, J.J.; Adey, B.T. Automated Construction Progress and Quality Monitoring for Commercial Buildings with Unmanned Aerial Systems: An Application Study from Switzerland. *Infrastructures* **2020**, *5*, 98.

85. Meshram, K.; Reddy, N.G. Development of a machine learning-based drone system for management of construction sites. In *Advances in Sustainable Materials and Resilient Infrastructure*; Springer: Singapore, 2022; pp. 77–78.

86. Irizarry, J.; Costa, D.B. Exploratory study of potential applications of unmanned aerial systems for construction management tasks. *J. Manag. Eng.* **2016**, *32*, 05016001.

87. Ngadiman, N.; Kaamin, M.; Nizam, M.A.H.M.; Johar, M.A.H.; Roslin, M.A. Unmanned aerial vehicle (UAV) visual monitoring in construction. *Ann. Rom. Soc. Cell Biol.* **2021**, *25*, 3097–3104.

88. Keyvanfar, A.; Shafaghat, A.; Awanghamat, A. Optimization and Trajectory Analysis of Drone's Flying and Environmental Variables for 3D Modelling the Construction Progress Monitoring. *Int. J. Civ. Eng.* **2021**, *20*, 363–388.

89. Qu, T.; Zang, W.; Peng, Z.; Liu, J.; Li, W.; Zhu, Y.; Zhang, B.; Wang, Y. Construction Site Monitoring using UAV Oblique Phtogrammetry and BIM Technologies. In Proceedings of the 22nd CAADRIA Conference, Suzhou, China, 5–8 April 2017.

90. Cheng, T.; Teizer, J. Real-time resource location data collection and visualization technology for construction safety and activity monitoring applications. *Autom. Constr.* **2013**, *34*, 3–15.

91. Yang, J.; Park, M.W.; Vela, P.A.; Golparvar-Fard, M. Construction Performance Monitoring via Still Images, Time-Lapse Photos, and Video Streams: Now, Tomorrow, and the Future. *Adv. Eng. Inform.* **2015**, *29*, 211–224.

92. Martinez, J.G.; Gheisari, M.; Alarcón, L.F. UAV integration in current construction safety planning and monitoring process: Case study of a high-rise building construction project in Chile. *J. Manag. Eng.* **2020**, *36*, 05020005.

93. Mavroulis, S.; Andreadakis, E.; Spyrou, N.I.; Antoniou, V.; Skourtsos, E.; Papadimitriou, P.; Kasssaras, I.; Kaviris, G.; Tselentis, G.A.; Voulgaris, N.; et al. UAV and GIS based rapid earthquake-induced building damage assessment and methodology for EMS-98 isoseismal map drawing: The June 12, 2017 mw 6.3 lesvos (northeastern aegean, greece) earthquake. *Int. J. Disaster Risk Reduct.* **2019**, *37*, 20.

94. Candigliota, E.; Immordino, F. Low Altitude Remote Sensing by UAV for monitoring and emergency management on historical heritage. In Proceedings of the ANIDIS Congress, Padova, Italy, 30 June–4 July 2013; Volume 30.

95. Feroz, S.; Abu Dabous, S. UAV-Based Remote Sensing Applications for Bridge Condition Assessment. *Remote Sens.* **2021**, *13*, 1809. [CrossRef]

96. Duque, L.; Seo, J.; Wacker, J. Bridge Deterioration Quantification Protocol Using UAV. *J. Bridge Eng.* **2018**, *23*, 04018080. [CrossRef]

97. Aliyari, M.; Ashrafi, B.; Ayele, Y.Z. Hazards identification and risk assessment for UAV–assisted bridge inspections. *Struct. Infrastruct. Eng.* **2022**, *18*, 412–428. [CrossRef]

98. Wackwitz, K.; Boedecker, H. *Safety Risk Assessment for UAV Operation*; Drone Industry Insights: Hamburg, Germany, 2015.

99. Namian, M.; Khalid, M.; Wang, G.; Turkan, Y. Revealing safety risks of unmanned aerial vehicles in construction. *Transp. Res. Rec.* **2021**, *2675*, 03611981211017134. [CrossRef]

100. Calamoneri, T.; Corò, F.; Mancini, S. A Realistic Model to Support Rescue Operations After an Earthquake via UAVs. *IEEE Access* **2022**, *10*, 6109–6125. [CrossRef]

101. Oubbati, O.S.; Badis, H.; Rachedi, A.; Lakas, A.; Lorenz, P. Multi-UAV Assisted Network Coverage Optimization for Rescue Operations using Reinforcement Learning. In Proceedings of the 2023 IEEE 20th Consumer Communications & Networking Conference (CCNC), Las Vegas, NV, USA, 8–11 January 2023; pp. 1003–1008.

102. Horyna, J.; Baca, T.; Walter, V.; Albani, D.; Hert, D.; Ferrante, E.; Saska, M. Decentralized swarms of unmanned aerial vehicles for search and rescue operations without explicit communication. *Auton. Robot.* **2023**, *47*, 77–93. [CrossRef]

103. Du, Y. Multi-UAV Search and Rescue with Enhanced A Algorithm Path Planning in 3D Environment. *Int. J. Aerosp. Eng.* **2023**, *2023*, 8614117.

104. Tu´snio, N.; Wróblewski, W. The Efficiency of Drones Usage for Safety and Rescue Operations in an Open Area: A Case from Poland. *Sustainability* **2022**, *14*, 327.

105. Lygouras, E.; Santavas, N.; Taitzoglou, A.; Tarchanidis, K.; Mitropoulos, A.; Gasteratos, A. Unsupervised Human Detection withan Embedded Vision System on a Fully Autonomous UAV for Search and Rescue Operations. *Sensors* **2019**, *19*, 3542.

106. Feraru, V.A.; Andersen, R.E.; Boukas, E. Towards an Autonomous UAV-Based System to Assist Search and Rescue Operations in Man Overboard Incidents. In Proceedings of the 2020 IEEE International Symposium on Safety, Security, and Rescue Robotics (SSRR), Abu Dhabi, United Arab Emirates, 4–6 November 2020; pp. 57–64.

107. Dong, J.; Ota, K.; Dong, M. UAV-Based Real-Time Survivor Detection System in Post-Disaster Search and Rescue Operations. *IEEE J. Miniaturization Air Space Syst.* **2021**, *2*, 209–219. [CrossRef]

108. Martinez-Alpiste, I.; Golcarenarenji, G.; Wang, Q.; Alcaraz-Calero, J.M. Search and rescue operation using UAVs: A case study. *Expert Syst. Appl.* **2021**, *178*, 114937.

109. Goodrich, M.A.; Cooper, J.L.; Adams, J.A.; Humphrey, C.; Zeeman, R.; Buss, B.G. Using a mini-uav to support wilderness search and rescue: Practices for human-robot teaming. In Proceedings of the 2007 IEEE International Workshop on Safety, Security and Rescue Robotics, Rome, Italy, 27–29 September 2007.

110. Koubaa, A.; Ammar, A.; Abdelkader, M.; Alhabashi, Y.; Ghouti, L. AERO: AI-Enabled Remote Sensing Observation with OnboardEdge Computing in UAVs. *Remote Sens.* **2023**, *15*, 1873.

111. Anhammer, A.; Lundeberg, H. Autonomous UAV Path Planning Using RSS Signals in Search and Rescue Operations. Master's Thesis, Linköping University, Linköping, Sweden, 2022.

112. Kalatzis, N.; Avgeris, M.; Dechouniotis, D.; Papadakis-Vlachopapadopoulos, K.; Roussaki, I.; Papavassiliou, S. Edge computing in IoT ecosystems for UAV-enabled early fire detection. In Proceedings of the 2018 IEEE International Conference on Smart Computing (SMARTCOMP), Taormina, Italy, 18–20 June 2018; pp. 106–114.

113. Van Persie, M.; Oostdijk, A.; Fix, J.; Van Sijl, M.C.; Edgardh, L. Real-time UAV based geospatial video integrated into the fire brigades crisis management GIS system. *Int. Arch. Photogramm. Remote Sens. Spat. Inf. Sci.* **2012**, *38*, 173–175. [CrossRef]

114. Daniel, K.; Wietfeld, C. *Using Public Network Infrastructures for UAV Remote Sensing in Civilian Security Operations*; Technical University Dortmund: Dortmund, Germany, 2011.

115. Moumgiakmas, S.S.; Samatas, G.G.; Papakostas, G.A. Computer Vision for Fire Detection on UAVs—From Software to Hardware. *Future Internet* **2021**, *13*, 200. [CrossRef]

116. Merino, L.; Caballero, F.; Martínez-de Dios, J.; Ollero, A. Cooperative fire detection using Unmanned Aerial Vehicles. In Proceedings of the 2005 IEEE International Conference on Robotics and Automation, Barcelona, Spain, 18–22 April 2005; pp. 1884–1889.

117. Seo, S.H.; Choi, J.I.; Song, J. Secure Utilization of Beacons and UAVs in Emergency Response Systems for Building Fire Hazard. *Sensors* **2017**, *17*, 2200. [PubMed]

118. Kumarapu, K.; Shashi, M.; Keesara, V.R. UAV in Construction Site Monitoring and Concrete Strength Estimation. *J. Indian Soc. Remote Sens.* **2020**, *49*, 619–627. [CrossRef]

119. Anwar, N.; Najam, F.A.; Izhar, M.A. Construction monitoring and reporting using drones and unmanned aerial vehicles (UAVs). In Proceedings of the Tenth International Conference on Construction in the 21st Century (CITC-10), Colombo, Sri lanka, 2–4 July 2018.

120. Jo, D.; Kwon, Y. Development of Rescue Material Transport UAV (Unmanned Aerial Vehicle). *World J. Eng. Technol.* **2017**, *5*, 720–729. [CrossRef]

121. Huang, Y.; Han, H.; Zhang, B.; Su, X.; Gong, Z. Supply distribution center planning in UAV-based logistics networks for post-disaster supply delivery. In Proceedings of the 2020 IEEE International Conference on E-health Networking, Application & Services (HEALTHCOM), Shenzhen, China, 1–2 March 2021; pp. 1–6.

122. Wenjian, Z.; Sidong, Z.; RongJie, C.; Jingchang, X.; Yeqian, L.; Huiru, C. Design of a relief materials delivery system based on UAV. *IOP Conf. Ser. Mater. Sci. Eng.* **2020**, *715*, 012049. [CrossRef]

123. Zhang, J.; Zhu, Y.; Wang, T.; Wang, W.; Wang, R.; Li, X. An Improved Intelligent Auction Mechanism for Emergency Material Delivery. *Mathematics* **2022**, *10*, 2184. [CrossRef]

124. Zhang, R.; Dou, L.; Xin, B.; Chen, C.; Deng, F.; Chen, J. A Review on the Truck and Drone Cooperative Delivery Problem. *Unmanned Syst.* **2023**, *12*, 1–25.

125. Amicone, D.; Cannas, A.; Marci, A.; Tortora, G. A Smart Capsule Equipped with Artificial Intelligence for Autonomous Delivery of Medical Material through Drones. *Appl. Sci.* **2021**, *11*, 7976.

126. Kim, J.; Moon, H.; Jung, H. Drone-Based Parcel Delivery Using the Rooftops of City Buildings: Model and Solution. *Appl. Sci.* **2020**, *10*, 4362.

127. Saponi, M.; Borboni, A.; Adamini, R.; Faglia, R.; Amici, C. Embedded Payload Solutions in UAVs for Medium and Small Package Delivery. *Machines* **2022**, *10*, 737. [CrossRef]

128. Ragauskas, U.; Bručas, D.; Sužiedelytė Visockienė, J. Research of remotely piloted vehicles for cargo transportation. *Aviation* **2016**, *20*, 14–20. [CrossRef]

129. Olsen, M.J.; Parrish, C.; Che, E.; Jung, J.; Greenwood, J. *Lidar for Maintenance of Pavement Reflective Markings and Retroreflective Signs (No. FHWA-OR-RD-19-01)*; Oregon Department of Transportation: Salem, OR, USA, 2018.

130. Escobar-Wolf, R.; Oommen, T.; Brooks, C.N.; Dobson, R.J.; Ahlborn, T.M. Unmanned Aerial Vehicle (UAV)-Based Assessment of Concrete Bridge Deck Delamination Using Thermal and Visible Camera Sensors: A Preliminary Analysis. *Res. Nondestruct. Eval.* **2017**, *29*, 183–198. [CrossRef]

131. Bolourian, N.; Soltani, M.M.; Albahria, A.H.; Hammad, A. High level framework for bridge inspection using LiDAR-equipped UAV. In Proceedings of the International Symposium on Automation and Robotics in Construction (ISARC), Taipei, Taiwan, 28 June–1 July 2017; Volume 34.

132. Bolourian, N.; Hammad, A. Li-equipped UAV path planning considering potential locations of defects for bridge inspection. *Autom. Constr.* **2020**, *117*, 103250.

133. Seo, J.; Duque, L.; Wacker, J.P. Field Application of UAS-Based Bridge Inspection. *Transp. Res. Rec. J. Transp. Res. Board* **2018**, *2672*, 72–81.

134. Tomiczek, A.P.; Whitley, T.J.; Bridge, J.A.; Ifju, P.G. Bridge Inspections with Small Unmanned Aircraft Systems: Case Studies. *J. Bridge Eng.* **2019**, *24*, 05019003.

135. Ellenberg, A.; Kontsos, A.; Moon, F.; Bartoli, I. Bridge deck delamination identification from unmanned aerial vehicle infrared imagery. *Autom. Constr.* **2016**, *72*, 155–165. [CrossRef]

136. Lei, B.; Wang, N.; Xu, P.; Song, G. New crack detection method for bridge inspection using UAV incorporating image processing. *J. Aerosp. Eng.* **2018**, *31*, 04018058.

137. Kim, I.-H.; Jeon, H.; Baek, S.-C.; Hong, W.-H.; Jung, H.-J. Application of crack identification techniques for an aging concrete bridge inspection using an unmanned aerial vehicle. *Sensors* **2018**, *18*, 1881. [CrossRef] [PubMed]

138. Pi, H.; Li, X.; Yuan, C.; Yang, Z.; Wei, L.; Lian, Z. Application of multi-rotor UAV patrol system in safety and quality management of power grid construction projects. In Proceedings of the 2020 International Conference on Artificial Intelligence and Electromechanical Automation (AIEA), Tianjin, China, 26–28 June 2020; pp. 416–419.
139. Xiong, D.; Wei, W.; Zhang, W.; Huang, K.; Li, B.; Zhang, L.; Bian, L.; Liu, J. UAV Inspection Monitoring and Management System based on Wireless Sensing Technology. In Proceedings of the 2022 IEEE 2nd International Conference on Electronic Technology, Communication and Information (ICETCI), Changchun, China, 27–29 May 2022; pp. 958–961.
140. Omar, T.; Nehdi, M.L. Remote sensing of concrete bridge decks using unmanned aerial vehicle infrared thermography. *Autom. Constr.* **2017**, *83*, 360–371. [CrossRef]
141. Yamazaki, F.; Kubo, K.; Tanabe, R.; Liu, W. Damage assessment and 3d modeling by UAV flights after the 2016 Kumamoto, Japan earthquake. In Proceedings of the 2017 IEEE International Geoscience and Remote Sensing Symposium (IGARSS), Fort Worth, TX, USA, 23–28 July 2017; pp. 3182–3185.
142. Parra-Peñuela, H.; Angulo-Morales, V.; Gaona-Garcia, E. Seismic Analysis on Historical Bridge Using Photogrammetry and Finite Elements. In Proceedings of the IGARSS 2020–2020 IEEE International Geoscience and Remote Sensing Symposium, Waikoloa, HI, USA, 26 September–2 October 2020; pp. 6627–6630.
143. Hirose, M.; Xiao, Y.; Zuo, Z.; Kamat, V.R.; Zekkos, D.; Lynch, J. Implementation of UAV localization methods for a mobile post-earthquake monitoring system. In Proceedings of the IEEE Workshop on Environmental, Energy, and Structural Monitoring Systems (EESMS 2015), Trento, Italy, 9–10 July 2015.
144. Kim, H.; Ahn, E.; Shin, M.; Sim, S.-H. Crack and non-crack classification from concrete surface images using machine learning. *Struct. Health Monit.* **2019**, *18*, 725–738. [CrossRef]
145. Oskouie, P.; Becerik-Gerber, B.; Soibelman, L. A data quality-driven framework for asset condition assessment using LiDAR and image data. In Proceedings of the 2015 International Workshop on Computing in Civil Engineering, Austin, TX, USA, 21–23 June 2015; pp. 240–248.
146. Kwon, S.; Park, J.W.; Moon, D.; Jung, S.; Park, H. Smart Merging Method for Hybrid Point Cloud Data Using UAV and LIDAR in Earthwork Construction. *Procedia Eng.* **2017**, *196*, 21–28. [CrossRef]
147. Room, M.H.M.; Ahmad, A. Fusion of Uav-Based LIDAR and Mobile Laser Scanning Data for Construction of 3d Building Model. *Int. Arch. Photogramm. Remote Sens. Spat. Inf. Sci.* **2023**, *48*, 297–302. [CrossRef]
148. Wang, J.; Sun, W.; Shou, W.; Wang, X.; Wu, C.; Chong, H.Y.; Liu, Y.; Sun, C. Integrating BIM and LiDAR for real-time construction quality control. *J. Intell. Robot. Syst. Theory Appl.* **2015**, *79*, 417–432.
149. Kashani, A.G.; Graettinger, A.J.; Dao, T. Lidar-Based Methodology to Evaluate Fragility Models for Tornado-Induced Roof Damage. *Nat. Hazards Rev.* **2016**, *17*, 04016006. [CrossRef]
150. Chen, G.; Wiede, C.; Kokozinski, R. Data Processing Approaches on SPAD-Based d-TOF LiDAR Systems: A Review. *IEEE Sens. J.* **2021**, *21*, 5656–5667. [CrossRef]
151. Chauve, A.; Bretar, F.; Durrieu, S.; Pierrot-Deseilligny, M.; Puech, W. FullAnalyze: A research tool for handling, processing and analyzing full-waveform LiDAR data. In Proceedings of the IEEE International Geoscience and Remote Sensing Symposium, Cape Town, South Africa, 12–17 July 2009.
152. Lyu, Y.; Bai, L.; Huang, X. Real-Time Road Segmentation Using LiDAR Data Processing on an FPGA. In Proceedings of the 2018 IEEE International Symposium on Circuits and Systems (ISCAS), Florence, Italy, 27–30 May 2018; pp. 1–5.
153. Venugopal, V.; Kannan, S. Accelerating real-time LiDAR data processing using GPUs. In Proceedings of the 2013 IEEE 56th International Midwest Symposium on Circuits and Systems (MWSCAS), Columbus, OH, USA, 4–7 August 2013; pp. 1168–1171.
154. Cao, V.H.; Chu, K.X.; Le-Khac, N.A.; Kechadi, M.T.; Laefer, D.; Truong-Hong, L. Toward a new approach for massive LiDAR data processing. In Proceedings of the 2nd IEEE International Conference on Spatial Data Mining and Geographical Knowledge Services, ICSDM 2015, Fuzhou, China, 8–10 July 2015; pp. 135–140.
155. Du, R.; Lee, H.J. A novel compression algorithm for LiDAR data. In Proceedings of the 2012 5th International Congress on Image and Signal Processing, Chongqing, China, 16–18 October 2012; pp. 987–991.
156. Lu, Y.; Xue, Z.; Xia, G.S.; Zhang, L. A survey on vision-based UAV navigation. *Geo-Spat. Inf. Sci.* **2018**, *21*, 21–32.
157. Conte, G.; Doherty, P. An Integrated UAV Navigation System Based on Aerial Image Matching. In Proceedings of the 2008 IEEE Aerospace Conference, Big Sky, MT, USA, 1–8 March 2008; pp. 1–10.
158. Nikolos, I.; Valavanis, K.; Tsourveloudis, N.; Kostaras, A. Evolutionary algorithm based offline/online path planner for uav navigation. *IEEE Trans. Syst. Man Cybern. Part B Cybern.* **2003**, *33*, 898–912. [CrossRef]
159. Devos, A.; Ebeid, E.; Manoonpong, P. Development of autonomous drones for adaptive obstacle avoidance in real world environments. In Proceedings of the 21st Euromicro Conference on Digital System Design, Prague, Czech Republic, 15–21 June 2018.
160. Aguilar, W.G.; Casaliglla, V.P.; Pólit, J.L. Obstacle avoidance based-visual navigation for micro aerial vehicles. *Electronics* **2017**, *6*, 10. [CrossRef]
161. Zheng, H.; Zhong, X.; Yan, J.; Zhao, L.; Wang, X. A Thermal Performance Detection Method for Building Envelope Based on 3D Model Generated by UAV Thermal Imagery. *Energies* **2020**, *13*, 6677. [CrossRef]
162. Quater, P.B.; Grimaccia, F.; Leva, S.; Mussetta, M.; Aghaei, M. Light Unmanned Aerial Vehicles (UAVs) for cooperative inspection of PV plants. *IEEE J. Photovolt.* **2014**, *4*, 1107–1113. [CrossRef]

163. Wang, Z.F.; Yu, Y.F.; Wang, J.; Zhang, J.Q.; Zhu, H.L.; Li, P.; Xu, L.; Jiang, H.-N.; Sui, Q.-M.; Jia, L.; et al. Convolutional neural-network-based automatic dam-surface seepage defect identification from thermograms collected from UAV-mounted thermal imaging camera. *Constr. Build. Mater.* **2022**, *323*, 126416. [CrossRef]

164. Citroni, R.; Di Paolo, F.; Livreri, P. A novel energy harvester for powering small UAVs: Performance analysis, model validation and flight results. *Sensors* **2019**, *19*, 1771. [CrossRef] [PubMed]

165. Hossain, E.; Faruque, H.M.R.; Sunny, M.S.H.; Mohammad, N.; Nawar, N. A comprehensive review on energy storage systems: Types, comparison, current scenario, applications, barriers, and potential solutions, policies, and future prospects. *Energies* **2020**, *13*, 3651.

Article

A Novel UAV Visual Positioning Algorithm Based on A-YOLOX

Ying Xu, Dongsheng Zhong, Jianhong Zhou, Ziyi Jiang, Yikui Zhai * and Zilu Ying

Department of Intelligent Manufacturing, Wuyi University, Jiangmen 529020, China
* Correspondence: yikuizhai@163.com; Tel.: +86-1802-298-7593

Abstract: The application of UAVs is becoming increasingly extensive. However, high-precision autonomous landing is still a major industry difficulty. The current algorithm is not well-adapted to light changes, scale transformations, complex backgrounds, etc. To address the above difficulties, a deep learning method was here introduced into target detection and an attention mechanism was incorporated into YOLOX; thus, a UAV positioning algorithm called attention-based YOLOX (A-YOLOX) is proposed. Firstly, a novel visual positioning pattern was designed to facilitate the algorithm's use for detection and localization; then, a UAV visual positioning database (UAV-VPD) was built through actual data collection and data augmentation and the A-YOLOX model detector developed; finally, corresponding high- and low-altitude visual positioning algorithms were designed for high- and low-altitude positioning logics. The experimental results in the actual environment showed that the AP50 of the proposed algorithm could reach 95.5%, the detection speed was 53.7 frames per second, and the actual landing error was within 5 cm, which meets the practical application requirements for automatic UAV landing.

Keywords: deep learning; data synthesis; A-YOLOX; visual positioning

1. Introduction

Public security, a critical field of national security, correlates strongly with personal interests and property safety. With its national economic development and modernization, China has been assigning more and more importance to public security. With the advantages of high flexibility, maneuverability, stealth, independence from the geographical environment, being low cost, and having the ability to carry different processing equipment, UAVs have been used for identification and detection in such areas as urban inspection [1], fire monitoring [2–5], criminal investigation and counter-terrorism [6], normal security patrolling [7], epidemic prevention and control [8], post-disaster rescue [9,10], agricultural inspection [11,12], and power inspection [13,14]. For example, UAVs can be used in agriculture for mapping farmland, spraying pesticides, seed sowing, monitoring crop growth, irrigation, pest diagnosis, artificial pollination, and much more. The use of UAVs greatly reduces working time and increases production efficiency, thus promoting the development of intelligent agriculture [15,16]. As UAVs are widely used in military and civil fields, their intelligent application has become a development trend, and autonomous positioning landing is the basis for realizing intelligent UAVs. With the efforts of researchers in recent years, UAV landing technology has made significant progress, but there are still some limitations. For example, GPS-based methods fail in places where there is no GPS signal [17], and traditional image recognition-based methods have poor recognition effects and poor stability in environments with changing light and complex backgrounds [18]. Therefore, research on a visual positioning algorithm for UAVs has important application value and diverse application scenarios.

In this study, we started from the above problems and strove to find relevant solutions to achieve accurate landing with UAVs. Compared to traditional methods, our method offers several advantages. First, the detection model uses an anchor-free target detection algorithm, which is much faster. The FPS can reach 53.7, which meets the requirements of

Citation: Xu, Y.; Zhong, D.; Zhou, J.; Jiang, Z.; Zhai, Y.; Ying, Z. A Novel UAV Visual Positioning Algorithm Based on A-YOLOX. *Drones* **2022**, *6*, 362. https://doi.org/10.3390/drones6110362

Academic Editor: Seokwon Yeom

Received: 12 October 2022
Accepted: 14 November 2022
Published: 18 November 2022

real-time detection. Second, in comparison with previous methods [19–21], our method possesses much higher actual landing accuracy. Third, we introduce deep-learning methods into the UAV landing process, which are characterized by powerful feature extraction and characterization capabilities. This significantly improves the detection performance of the model, which is able to undertake detection accurately despite light changes, scale changes, and wind impacts and shows better robustness. In summary, this paper makes the following contributions:

- During the process of UAV landing, when moving from high to low altitudes, the visual imaging constantly changes, and the pixel area of the target pattern gradually increases, which poses a great challenge for target detection. Therefore, we developed high- and low-altitude visual positioning algorithms to achieve stable detection with UAVs throughout the process of moving from high to low altitudes;
- To solve the problem of poor detection of small- and medium-sized targets with the model, we supplemented the YOLOX algorithm [22] with an attention mechanism and proposed the attention-based YOLOX (A-YOLOX) detection algorithm, which improves the detection performance of the model for small- and medium-sized targets;
- We collected 6541 actual images under different conditions and expanded the data with data synthesis techniques in order to compile the UAV Visual Positioning Database (UAV-VPD), a database applicable for UAV landing scenarios;
- Extensive experiments were carried out with the newly created database and in the real environment, and our model proved to be robust. Our model achieved an actual landing accuracy within 5 cm, and the FPS reached 53.7, which meets the requirements of real-time detection.

The organization of the remaining sections is as follows: Section 2 concerns related work, describing current approaches to autonomous positioning and the existing problems; Section 3 describes the visual positioning algorithm proposed in this paper in detail; Section 4 presents the experiments and discussion; and Section 5 is devoted to conclusions and future work.

2. Related Work

Autonomous positioning landing is generally divided into visual positioning landing [23] and satellite navigation landing [24]. Satellite navigation landing is a traditional UAV positioning technique that uses the Global Positioning System (GPS) for positioning, and it is suitable for long-duration tasks [25,26]. However, there are some limitations in satellite navigation landing, such as easy signal loss in scenes with more occlusions, the lack of a guarantee of stability, and low accuracy [27,28], meaning that it cannot meet the requirement for centimeter-level error.

UAV visual positioning landing mainly relies on image sensors and uses image processing technology to achieve an accurate landing, and this is a research hotspot for scholars in China and abroad. Sharp et al. [19] proposed a precision landing system for the autonomous landing of multi-rotor UAVs. This system uses a large square and five small squares as landmark patterns. The landing process starts with initial recognition through the large square and then combines image processing techniques, such as feature point extraction, area segmentation, and motion estimation, to guide the UAV to land. Lange et al. [29] put forward a method for UAV landing based on a moving target plate with a landmark pattern consisting of a black, square hexagon and four white concentric circles, using optical flow sensors to acquire the velocity of the moving target and, thus, track the moving target, while the flight altitude of the UAV is acquired from the size of the landmark pattern imaging. Marut et al. [20] introduced a simple and low-cost visual landing system. The system uses Aruco markers and obtains candidate marker points by extracting contours, filtering, and other image processing techniques and then compares them with a marker dictionary to determine the location of the markers. Yuan et al. [21] proposed a hierarchical vision-based open landing and positioning method for rotary wing UAVs. This method defines the landing of UAVs as "Approaching", "Adjustment", and

"Touchdown" and develops the corresponding detection and positioning systems for these three phases. In addition, a federated Extended Kalman Filter (EKF) is designed to evaluate the attitude of UAVs. Zhou et al. [23] designed a monocular camera-based AprilTages visual positioning algorithm for UAV positioning and state estimation. They design a number of different sizes of labels to enable UAVs to position themselves at different altitudes. Xiu et al. [30] proposed a tilt-rotor quadrotor model for autonomous landing, which controlled the motor direction by using four servos so as to control UAVs' positions and attitudes, achieving tilt-rotor parking and tilting flight. This model controls UAVs' positions and attitudes more precisely, which enables a more effective landing for UAVs. Sefidgar et al. [31] designed a landing system with sensors that consisted of four ToF sensors and a monocular camera. First, the features of the AprilTag pattern are extracted by the designed algorithm to find the center point and calibrate it. Then, the sensor contacts are used to set up coordinate equations, and focal lengths in X and Y directions are solved to derive the coordinates of the ground pattern. With the continuous research and exploration of researchers, traditional image processing algorithms have undoubtedly made great efforts to improve the accuracy of UAV landing. However, their good performance depends on a good imaging environment, and the algorithm's performance will be significantly degraded under the situations of insufficient light, complex background, occlusion, scale transformation, etc. It is difficult to meet the actual demand for centimeter-level landing errors for UAVs in different scenes. Table 1 shows the comparison of different visual landing methods.

Table 1. The comparison of different visual landing methods.

	Methods	Landmark Pattern Type	Landing Accuracy	Test Type
[26]	Feature point extraction, area segmentation, and motion estimation.	Square	Position 5 cm Pose 5°	Landing test
[27]	Optical flow sensor, fixed threshold, segmentation, and contour detection.	Orthohexagon and circular	Position 3.8 cm	Landing test
[28]	Contour extraction and filtering.	ArUco	10% error rate	Landing test
[29]	Optical flow sensors and extended Kalman filter	Square	Position 6.4 cm pose 0.08°	Landing test
[30]	Histogram of oriented gradients (HOG) and normalized cross-correlation (NCC).	AprilTag	Landing error within (−20 cm, +50 cm)	Landing test
[31]	Canny, Adaptive thresholding and Levenberg–Marquardt (LM).	Combination patterns	Position < 10 cm	Simulation
[32]	Contour extraction and 3D rigid body transformation.	AprilTag	X: 0.47 cm Y: 0.42 cm	Simulation

Deep learning methods have been developed rapidly in recent years [32,33]. In 2014, Girshick et al. [34] proposed RCNN (Region-based Convolutional Neural Networks, RCNN) and introduced deep learning into target detection for the first time, opening a new chapter for target detection. Deep-learning-based target detection algorithms have also become a hot topic for scholars in recent years. Many scholars have successively proposed two-stage detection networks such as SPPNet (Spatial Pyramid Pooling in Deep Convolutional Networks, SPPNet) [35], FastR-CNN, and FasterR-CNN [36], which use RPN (Region Proposal Network, RPN) [37] to generate a large number of candidate frames to improve the recall rate, and the confidence of these candidate frames is not utilized in the inference stage, which reduces the inference speed. In 2016, Redmon et al. [38] proposed the first version of the YOLO (You Only Look Once, YOLO) series of single-stage networks, YOLOv1, which surpassed the detection speed of two-stage detectors. Moreover, its accuracy is continuously improveed by subsequent researchers, which is comparable to that of two-stage detection networks, meeting the requirements of most industrial scenarios. As a result, the YOLO series has also become the mainstream target detection algorithm in the industry. Deep learning models have powerful learning and characterization capabilities, and their utility and generalization capabilities are stronger [39,40]. Therefore, they are considered to be introduced into the autonomous landing process of UAVs in order to solve the various environmental interference problems mentioned and to further improve the detection speed.

3. Methods

According to the different visual imaging of UAVs at different altitudes, this paper designs a high-altitude visual positioning algorithm and a low-altitude visual positioning algorithm to guide UAVs to land accurately. When UAVs return to the vicinity of the target point, they automatically adjust the direction of the camera, return the video captured by the camera, detect it by a trained detector, and output the number and coordinates of special patterns of the image. The visual positioning algorithm automatically selects a high-altitude positioning algorithm or a low-altitude positioning algorithm by calculating the area and number of patterns. The algorithm flow is shown in Figure 1.

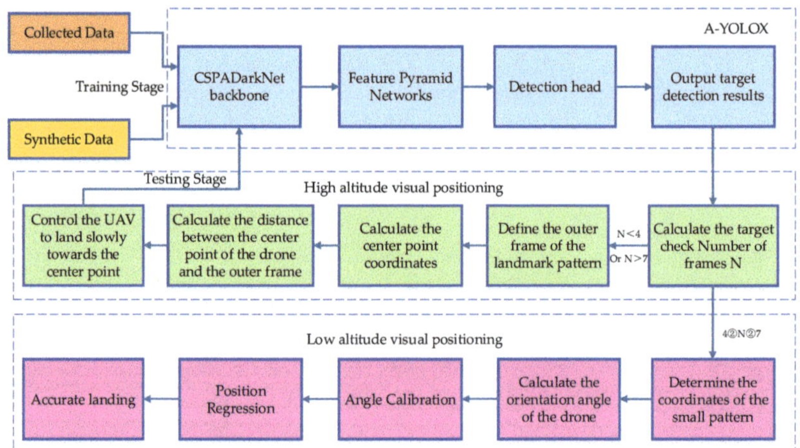

Figure 1. Flow chart of the proposed UAV visual positioning algorithm based on A-YOLOX. Firstly, the actual acquisition data and the synthesized data are used to train the A-YOLOX model to obtain a detector with good accuracy and robustness; then, the detector is used for target detection during UAV landing, and the high-altitude visual positioning algorithm is called when the number of detected target frames is N < 4 or N > 7, and the low-altitude visual localization algorithm is called when the number of target frames is $4 \leq N \leq 7$.

3.1. The Construction of UAV-VPD

A pattern that facilitates fast recognition for a detection algorithm is an important condition for UAVs to land accurately at the designated location. When the UAV flies over the landing point, it obtains ground information through the camera and then adjusts its orientation and lands toward the target point after valid information is detected. The design of the visual positioning pattern mainly follows two principles: feature discriminability and visual imaging adaptability. Feature discriminability: in order to make the model easy to recognize, the basic circular and square patterns are used in designing visual positioning patterns. However, the single basic pattern is not conducive to feature discrimination, so the circles and squares are combined inline to improve feature discriminability; visual imaging adaptability: there are two stages in the landing process of the UVA: high-altitude phase and low-altitude phase. Since the visual imaging of the UAV changes continuously during the landing process and the pixel area of the target pattern gradually increases, in order to avoid the loss of the visual pattern due to the narrow field of view of the UAV at low altitude, the designed positioning pattern adopts the mutual fusion of large and small patterns. In other words, six small patterns are added inside a large pattern, whose structure is similar to the large pattern. It is worth noting that with such a design, end-to-end high- and low-altitude visual positioning can be achieved with only one model so as to complete an efficient UAV parking and landing process. In actual application scenarios, for the weak GPS signal, UAVs will pre-bind QR codes on the apron to assist themselves in finding the location of visual positioning patterns. The visual positioning pattern is shown in Figure 2.

Figure 2. The visual positioning pattern designed in this paper.

To construct the UAV-VPD, we printed the designed patterns onto KT plate and then manually operated the UAV to fly and shoot the patterns. During the shooting process, since the adjacent frames of the video have extremely high similarity, we tried to put the KT plate in different positions of the frame during the acquisition process instead of simply having the KT plate presented in the center of the video frame. Furthermore, for the sake of improving the fit of the data in actual application scenarios, videos of different scenes, different periods, different heights and angles were collected during data collection; then, a total of 6541 images were obtained after video split-frame processing and data cleaning; finally, the visual positioning patterns were labeled by the open source labeling software. Remarkably, when the UAV returns to a relatively high position over the target location, only the outer frame of the visual positioning pattern needs to be marked. When the UAV is at low altitude, the six small marker patterns on the visual positioning pattern can be clearly seen, so all the markers within the field of view need to be marked. Some of the collected data are shown in Figure 3.

Since the scenes of UAV inspection are varied, simply relying on manual acquisition and labeling requires a lot of labor and material costs, and the scenes are also relatively single, which cannot well fit the real situation of different scenes during UAV inspection. Therefore, we use data synthesis technology to expand the data of scenes that are difficult to collect and use the synthesized data together with the real data for model training so as to improve the accuracy and strengthen the robustness and generalization ability of the model. The data synthesis uses the copy-paste method [41]. Firstly, 970 background images with high semantic similarity from different scenes are collected from the Internet using crawler technology, and then the visual localization patterns are randomly copied and pasted onto these background images, and the corresponding pattern coordinate information is extracted, which no longer needs to be re-labeled manually. When UAVs are at a low altitude and an ultra-low altitude, their imaging pictures only have recognition patterns and no other objects, and only at a high altitude will other objects be recognized, so we only need to synthesize the high-altitude data. Some of the synthesized data are shown in Figure 4.

Figure 3. Example of UAV-VPD database samples.

Figure 4. Synthetic data.

The data of 6541 images collected from real scenes are combined with the synthetic data of 970 images to form the UAV-VPD. A large number of images are needed for testing to achieve an efficient model performance evaluation, so the training set, validation set, and test set are divided in the ratio of 2:2:6. The details are shown in Table 2.

Table 2. Database composition and division.

Data Division	Training Set	Validation Set	Test Set
Training set:Validation set:Test set = 2:2:6 High:Low:Ultra Low = 4:2:4	1557	1769	4185

3.2. Object Detection Algorithm A-YOLOX

The proposed target detection algorithm A-YOLOX is based on YOLOX with the addition of CBAM (Convolutional Block Attention Module) [42], allowing CBAM to be used throughout the backbone network part of the depth model. The CBAM contains two separate submodules, a channel attention module, and a spatial attention module, which perform "attention" on the channel and space, respectively. The module structure is shown in Figure 5.

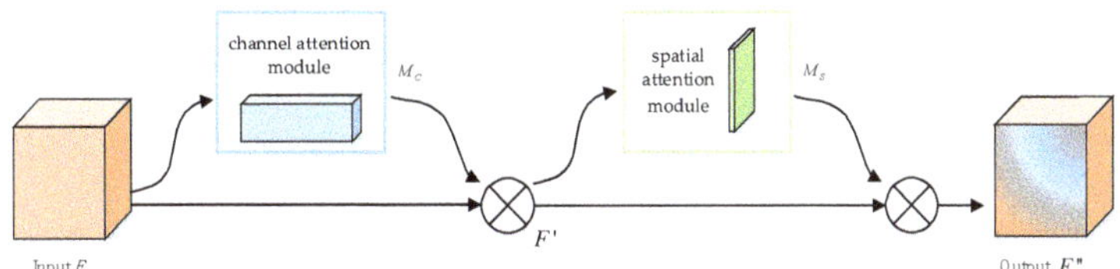

Figure 5. CBAM. It consists of a channel attention module and a spatial attention module and operates sequentially on the channel and in space.

Assuming an intermediate feature graph F is input, $F \in R^{C \times H \times W}$, CBAM first performs global maximum pooling and average pooling of F by channel, sends the pooled two one-dimensional vectors into the fully connected layer operation and sums them to generate one-dimensional channel attention $M_C \in R^{C \times 1 \times 1}$; then, multiply M_C with the input element F to obtain the channel attention-adjusted feature graph F'. Secondly, F' is conducted global maximum pooling and average pooling by space, and the two two-dimensional vectors generated by pooling are stitched together and subjected to convolution operation to eventually generate two-dimensional spatial attention $M_S \in R^{1 \times H \times W}$. Finally, the output feature F'' is obtained by multiplying M_S with F' by elements. The CBAM generation attention process can be described by Equations (1) and (2), where $\otimes$ denotes the corresponding element multiplication.

$$F' = M_c(F) \otimes F \tag{1}$$

$$F'' = M_S(F') \otimes F' \tag{2}$$

The A-YOLOX network is mainly composed of three parts, which are CSPADarkNet (Cross Stage Partial Attention DarkNet, CSPADarkNet) backbone network, FPN (Feature Pyramid Networks, FPN) [43] feature fusion network, and detection heads, as shown in Figure 6. DarkNet is a classical deep framework, which is often used as the backbone network for feature extraction in the YOLO series. Its design process borrows the idea from residual network ResNet [44] to prevent the gradient from disappearing during the deepening of the network by adding the residual module to the network, which is beneficial to the fast convergence of model training. In this paper, CSPDarkNet [45] is used and combined with CBAM attention mechanism to form CSPADarkNet backbone feature extraction network, the output of which is three effective feature layers. The three effective feature layers are then fused by the FPN network, and finally, three different scales of features: $20 \times 20 \times 512$, $40 \times 40 \times 256$, and $80 \times 80 \times 128$ are output for target classification and localization. The detection head of A-YOLOX determines whether there is an object corresponding to it at the feature point by the three feature graphs output by FPN. The structure of CSPADarknet and the refinement network structure of the detection head are shown in Figures 7 and 8.

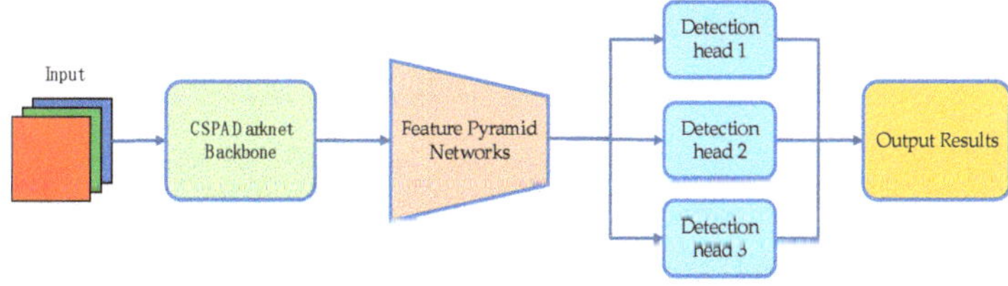

Figure 6. The network structure of A-YOLOX. Features of the input image are extracted by our modified CSPADarknet backbone network, and the three extracted effective feature layers are fused by the FPN network. Then the target is detected and determined by the detection head, and finally, the prediction result is output.

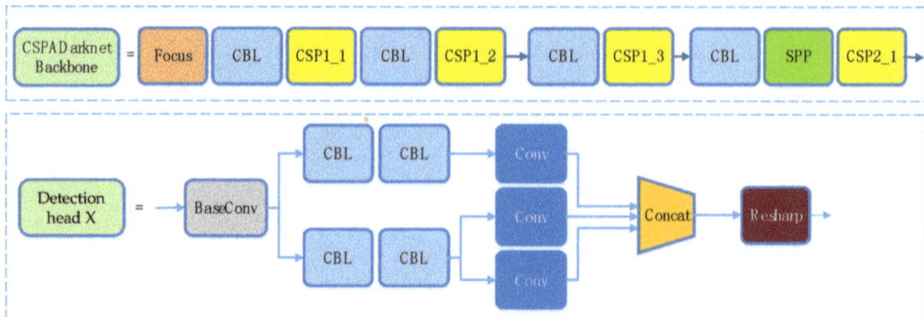

Figure 7. The network structure of CSPADarknet and detection head, which is a refined network structure diagram of the modules in Figure 6.

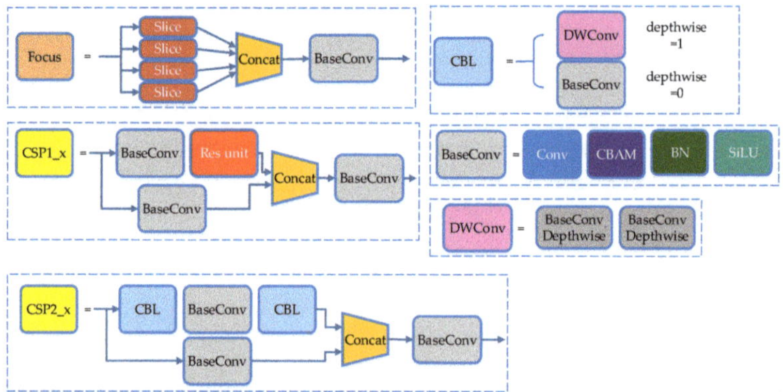

Figure 8. The refinement network structure of the CSPADarknet and the detection head, including the basic composition and order of the sub-modules. CBAM is inserted in the BaseConv between the Conv layer and the BN layer.

Since the detection head of the network respectively predicts the category, location, and object boundary frame, the loss function of the network consists of three parts: the category loss L_{cls}, the location loss L_{reg}, and the object boundary frame loss L_{obj}. L_{cls} and L_{obj} adopt the cross-entropy loss, and L_{reg} adopts the IoU loss. The formula for calculating the total loss is shown in Equation (3).

$$L = \frac{L_{cls} + \lambda L_{reg} + L_{obj}}{N_{pos}} \tag{3}$$

In the above equation, λ refers to the balance coefficient of the location loss and N_{pos} refers to the positive sample number. The A-YOLOX algorithm employs several training strategies during the training process, such as Exponential Moving Average (EMA), cosine annealing learning rate, and IOU loss, and uses the means of data enhancement such as mosaic, horizontal random rotation, and color change.

3.3. The Design of High- and Low-Altitude Visual Positioning Algorithm

3.3.1. High-Altitude Visual Positioning Algorithm

As GPS navigation technology is relatively stable in wide-open areas, when UAVs receive return instructions at a high altitude, we first use GPS positioning technology to return the UAVs to a high-altitude position a few dozen meters from the marker pattern. Then, the landing position can be determined only by recognizing the outer frame of the visual positioning pattern designed in this paper. The detection algorithm is called in real-time for recognition, and when the algorithm recognizes the target object, it outputs information such as the number and coordinates of the target frame in real-time. The high-altitude positioning algorithm calculates the target frame area A, confidence degree P, and comprehensive score S from the information output by the detector, and the target frame with the highest comprehensive score is the landing position for the target. After the target frame is determined, the relative distances dx and dy between the central point of the UAVs and the central point of the target frame can be calculated so as to call UAV flight control module to allow UAVs to descend slowly toward the center of the pattern. The UAV flight control adopts PID control mode, and its control quantity is calculated according to Equation (4).

$$u(k) = K_P \cdot e(k) + K_I \cdot \sum_{i=0} e(i) + K_D \cdot [e(k) - e(k-1)] \tag{4}$$

In the above equation, K_P refers to the proportional coefficient, K_I refers to the integral time constant, and K_D refers to the differential time constant. The target detection algorithm continuously detects and updates dx and dy, and the flight control algorithm updates $e(k)$, the deviation distance between the current position and the target position, according to dx and dy to obtain the control quantity $u(k)$ output by the PID controller, so that UAVs can quickly and steadily approach the target position until they reach the ideal position. The flow of the high-altitude visual positioning algorithm is shown in Algorithm 1.

Algorithm 1. High-Altitude Visual Positioning Algorithm.

Step 1	Call the target detection algorithm for the first detection when UVAs return over the landing site to obtain the position of center point of the camera carried by UAVs as (x_c, y_c).
Step 2	a. Take the detected target as the target landing point when only one outer frame of the visual positioning pattern is detected. b. Calculate the confidence degree of each detected target when two or more outer frames of the visual positioning pattern are detected, and score the target frame with the higher confidence level by the equation $S = 0.5 \times A + P$, thus the target frame with the highest score being the landing point.
Step 3	Calculate the relative distance between the center point of the camera and the center point of the visual positioning pattern by equations $dx = x_c - width/2$, and $dy = y_c - heigth/2$.
Step 4	Calculate the control quantity according to $u(k) = K_P \cdot e(k) + K_I \cdot \sum_{i=0} e(i) + K_D \cdot [e(k) - e(k-1)]$ to lead UAVs closer to the target position.
Step 5	Repeat steps 2 to 4.

3.3.2. Low-Altitude Visual Positioning Algorithm

During the slow descent of UAVs, the visual field of the camera will slowly become narrower, and the outer frame of the pattern will slowly disappear on the imaging. Therefore, this paper designs a low-altitude positioning logic algorithm according to the actual situation. The algorithm mainly post-processes the identification results of the six small logo patterns given by the detector.

When UAVs land at a low-altitude position, firstly, the angle should be calibrated. Assuming the visual imaging of UAVs is shown in Figure 9, set the coordinates of point A as (x_1, y_1), point B as (x_2, y_2), point C as (x_3, y_3), and point O as (x, y). Calculate vector $\left|\overrightarrow{AB}\right|$, $\left|\overrightarrow{AC}\right|$, and $\left|\overrightarrow{BC}\right|$ to determine whether the triangle is isosceles triangle, and find the vertex (assumed to be A) and the bottom side of the isosceles triangle, and the midpoint D (x_4, y_4) of the bottom side; then, the vector $\left|\overrightarrow{AD}\right|$ is the most optimal direction for UAV's landing. The midpoints of the upper and lower edges of the screen are P_1 and P_2, respectively, and the vector $\left|\overrightarrow{P_1P_2}\right|$ is the current orientation for UAVs. Therefore, our main task is to control UAVs to make attitude adjustments so that the pinch angle θ tends to 0. θ is calculated as shown in Equation (5).

$$\theta = \arg\cos \frac{(\overrightarrow{AD}, \overrightarrow{P_1P_2})}{\left|\overrightarrow{AD}\right| \cdot \left|\overrightarrow{P_1P_2}\right|} \tag{5}$$

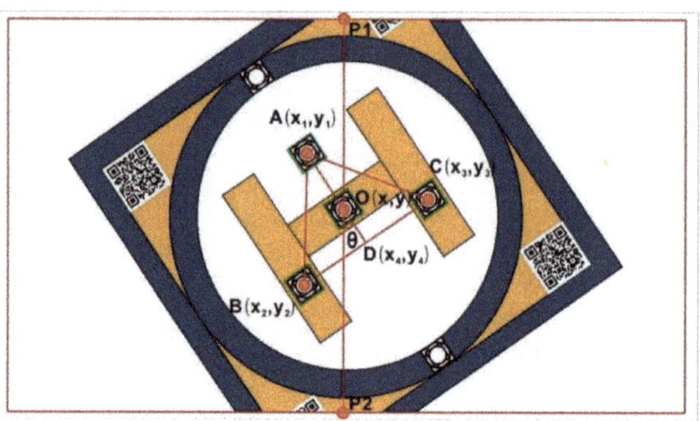

Figure 9. The visual imaging of the UAV.

After UAVs have completed angle correction, they need to perform position regression, which is to find the coordinates of the best landing point. As can be seen from Figure 9, in the case that the angle correction has been completed, the detection frame at the top of the isosceles triangle is the point we need.

4. Results and Discussion

This experiment is conducted on the ubuntu 18.04 system with Intel Xeon(R) E3-1241 v3@3.50 GHz processor. Its running memory is 24 Gb, the graphics card is NVIDIA GTX1080, the video memory is 8 Gb, and the parallel computing framework version is cuda10.2.

To evaluate the model performance, we use the target detection evaluation metrics of the COCO database as our evaluation metrics in this paper. AP_{50} and AP_{75} are the average accuracy at IoU = 0.5 and IoU = 0.75, respectively. mAP is the average of the average accuracy of the IoU from 0.5 to 0.95, in a step length of 0.05. The detection speed is evaluated by Frames Per Second (FPS).

4.1. Experiment to Verify the Validity of Synthetic Data

To improve the accuracy and generalization of the model, we start with the data and fit as many real-world application scenarios as possible. We also increase the number and complexity of the training set. However, it is not easy to obtain the data in the actual scenes, which are usually collected and labeled manually, requiring extremely huge manpower, physical resources, and time costs, so we start from synthetic images and synthesize the data of different scenes similar to the actual scenes offline.

In this paper, there are 1557 images in the training sets, of which 970 are synthetic images and 587 are real images. Both the validation set and the test set are actual acquisition data.

As can be seen from Table 3, the AP50 is 51.8% when obtained by training with only 970 synthetic images. Since the synthetic images only include the high-altitude part of the scene, but the test set contains both high-altitude and low-altitude images, so the accuracy of the test is not very high, yet it is sufficient to show that the synthetic data is effective for the detection algorithm. Adding synthetic data and real data together to the training can reach an accuracy of 95.5%, which is nearly 3% higher than when training with only real data.

Table 3. Experimental comparison of synthetic data.

Train Set	Validation Set	Test Set	AP_{50} (%)	mAP (%)
970 (synthetic images)	1769	4145	51.8	33.6%
587 (real images)	1769	4145	92.8	76.3
1557 (real images + synthetic images)	1769	4145	95.5	77.3

4.2. Experiments on Attention Mechanism

The attention mechanism was introduced mainly to enhance the poor effectiveness of the model on small and medium targets. It is desirable to validate the detection performance of the model on small, medium, and large targets so as to exhaustively verify the effectiveness of our model.

As shown in Table 4, the attention mechanism is beneficial to improve the detection accuracy of small and medium targets by 0.5% and 2%, respectively, which is helpful for UAVs to accurately identify the visual localization pattern at a high altitude and accurately locate the small graphs in the visual localization pattern at a low altitude. Since there are more computational parameters after the introduction of the attention mechanism, the FPS is much lower, but the speed of 53.7 frames per second can still meet the requirements of real-time detection.

Table 4. Experimental comparison of attention mechanisms.

Attentional Mechanisms	mAP (%) (Small Targets)	mAP (%) (Medium Targets)	mAP (%) (Large Targets)	FPS
No	35.7	66.3	87.2	149.9
Yes	36.2	68.3	87.1	53.7

4.3. Performance Comparison Experiments of Target Detection Algorithm

This paper researches the target-detection-based UAV vision localization algorithm. The detector in the vision localization algorithm can be replaced with arbitrary target detection models. To demonstrate the superiority of A-YOLOX, experiments are conducted to compare it with the target detection algorithms commonly used today. The backbone network of each model in the experiments is DarkNet53; the Epoch is 300. The learning rate is set to 0.01, and the BatchSize is 8. To be fair, all parameters are used with the same hyperparameter.

As can be seen from Table 5, A-YOLOX has a distinct advantage in the AP_{50} and mAP metrics, and its accuracy rate exceeds that of other models, especially in the mAP metric, which reaches 77.3% more than 10 points higher than other models. RetinaNet's AP_{50} reached 93.4%, which is very similar to A-YOLOX's 95.5%, but its FPS is only 6.89 frames per second, which cannot satisfy the demands of real-time detection. Taken together, the A-YOLOX offers the performance of both detection efficiency and accuracy.

Table 5. Performance comparison experiments of target detection algorithms.

Model	FPS	mAP (%)	AP_{50} (%)	AP_{75} (%)
DETR [46]	4.93	43.1	76.4	46.9
YOLOR [47]	26.5	61.2	91.5	75.2
CenterNet2 [48]	10	62.3	85.8	75.8
Faster-rcnn [31]	11.45	64.1	88.9	77.1
RetinaNet [49]	6.89	62.6	93.4	76.6
A-YOLOX[OURS]	53.7	77.3	95.5	84.6

4.4. Drone Actual Landing Experiment

We deployed the trained model to the local server for the actual landing test. In order to measure the deviation size of the actual landing point of UAVs more intuitively, the coordinate system was established with the vertex of the isosceles triangle in the visual positioning pattern as the reference point, the direction of the UAV nose as the positive Y-axis direction and the 90° clockwise rotation as the positive X-axis direction. The data and pictures recorded during the return flight of the UAV are shown in Table 6 and Figure 10.

Table 6. Experimental data of actual landing.

Test Serial Number	X-Direction (Unit: cm)	Y-Direction (Unit: cm)	Image Number
1	5.5	2.1	01
2	1.0	2.4	02
3	2.7	4.1	03
4	1.0	3.1	04
5	4.0	2.5	05
6	2.5	4.5	06
7	6.3	0.8	07
8	2.0	0	08
9	0.5	0.5	09
10	0	0	10

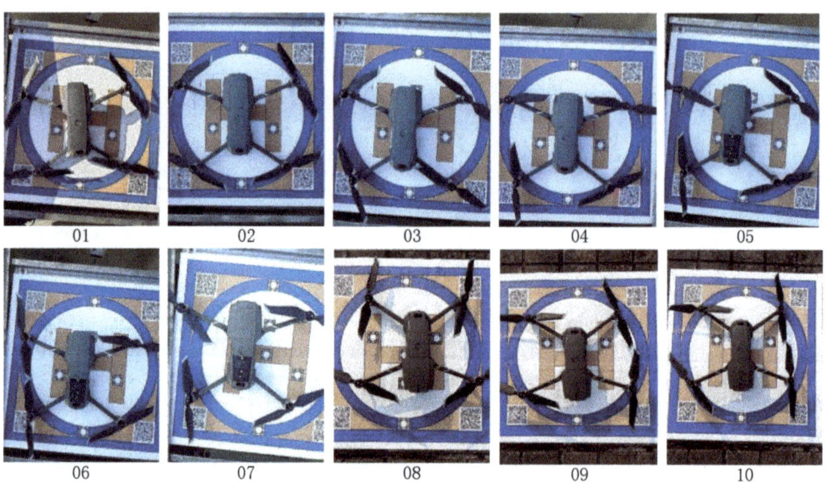

Figure 10. Actual landing pictures of drone. The pictures numbered 01, 07, 08, 09, and 10 are the landing results in strong light environment, while the rest are the landing results in non-strong light environment.

According to the above data, the average deviation value μ_x of the UAV in the X-axis direction can be calculated as 2.56 cm, the average deviation μ_y in the Y-axis direction as 2.0 cm, and the variance σ_x and σ_y in the X-axis direction and Y-axis direction are 4.07 and 2.40, respectively. Overall, our UAV's positioning algorithm achieves a centimeter-level landing error, which meets the industry precision positioning landing requirements. However, during the descent process, as it will inevitably be affected by the external airflow and its own wind field generated by the high-speed rotation of the UAV wings, the UAV tends to sway from side to side, so the landing position in the X-axis direction changes a bit more.

Actually, we conduct landing tests on sunny and cloudy days, as well as in the morning, noon and afternoon. As shown in Figure 10, the pictures numbered 01, 07, 08, 09, and 10 are the landing results in a strong light environment, while the rest are the landing results in a non-strong light environment. The tests are carried out in different areas, such as car parks, intersections, and rooftops. After calculation, it can be seen that in the non-bright light environment, the average deviation μ_x in the X-axis direction is equal to 2.24 cm; the average deviation μ_y in the Y-axis direction is equal to 3.32 cm, the variance σ_x is equal to 1.29, and σ_y is equal to 1.88. In comparison, in the bright light environment, the mean deviation μ_x is equal to 2.86 cm in the x-axis direction, μ_y is equal to 0.68 cm in the y-axis direction, and the variance σ_x is equal to 6.66 and σ_y is equal to 0.6. The difference in average landing accuracy between the two conditions is small, but the stability of the UAV landing in non-bright light conditions is much better than in bright light conditions. However, in terms of overall landings, our UAV positioning algorithm achieves centimeter-level landing errors in both bright light and non-bright light conditions.

During the actual test process, we find that the ambient wind has an effect on the UAV landing. The landing time of UAVs becomes longer as the wind increases. The main reason is that the wind makes UAVs sway, so the flight control algorithm has to constantly adjust UAVs' position according to the target detection results in order to land UAVs in an accurate position. Therefore, the constant adjustment process will cause a longer landing time. Fortunately, the target detection algorithm is still able to accurately detect the target frame of the visual positioning pattern under these circumstances. Thus, the visual positioning algorithm still shows good robustness under the influence of light changes, scene changes, and ambient wind.

5. Conclusions and Future Work

The combined application of UAV technology and computer vision technology is of great value and research significance in both civilian and military fields. In this paper, in order to improve the accuracy of automatic landing for UAVs, based on the actual situation that the performance of traditional image processing algorithms is sensitive to environmental changes, we introduce deep learning methods into target detection, propose the A-YOLOX target detection algorithm, and improve the training model with data synthesis technology and an attention mechanism to enhance the accuracy and generalization of the detection network. The corresponding high- and low-altitude visual localization algorithms are designed for the height change and visual transformation of UAV landing, and the landing test is conducted in the actual scene. The experimental results show that the proposed algorithm can achieve a processing speed of 53.7 frames/second and an accuracy rate of 95.5%, and the actual landing error is within 5cm, which effectively solves the problem of low landing accuracy under changing light, scale change, and complex background, thus realizing the high-precision autonomous landing for UAVs.

Although our model performs relatively better, we have to admit that it still has some limitations. For example, there is hovering in complex scenes. Although UAVs can still land in the expected position, the process consumes a certain amount of time. There may be some safety risks in low-power situations. In addition, the stability of UAVs when landing is slightly poor in strong ambient wind conditions, and they tend to swing. Therefore, there is room for further improvement of the control algorithm. In the future, we will continue to work on the basis of the current research to continuously improve these deficient aspects and achieve a more efficient and accurate autonomous landing for UAVs.

Author Contributions: Y.X. and D.Z. wrote the code and paper. Y.X. and Y.Z. conceived of and designed the experiments. D.Z. and J.Z. performed the experiments. J.Z. and Z.J. collected the data. Y.Z. and Z.Y. revised the paper. All authors have read and agreed to the published version of the manuscript.

Funding: This work was supported by Key Research Projects for the Universities of Guangdong Provincial Education Department (No. 2019KZDZX1017, No. 2020ZDZX3031); Guangdong Basic and Applied Basic Research Foundation (No. 2019A1515010716, No. 2021A1515011576, No. 2017KCXTD015); Guangdong, Hong Kong, Macao and the Greater Bay Area International Science and Technology Innovation Cooperation Project (No. 2021A050530080, No. 2021A0505060011). Jiangmen Basic and Applied Basic Research Key Project (2021030103230006670). Key Laboratory of Public Big Data in Guizhou Province (No. 2019BDKFJJ015).

Institutional Review Board Statement: Not applicable.

Informed Consent Statement: Not applicable.

Data Availability Statement: Not applicable.

Conflicts of Interest: The authors declare no conflict of interest.

References

1. Ming, Z.; Huang, H. A 3d vision cone based method for collision free navigation of a quadcopter UAV among moving obstacles. *Drones* **2021**, *5*, 134. [CrossRef]
2. Giuseppi, A.; Germanà, R.; Fiorini, F. UAV Patrolling for Wildfire Monitoring by a Dynamic Voronoi Tessellation on Satellite Data. *Drones* **2021**, *5*, 130. [CrossRef]
3. Ausonio, E.; Bagnerini, P.; Ghio, M. Drone swarms in fire suppression activities: A conceptual framework. *Drones* **2021**, *5*, 17. [CrossRef]
4. Akhloufi, M.A.; Couturier, A.; Castro, N.A. Unmanned aerial vehicles for wildland fires: Sensing, perception, cooperation and assistance. *Drones* **2021**, *5*, 15. [CrossRef]
5. Aydin, B.; Selvi, E.; Tao, J. Use of fire-extinguishing balls for a conceptual system of drone-assisted wildfire fighting. *Drones* **2019**, *3*, 17. [CrossRef]
6. Zhang, J.; Huang, H. Occlusion-aware UAV path planning for reconnaissance and surveillance. *Drones* **2021**, *5*, 98. [CrossRef]
7. Khan, A.; Rinner, B.; Cavallaro, A. Cooperative Robots to Observe Moving Targets: Review. *IEEE Trans. Cybern.* **2018**, *48*, 187–198. [CrossRef]
8. Fan, J.; Yang, X.; Lu, R. Design and implementation of intelligent inspection and alarm flight system for epidemic prevention. *Drones* **2021**, *5*, 68. [CrossRef]
9. Alsamhi, S.H.; Shvetsov, A.V.; Kumar, S. UAV computing-assisted search and rescue mission framework for disaster and harsh environment mitigation. *Drones* **2022**, *6*, 154. [CrossRef]
10. Ding, J.; Zhang, J.; Zhan, Z. A Precision Efficient Method for Collapsed Building Detection in Post-Earthquake UAV Images Based on the Improved NMS Algorithm and Faster R-CNN. *Remote Sens.* **2022**, *14*, 663. [CrossRef]
11. Jumaah, H.J.; Kalantar, B.; Halin, A.A. Development of UAV-based PM2. 5 monitoring system. *Drones* **2021**, *5*, 60. [CrossRef]
12. Krul, S.; Pantos, C.; Frangulea, M. Visual SLAM for indoor livestock and farming using a small drone with a monocular camera: A feasibility study. *Drones* **2021**, *5*, 41. [CrossRef]
13. Zhao, W.; Dong, Q.; Zuo, Z. A Method Combining Line Detection and Semantic Segmentation for Power Line Extraction from Unmanned Aerial Vehicle Images. *Remote Sens.* **2022**, *14*, 1367. [CrossRef]
14. Ben, M.B. Power Line Charging Mechanism for Drones. *Drones* **2021**, *5*, 108. [CrossRef]
15. Aslan, M.F.; Durdu, A.; Sabanci, K. A comprehensive survey of the recent studies with UAV for precision agriculture in open fields and greenhouses. *Appl. Sci.* **2022**, *12*, 1047. [CrossRef]
16. Kim, J.; Kim, S.; Ju, C. Unmanned aerial vehicles in agriculture: A review of perspective of platform, control, and applications. *IEEE Access* **2019**, *7*, 105100–105115. [CrossRef]
17. Bassolillo, S.R.; D'Amato, E.; Notaro, I. Enhanced Attitude and Altitude Estimation for Indoor Autonomous UAVs. *Drones* **2022**, *6*, 18. [CrossRef]
18. Xin, L.; Tang, Z.; Gai, W. Vision-Based Autonomous Landing for the UAV: A Review. *Aerospace* **2022**, *9*, 634. [CrossRef]
19. Sharp, C.S.; Shakernia, O.; Sastry, S.S. A vision system for landing an unmanned aerial vehicle. In Proceedings of the 2001 IEEE International Conference on Robotics and Automation, (ICRA), Seoul, Korea, 21–26 May 2001. [CrossRef]
20. Marut, A.; Wojtowicz, K.; Falkowski, K. ArUco markers pose estimation in UAV landing aid system. In Proceedings of the 2019 IEEE 5th International Workshop on Metrology for AeroSpace (MetroAeroSpace), Torino, Italy, 19–21 June 2019. [CrossRef]
21. Yuan, H.; Xiao, C.; Xiu, S. A hierarchical vision-based UAV localization for an open landing. *Electronics* **2018**, *7*, 68. [CrossRef]
22. Ge, Z.; Liu, S.; Wang, F. Yolox: Exceeding yolo series in 2021. *arXiv* **2021**, arXiv:2107.08430.
23. Li, Z.; Chen, Y.; Lu, H. UAV autonomous landing technology based on AprilTags vision positioning algorithm. In Proceedings of the 2019 Chinese Control Conference (CCC), Guangzhou, China, 27–30 July 2019. [CrossRef]
24. Al-Radaideh, A.; Sun, L. Self-Localization of Tethered Drones without a Cable Force Sensor in GPS-Denied Environments. *Drones* **2021**, *5*, 135. [CrossRef]
25. Kwak, J.; Sung, Y. Autonomous UAV flight control for GPS-based navigation. *IEEE Access* **2018**, *6*, 37947–37955. [CrossRef]

26. Abdelkrim, N.; Aouf, N.; Tsourdos, A. Robust nonlinear filtering for INS/GPS UAV localization. In Proceedings of the 2008 16th Mediterranean Conference on Control and Automation, Ajaccio, France, 25–27 June 2008. [CrossRef]

27. Vanegas, F.; Gaston, K.J.; Roberts, J. A framework for UAV navigation and exploration in GPS-denied environments. In Proceedings of the 2019 IEEE Aerospace Conference, Big Sky, MT, USA, 2–9 March 2019. [CrossRef]

28. Wubben, J.; Fabra, F.; Calafate, C.T. Accurate landing of unmanned aerial vehicles using ground pattern recognition. *Electronics* **2019**, *8*, 1532. [CrossRef]

29. Lange, S.; Sunderhauf, N.; Protzel, P. A vision based on board approach for landing and position control of an autonomous multirotor UAV in GPS-denied environments. In Proceedings of the 14th International Conference on Advanced Robotics (ICAR), Munich, Germany, 22–26 June 2009.

30. Xiu, S.; Wen, Y.; Xiao, C. Design and Simulation on Autonomous Landing of a Quad Tilt Rotor. *Syst. Simul.* **2020**, *32*, 1676. [CrossRef]

31. Sefidgar, M.; Landry, J.R. Unstable landing platform pose estimation based on Camera and Range Sensor Homogeneous Fusion (CRHF). *Drones* **2022**, *6*, 60. [CrossRef]

32. Zhao, Z.Q.; Zheng, P.; Xu, S. Object detection with deep learning: A review. *IEEE Trans. Neural Netw. Learn. Syst.* **2019**, *30*, 3212–3232. [CrossRef]

33. Xiao, Y.; Tian, Z.; Yu, J. A review of object detection based on deep learning. *Multimed. Tools Appl.* **2020**, *79*, 23729–23791. [CrossRef]

34. Girshick, R.; Donahue, J.; Darrell, T. Rich Feature Hierarchies for Accurate Object Detection and Semantic Segmentation. In Proceedings of the 2014 IEEE Computer Society Conference on Computer Vision and Pattern Recognition (CVPR), Columbus, OH, USA, 23–28 June 2014.

35. He, K.; Zhang, X.; Ren, S. Spatial pyramid pooling in deep convolutional networks for visual recognition. *IEEE Trans. Pattern Anal. Mach. Intell.* **2015**, *37*, 1904–1916. [CrossRef]

36. Ren, S.; He, K.; Girshick, R. Faster R-CNN: Towards Real-Time Object Detection with Region Proposal Networks. *IEEE Trans. Pattern Anal. Mach. Intell.* **2017**, *39*, 1137–1149. [CrossRef]

37. Fan, Q.; Zhuo, W.; Tang, C.K. Few-shot object detection with attention-RPN and multi-relation detector. In Proceedings of the IEEE/CVF Conference on Computer Vision and Pattern Recognition, Seattle, WA, USA, 13–19 June 2020.

38. Redmon, J.; Divvala, S.; Girshick, R. You Only Look Once: Unified, Real-Time Object Detection. In Proceedings of the 2016 IEEE Conference on Computer Vision and Pattern Recognition (CVPR), Las Vegas, NV, USA, 27–30 June 2016.

39. Sun, C.; Shrivastava, A.; Singh, S. Revisiting unreasonable effectiveness of data in deep learning era. In Proceedings of the IEEE International Conference on Computer Vision (ICCV), Venice, Italy, 22–29 October 2017.

40. Karg, B.; Lucia, S. Efficient representation and approximation of model predictive control laws via deep learning. *IEEE Trans. Cybern.* **2020**, *50*, 3866–3878. [CrossRef]

41. Chiu, M.C.; Chen, T.M. Applying data augmentation and mask R-CNN-based instance segmentation method for mixed-type wafer maps defect patterns classification. *IEEE Trans. Semicond. Manuf.* **2021**, *34*, 455–463. [CrossRef]

42. Wang, W.; Tan, X.; Zhang, P. A CBAM Based Multiscale Transformer Fusion Approach for Remote Sensing Image Change Detection. *IEEE J. Sel. Top. Appl. Earth Obs. Remote Sens.* **2022**, *15*, 6817–6825. [CrossRef]

43. Zhang, Y.; Chen, G.; Cai, Z. Small Target Detection Based on Squared Cross Entropy and Dense Feature Pyramid Networks. *IEEE Access* **2021**, *9*, 55179–55190. [CrossRef]

44. He, K.M.; Zhang, X.; Ren, S. Deep residual learning for image recognition. In Proceedings of the 2016 IEEE Conference on Computer Vision and Pattern Recognition (CVPR), Las Vegas, NV, USA, 27–30 June 2016.

45. Wang, C.Y.; Liao, H.Y.M.; Wu, Y.H. CSPNet: A new backbone that can enhance learning capability of CNN. In Proceedings of the IEEE/CVF Conference on Computer vision and Pattern Recognition Workshops, Seattle, WA, USA, 14–19 June 2020.

46. Carion, N.; Massa, F.; Synnaeve, G. End-to-end object detection with transformers. In *European Conference on Computer Vision*; Springer: Cham, Switzerland, 2020.

47. Wang, C.Y.; Yeh, I.H.; Liao, H.Y.M. You only learn one representation: Unified network for multiple tasks. *arXiv* **2021**, arXiv:2105.04206.

48. Zhou, X.; Koltun, V.; Krähenbühl, P. Probabilistic two-stage detection. *arXiv* **2021**, arXiv:2103.07461.

49. Lin, T.Y.; Goyal, P.; Girshick, R. Focal loss for dense object detection. In Proceedings of the IEEE International Conference on Computer Vision (ICCV), Venice, Italy, 22–29 October 2017.

Article

Non-Linear Signal Processing Methods for UAV Detections from a Multi-Function X-Band Radar

Mohit Kumar * and P. Keith Kelly

Agile RF Systems LLC, Berthoud, CO 80513, USA
* Correspondence: mkumar@agilerfsystems.com

Abstract: This article develops the applicability of non-linear processing techniques such as Compressed Sensing (CS), Principal Component Analysis (PCA), Iterative Adaptive Approach (IAA), and Multiple-input-multiple-output (MIMO) for the purpose of enhanced UAV detections using portable radar systems. The combined scheme has many advantages and the potential for better detection and classification accuracy. Some of the benefits are discussed here with a phased array platform in mind, the novel portable phased array Radar (PWR) by Agile RF Systems (ARS), which offers quadrant outputs. CS and IAA both show promising results when applied to micro-Doppler processing of radar returns owing to the sparse nature of the target Doppler frequencies. This shows promise in reducing the dwell time and increases the rate at which a volume can be interrogated. Real-time processing of target information with iterative and non-linear solutions is possible now with the advent of GPU-based graphics processing hardware. Simulations show promising results.

Keywords: compressed sensing radar processing; iterative adaptive algorithm; principal component analysis; X-band phased array radars; UAV

1. Introduction

The main goal of CS is to use optimization methods to recover a sparse signal from a small number of non-adaptive measurements. The radar measurements can be viewed as sparse in both time and Doppler space and are possibly sampled at sub-Nyquist rates, which breaks the relationship between the number of samples acquired and the perfect recovery of radar parameters like delay, velocity, and target angle. The recovery of essential micro-Doppler signatures from the UAV target through the sparse representation of the signal in the frequency domain and following optimization of the sparse signal's l_1 norm using CS can improve the classification accuracy of various UAV targets. Additionally, MIMO-based virtual aperture formation can impart a better spatial resolution for the small spatial footprint UAV targets. IAA is another Doppler resolution enhancement technique that is considered in this article and it shows a promising application for UAV detections with few pulses. Prior to CS, filtering is accomplished using PCA-based decomposition into eigen sub-spaces to get rid of clutter contamination principally due to sidelobes pointing towards the ground. A unified theory is developed for the applicability of these non-linear processing methods and shows their enhancements for better UAV detection using simulations. A common theoretical framework is developed for ease of understanding and applicability of these techniques.

For the US Air Force, Agile RF Systems (ARS) has finished developing a portable weather radar (PWR) system built on phased arrays and a four-quadrant architecture. It can be mounted on a roof or tower. It has a sealed radome that provides wind, rain, snow, hail, and sand protection. The CS, IAA, and PCA methods elaborated in this article are with reference to this phased array design. Figure 1 depicts the conceptual representation of the various sub-sections of this phased array radar. The data from the quadrant-based four-phased array centers can be processed by the signal processor and backend processing

Citation: Kumar, M.; Kelly, P.K. Non-Linear Signal Processing Methods for UAV Detections from a Multi-Function X-Band Radar. *Drones* **2023**, *7*, 251. https://doi.org/10.3390/drones7040251

Academic Editor: Seokwon Yeom

Received: 28 February 2023
Revised: 22 March 2023
Accepted: 30 March 2023
Published: 6 April 2023

algorithms implemented in servers. This radar is based on quadrant-level processing to implement a 4-channel MIMO architecture. This article makes use of this hardware platform to demonstrate non-linear processing which also has a quadrant-wise aperture for MIMO-related enhancements.

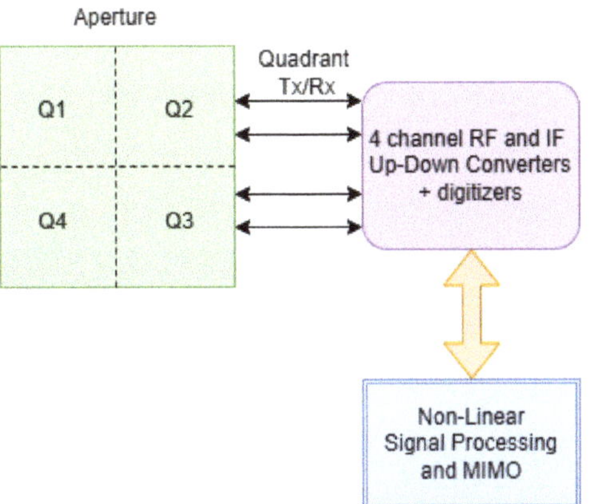

Figure 1. A conceptual representation of a MIMO quadrant phased array for PWR Weather radar system.

Recent advances in computational methods and increased computing capacity for real-time radar operations have greatly increased the use of non-linear processing in radars and communication. For radar applications, the design complexity is typically higher, and a variety of computational techniques can be used to achieve the desired properties [1]. Today's modern phased array radars are able to switch beams faster and are based on inertia-less electronic phase programmability for observing different directions. With such a rapid observation capacity needed, a software framework that can extract information from the least number of acquired samples and pulses, aids in reducing the dwell time of radar in a specific direction. This ultimately results in an overall increase in the rate at which targets can be revisited or surveilled. This article combines the power of non-linear CS, IAA, and PCA to make this advancement into the next generation of radar processing and develops a theoretical understanding with the modeling of signal, clutter, and noise spaces for non-linear processing. The key idea behind CS is sparsity. A signal is considered sparse if most of its information is contained within a few non-zero samples. Consequently, a CS-based signal reconstruction algorithm has to find a sparse vector that best represents the measured signal. Sparsity may be in the time domain or in the transformed frequency domain signal. As will be shown later, only a few micro-Doppler non-zero components are present in the frequency domain response of a drone echo, therefore the frequency domain can be considered sparse.

Taking a peek into CS, if $\mathbf{x}$ is a sparse vector, it can be recovered from the knowledge of the observation vector $\mathbf{y}$ by solving the following optimization problem:

$$arg \min_{x} ||\mathbf{x}||_0 \quad subject\ to\ \mathbf{y} = \Theta\mathbf{x} \tag{1}$$

This search is, however, NP-hard and can be replaced by its closest convex norm, the l_1 norm [2]. The equation above can thus be reformulated as:

$$arg \min_{x} ||\mathbf{x}||_1 \quad subject\ to\ \mathbf{y} = \Theta\mathbf{x} \tag{2}$$

where Θ is the reconstruction matrix. This condition is influenced by the incoherence of the matrix (the sensing matrix), as well as the sparsity of the initial vector $\mathbf{x}$ [2]. The literature offers a number of solutions to this optimization issue. To locate the sparse approximation of the incoming signal $\mathbf{x}$ in a dictionary or matrix ψ, basis pursuit is used in CS. The

Dantzig selector, basis pursuit denoising (BPDN), total variation (TV) minimization-based denoising, etc. are additional commonly used formulations for reliable data recovery from noisy measurements [3]. The squared l_2-norm of the error between the reconstructed signal $\mathbf{y}$ and the sparse signal $\hat{x}$ in the case of BPDN should be less than or equal to ϵ for the obtained solution.

$$arg \min_{x} ||\mathbf{x}||_1 \quad subject\ to\ ||\mathbf{y} - \Theta\mathbf{x}||_2^2 \leq \epsilon \qquad (3)$$

We can also solve BPDN in its Lagrangian form, which is an unconstrained optimization problem and can be rewritten as:

$$\hat{x} = arg \min_{x} \lambda||\mathbf{x}||_1 + ||\mathbf{y} - \Theta\mathbf{x}||_2^2. \qquad (4)$$

The primal-dual interior-point technique and fixed-point continuation are two well-known algorithms that have been applied to the aforementioned equation. Algorithms for linear programming, such as the simplex algorithm known as BP-simplex and the interior-point algorithm known as BP-interior, can also be used to solve the optimization problem in Equation (3). These are solvers for convex problems.

In this article, formulations are developed of the MIMO, CS, and PCA operations with PWR as the platform for UAV detections which is a novel combination of non-linear processing techniques not explored before in literature. The data from the four sub-aperture channels are processed by the signal processor and backend processing algorithms implemented in servers kept in an enclosed thermally controlled chamber as part of PWR radar hardware. To enable MIMO and CS-related enhancements, the signal processor implemented in the radar server can incorporate MIMO, CS, and PCA-related data processing. These non-linear processing methods will eventually lead to the development of low-cost, power-efficient, and small-size radar systems that can scan faster and acquire larger volumes than traditional systems. The evolution of these methods is presented briefly next. Many previous works on CS methods allow recovery of sparse, under-sampled signals from random linear measurements [4]. In [5], authors present Xampling as a sub-Nyquist framework for signal acquisition and processing of signals in a union of subspaces. However, Xampling is not utilized for analog-to-digital conversion in this article. All processing techniques are after the Nyquist rate ADC conversions in a fast time. Ref. [6] used CS to enhance micro-Doppler signatures of drones; however, what is lacking is a common framework for understanding and evaluating other non-linear methods like IAA that is presented here and how IAA compares against CS in terms of performance. In [7], an optimal dwell time is evaluated for effectiveness to capture at least one full rotation of the blades. Comments are made on the total dwell time required but no discussion is included on the sampling rate requirement over dwell time. The article [4] serves as a good introduction to and a survey about compressed sensing. In [8] authors analyze the number of samples required for perfect recovery under noiseless conditions. A good theoretical framework is devised which has been extended here to PCA and IAA under clutter and noise conditions. In [3], authors have summarized a whole set of optimization routines that can be used to reconstruct a signal using CS. Authors in [2] developed the beginning of a mathematical theory of super-resolution. They illustrated that point sources can be super-resolved with infinite precision i.e. recover the exact locations and amplitudes by solving a simple convex optimization problem, which can essentially be reformulated as a semi-definite program. This holds provided the distance between the sources meets certain criteria. The article [9] talks about a method that exploits the difference in the statistics of the returns from sea clutter and the target to improve detection performance. Contrary to this, PCA is used here as the dominant approach to remove clutter echoes by suppressing clutter eigenvectors and also removing a few noise eigenvectors to enhance SNR. In [10], authors discuss Subspace space–time adaptive processing (STAP) algorithms to eliminate clutter. A few interesting research articles blending deep learning methods with CS in the last three years have been [11–13].

Prior Work in Drone Detection

In [14] authors discussed drone detection based on FMCW radars, however, our approach is based on the pulsed system because the leaked transmitter signal can overwhelm the receiver in case of large average powers and co-located transmit and receive systems. Authors in [15] discuss the detection of small RCS drone targets using X-band radar. Unfortunately, they do not address concerns for resolutions needed for micro-Doppler observations and subsequent classification problems. Also, authors in [16] have used a small phased array X-band radar using an AD9361 transceiver chip and showed detections and tracking drones up to 5 km. They have not dealt with micro-Doppler detection and classification in their article. Another interesting article is [17], in which authors have tried to fuse features of micro-Doppler echoes from a dual-band radar and since there are twice the number of features available, they claim to have obtained a better classification than a single-band sensor. It is not sure, however, how much the improvement gained and no quantitative comparisons have been made.

This article explores using CS and IAA-based reconstruction of micro-Doppler for small UAV targets from fewer pulses, such that we do not lose micro-Doppler characteristics for the detection and classification of these targets. Traditionally using Fourier transform on these fewer pulses will degrade the resolution to such an extent that the nearby micro-Doppler features cannot be identified. This aspect is simulated using CS and IAA performance versus FFT-based reconstruction and the benefits can be readily observed. The MIMO formulation is also presented which aids in better spatial resolution indeed needed to support the accurate localization of these small targets.

Figure 2 gives the basic conceptual processing steps needed for building up this system. As would be evident later that for CS-based recovery from a minimum number of pulses, the pulses must be randomly transmitted in different elevation states (in the case of PWR radar) thus a random ray (direction) selector is needed to send out a pulse. At the receiver, all the pulses can be segregated together for a ray and processed along the slow time (pulse) axis for Doppler super-resolution. IAA however, doesn't have this requirement which can be one of its advantages as compared to CS. Uniform sampling, in the case of IAA, also aids in PCA-based clutter suppression which would not work for the non-uniformly sampled received echo. In that case, all the pulses would need to go out in all directions (rays) for PCA and CS to start. In the case of IAA, however, as soon as one ray (direction) echoes have been received, processing can start.

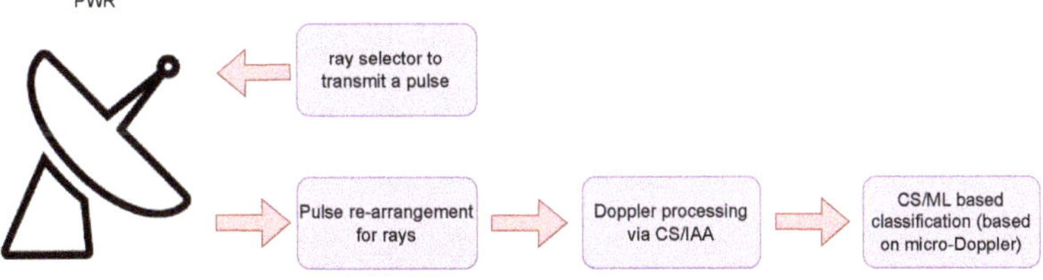

Figure 2. Simple illustration of the processing system.

There is an urgent need for faster scanning for a drone detection radar system because these small objects are highly agile and maneuvering. A really fast update is required to search and track the full space for drones and swarms of drones to keep an eye on their ever-changing activities and strategies. Counter UAS systems must be equipped with very fast scan strategies using very few pulses per direction and still being able to recover high-resolution Doppler features from detected drones. These non-linear processing techniques would aid PWR in achieving this goal.

The rest of the article is organized as follows: Section 2 discusses the important features of the PWR radar developed by ARS and how it is an ideal platform for these non-linear methods, Section 3 provides theoretical understanding and a common framework that encompasses CS, IAA, and PCA based application for a drone detection system,

Sections 4 and 5 summarizes the simulation setup and discusses results obtained. Finally, we conclude with Section 6.

2. Portable Weather Radar (PWR)

All the techniques discussed in this article are being developed with reference to the the ARS PWR weather radar sensor. PWR is a flexible and agile radar due to the phase spin architecture and central Radar System Controller (RSC) [18]. This radar is based on a phased array design and is inherently very different from parabolic dish antenna radars like D3R [19,20]. The radar located at the Greeley radar test facility is shown in Figure 3a. The phase gradient that the phased array controller uses are coordinated by the RSC, along with the motor control for azimuth positioning and rotation. The FPGA logic in the Software defined radio (SDR) has a programmable register interface that enables the RSC to change a broad range of operational radar parameters. The RSC uses alternate horizontal and vertical polarization to allow transmit pulses with a Linear Frequency Modulated Chirp (LFMC) waveform in the SDR. The Host Processor in the local cabinet receives the filtered radar returns from the multichannel receive hardware and processes them there.

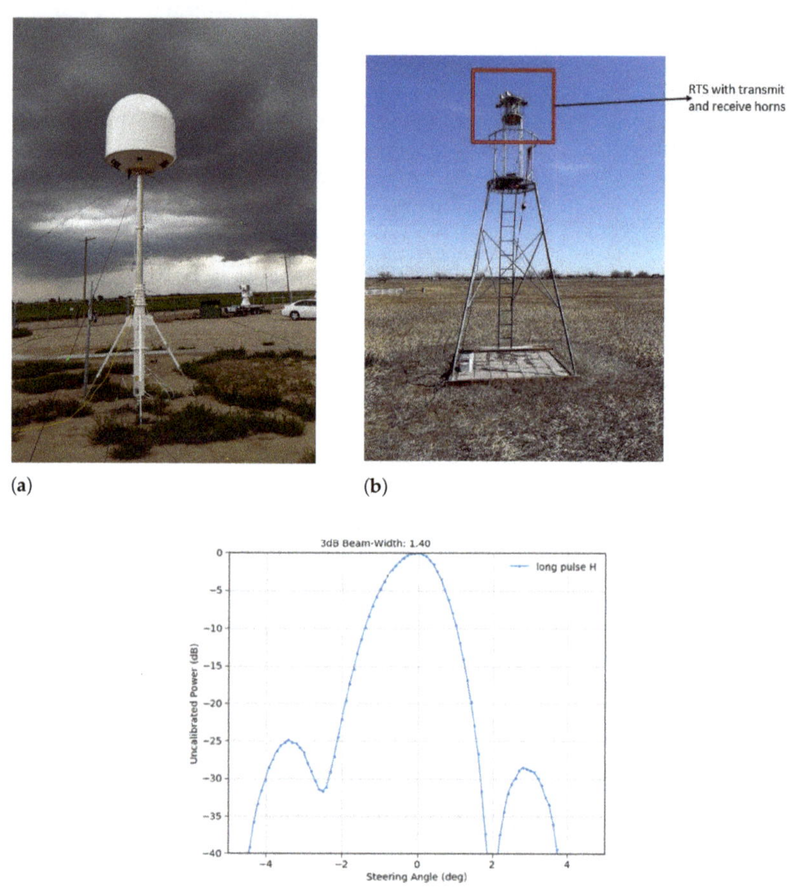

Figure 3. Radar and RTS Setup. (**a**) The Radar on its tower at the CSU-CHILL radar test facility in Greeley. (**b**) The RTS setup with horns on a tower pointing towards radar in a far field. (**c**) The obtained beam shape.

The beamforming network and phased array antenna for PWR underwent extensive testing. Array calibration and aperture beam pattern data were collected to confirm expected aperture performance. Radar Target Simulator (RTS) was created to test the complete functionality of the radar system with the calibrated aperture. The RTS was positioned for this test in the far field of the aperture (Figure 3b). The PWR waveform was received by RTS, a digital time delay was applied, and the result was transmitted back to the PWR while it was still in receive mode. An accurate estimate of the combined beam pattern and all four quadrants was confirmed by determining the peak value from the returned waveform in PWR at each elevation. The two-way combined H-pol antenna pattern is shown in Figure 3c measured with the help of RTS. The two-way pattern sidelobes are approximately 25 dB below the main lobe peak power and the 3 dB beamwidth measured is 1.4 deg confirming good phase/time alignment of all quadrant channels in the combined pattern. With a similar setup of RTS, by applying MIMO Coherent implementation, we expect to measure 0.9 to 1 deg of 3 dB beamwidth. This improvement is enabled by four-quadrant based MIMO signal processing.

To remove any spatial ambiguity, PWR was co-located with the CSU-CHILL radar and concurrently gathered weather observations. This was done to determine the PWR data products' level of quality for Single-Input-Single-Output (SISO) based radar operations. It has been described here so that it can serve as a standard by which to compare the efficacy of MIMO. Data comparisons between these two radars were carried out while PWR rotated at a constant quarter RPM and CHILL transmitted in the eastern region for the same 14 elevation states. Let's examine one of the light rain cases that both radars recorded on 31 May 2022. Figure 4a,d show the reflectivity field for CHILL and PWR radars respectively while Figure 4b,e shows the comparison of differential reflectivity between the two. Figure 4c,f shows the differential phase encountered going through the storm from these radars. The top plots are from the CHILL radar and the bottom ones are from PWR. The CHILL radar was scanning only the eastern sector while the PWR did the whole 360-degree coverage. Both radars observed 14 elevation states. The 2 deg elevation state is shown in the figures. Several of the bright thunderstorm features that both radars picked up in the southeast can be seen distinctly in these figures. All of the level 2 products were subjected to this comparison. With nine times larger antenna dimensions than PWR, we can easily see the CHILL radar's high spatial resolution. With MIMO Coherent processing, the spatial resolution of PWR is anticipated to improve without a physical size increase in an effort to resolve the weather storm features better.

PWR is easily configured for sensing both weather and drone targets. This is the hardware platform developed at ARS currently for weather sensing. Using a separate processing chain shown in Figure 2, its capabilities for drone target detection using non-linear processing can be easily expanded. It is fully capable of MIMO aperture extension because of its four quadrant transmit and receive channels and because of its software-defined capabilities in terms of beam agility and waveforms, PWR is an ideal platform for testing out non-linear techniques.

Spectrogram and smoothed pseudo-Wigner-Ville distribution are two time-frequency representation techniques that have been widely used to analyze drone micro-Doppler signatures. Furthermore, a number of classification methods based on micro-Doppler signatures have been reported for classifying drones of various sizes, types, and loads, as well as drones and people, dogs, and birds. The radar antennas in real-world ground-based surveillance radar systems must scan rapidly to cover a large spatial area of up to 360°. This implies that the radar beam's dwell period on any given target is quite short (which is, usually, a few tens of milliseconds). Thus, when adopting the conventional fast Fourier transform (FFT) for Doppler processing, the radar Doppler resolution is very poor and the accurate micro-Doppler signatures of drones are difficult to discriminate.

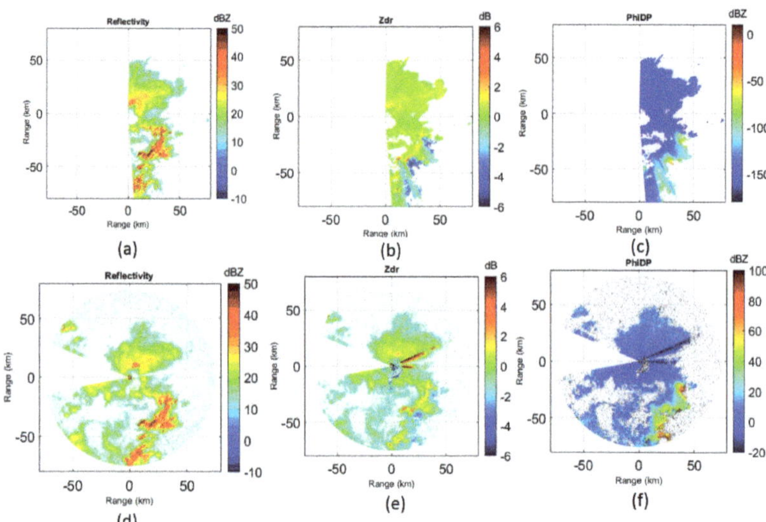

Figure 4. (**a–f**)The different polar products being generated by CHILL and PWR radars.

3. Methods

3.1. PCA and CS Formulation for Micro-Doppler Enhancement

Radar echoes from drones can be identified, categorized, and tracked using Micro-Doppler. A spinning blade is a feature of the majority of drones, including single-rotor, quadrotor, six-rotor, and even hybrid vertical takeoff and landing (VTOL) drones. They are typically active in low-altitude airspace, are small, and fly slowly [21,22]. The rotating movement of rotating blades can modulate the incident radar wave and produce an additional micro-Doppler on the base of the body Doppler contributed by the flying motion of the drone body. Micro-Doppler signals are thought to be quite useful signatures for radar-based drone detection and classification [7].

The importance of micro-Doppler for drone detections cannot be overstated. Using lengthy FFT sizes in traditional signal processing, drone detections can be more accurately resolved at higher Doppler resolutions. In general, greater Doppler resolution is associated with longer radar dwell times (sending out more pulses for longer FFTs). However, the maximum radar dwell period for a functional radar sensor applies. A practical radar system should be able to track targets more quickly and look quickly in all directions to search the entire volume. The secret to observing such a micro-Doppler is the radar dwell time. The dwell period should be sampled quickly enough to improve the Doppler resolution of our spectral analysis. CS and IAA-based non-linear processing can break this relation of linear dependence of resolution to the number of pulses required to observe micro-Doppler features of drones [8].

Prior to performing CS/IAA, PCA was used to get rid of clutter contamination of the drone echo. The cleaned-up signal can then go through spectral analysis.

3.2. PCA Decomposition of Clutter

Consider a radar transceiver, similar to [8] that transmits a pulse train:

$$\mathbf{x}_T(t) = \sum_{p=0}^{P-1} h(t - p\tau), \ \ 0 \leq t \leq P\tau. \tag{5}$$

consisting of P equally spaced pulses $h(t)$. The pulse-to-pulse delay τ is referred to as the PRI, and its reciprocal $1/\tau$ is the PRF. The entire span of the signal in Equation (5) is called the coherent processing interval (CPI). Let L Doppler-producing drone targets make a scene. The pulses travel back to the transceiver after reflecting off the L targets. Three parameters are used to describe each target: a complex amplitude α_l that is proportional to the target's radar cross-section (RCS), a Doppler radial frequency v_l that is proportional to the target-radar closing velocity, and a time delay τ_l that is proportional to the target's distance from the radar. We can write the received signal as:

$$x(t) = \sum_{p=0}^{P-1} \sum_{l=0}^{L-1} \alpha_l h(t - \tau_l - p\tau) e^{-jv_l p\tau}. \tag{6}$$

It might be convenient to express the signal as a sum of single frames:

$$x(t) = \sum_{p=0}^{P-1} x_p(t) \tag{7}$$

where

$$x_p(t) = \sum_{l=0}^{L-1} \alpha_l h(t - \tau_l - p\tau) e^{-jv_l p\tau}. \tag{8}$$

This is the case when the target can be characterized using a single velocity, v_l, however, in the case of micro-Doppler frequencies there exists a band of frequencies around the main body Doppler component comprising the micro-Doppler (Δv_l) as:

$$x_p(t) = \sum_{l=0}^{L-1} \sum_{i=0}^{I-1} \alpha_l h(t - \tau_l - p\tau) e^{-j(v_l + \Delta v_i) p\tau}. \tag{9}$$

I are the number of micro-Doppler components.

In practice, this signal is contaminated with noise and clutter:

$$x(t) = \sum_{p=0}^{P-1} [x_p(t) + \omega_p(t) + C_p(t)]. \tag{10}$$

where $\omega(t)$ is a zero mean wide-sense stationary random signal with auto-correlation $r_\omega(s) = \sigma^2 \delta(s)$ and $C(t)$ is the clutter component. A synonymous equation quantized in time would be:

$$x(n) = \sum_{p=0}^{P-1} [x_p(n) + \omega_p(n) + C_p(n)]. \tag{11}$$

In order to decrease its effect on micro-Doppler features, removing the clutter component is necessary. The Equation (11) can be thought to be composed of signal, noise, and clutter sub-spaces. Let the mean values of $x_p(n)$ be μ_p. Then the mean subtracted received signal can be written as:

$$x(n) = \sum_{p=0}^{P-1} (x_p(n) - \mu_p). \tag{12}$$

Forming the auto-correlation matrix $\mathbf{R}_{xx}$ of $x(n)$ and performing SVD decomposition on it yields,

$$\mathbf{R}_{xx} = \mathbf{U}\mathbf{S}\mathbf{V}^T. \tag{13}$$

Sorting out the eigenbasis vectors in $\mathbf{U}$ in descending order, we get the largest principal components in the received signal. If clutter is supposed to be the dominant return signal component, the corresponding eigenvector of the largest eigenvalue in $\mathbf{S}$ is set to zero. The rest of the received signal $x_p(t)$ is projected to remaining eigenvectors and sum them up as follows to reconstruct signal and noise:

$$x_{recons}(n) = \sum_{p=0}^{P-1}\sum_{d=0}^{D-1} u_d(n)x_p(n). \tag{14}$$

$u_d's$ span the eigenvector space comprising of signal and noise sub-spaces minus the clutter sub-space as that eigenvector is not part of this space. The signal and noise sub-spaces are orthogonal to each other. Noise power can be reduced by considering only a few noise eigenvectors and adding them up to the signal sub-space. This would improve SNR. A demonstration of this is part of the simulations section. A point worth noting is that clutter power can be estimated by averaging out the received signal whose mean will give an estimate of clutter power centered at DC. Based on this power estimation, the eigenvector nearest in power level to this DC power should be removed to nullify the clutter sub-space, if clutter is not the most dominant echo in the received signal.

Comparing PCA with MTI clutter filter, it can be observed that MTI removes the clutter component while low-frequency micro-Doppler components can also get completely suppressed, which would decrease the distinction of micro-Doppler features and finally influence the classification accuracy of drone targets. After the clutter signal has been suppressed, the CS step can be performed for enhancing micro-Doppler features. Micro-Doppler spectral lines can have better distinction when they are CS processed.

Apart from the primary signal Doppler, a few micro-Doppler lines, and clutter, which have high values, the majority of the entries in the spectral domain of drone targets are zeros or low values. Only the primary Doppler and a few high spectral lines caused by micro-Doppler may remain after removing clutter. In order to improve drone classification and identification using fewer pulse samples, CS may be able to provide high-resolution Doppler components for such a sparse signal. If we use fewer pulses to give the same resolution as with, say, 10 times the number of pulses, then we are effectively reducing the dwell time on the target and can potentially spin faster as in the case of PWR. This faster scanning radar can track and do multiple functions at the same time which may mean portable systems like PWR, is able to accomplish weather surveillance, seach and track UAVs.

3.3. CS-Based Enhancement of Doppler Space

The clutter-suppressed received signal can now be processed by CS to better resolve the micro-Doppler frequencies with relatively fewer random measurements. The premise is CS would be able to provide a higher resolution Doppler space with few samples as against conventional FFT processing which would need a sufficiently larger number of measurements or pulses to give the same resolution as CS. The first stage of CS is multiplying the random measurement matrix, $\psi(n)$ with $x(n)$:

$$y(n) = \sum_{p=0}^{P-1} \psi_p(n)[x_p(n) + \omega_p(n) + C_p(n)]. \tag{15}$$

where $\psi_p(n) \in \mathbb{R}^{MxN}$ *or* $\mathbb{C}^{MxN}$ and $y(n) \in \mathbb{R}^M$ *or* $\mathbb{C}^M$. The number of measurements taken is much lesser than the length of the input signal, i.e., $M << N$. To further reduce the number of measurements which are necessary for perfect reconstruction, the measurement matrix must be incoherent with the basis in which the signal is sparse. The inputs to the reconstruction algorithm are the measurement vector $y(n)$ and reconstruction matrix Θ where,

$$\Theta = \psi\zeta \in \mathbb{R}^{MxN} \text{ } or \text{ } \mathbb{C}^{MxN}. \tag{16}$$

ζ is the basis vector of the space where $x(n)$ is sparse. Thus $x(n)$ can be written as:

$$x(n) = \sum_{p=0}^{P-1} s_p(n)\zeta_p(n). \tag{17}$$

$s \in \mathbb{R}^N$ is the sparse coefficient vector of length N. The optimization problem expressed as l_1 norm (for reconstruction) can thus be expressed as:

$$\hat{s} = arg\ \min_{s} ||\mathbf{s}||_1 \quad subject\ to\ \Theta\mathbf{s} = \mathbf{y} \tag{18}$$

The estimate of x(n), i.e., $\hat{x}$ can be obtained from $\hat{s}$ by taking its inverse transform. Some of the other types of this same optimization problem with noise included and a lagrangian form of the above equation was discussed in the paragraphs preceding Equations (3) and (4).

It is shown in the literature [8] that for a noise-free case, the estimation of parameters $(\alpha_l, \tau_l, v_l)_{l=0}^{L-1}$ without micro-Doppler frequencies can be recovered using 3L samples using a Xampling framework and assuming Finite Rate of Innovation (FRI) samples. However, with micro-Doppler and the presence of noise, there are likely more samples required for perfect recovery. Simulations do confirm the fact that the number of slow time, pulse measurements required for a higher resolution Doppler reconstruction is sufficiently less so that either the radar can be made to scan faster or it can be made to accomplish multi-functions like weather detections and forecasting too. The software-defined phased array architecture of PWR is ideally suited for drone detection and weather surveillance.

SNR

The SNR is linked to the attenuation that the signal receives going through the link. However, noise power can be reduced in the prior step of PCA with fewer noise eigenvectors considered for the reconstruction of the received signal. Certainly, this can improve SNR and it is demonstrated in simulations too.

3.4. MIMO and CS Framework

The multi-function PWR radar is capable of MIMO because of its four-quadrant array structure. So quadrant-wise MIMO formulation can be used along with CS. This gives the benefit of virtual array formation without the addition of physical array elements, and also it is cost-effective since each element is not required to have an RF and IF hardware chain associated, and only four channels are sufficient to make use of quadrant MIMO benefits instead of hundreds if not thousands of channels for a full MIMO implementation.

The quadrant MIMO system is equivalent to the spatial convolution of the transmit and receive quadrant phase centers and the formation of virtual array elements beyond the physical aperture size. The virtual array dimensions are 1.5 times the physical array (in both axes) as evident from Figure 5. Equivalently, this would give the beamwidth reduction by the same factor and the spatial resolution will improve. The PWR system provides a very cost-effective MIMO radar system using a quadrant phased array structure. One of the main challenges of an element-wise MIMO radar is coping with complicated systems in terms of cost, high computational load, and complex implementation, which have been traded very well using quadrant MIMO in PWR radar hardware.

To demonstrate quadrant MIMO processing, transmissions were assumed using the same LFM waveforms from all the quadrants, however, there is a sequential transmission from quadrants in a time-multiplexed manner. The data cube received by the quadrants would have to be processed to form the virtual array data cube. Let the data collected when quadrant 1 transmits (from all four receive quadrants) given by:

$$\mathbf{Ph}_r = \begin{bmatrix} \mathbf{M}_{11} & \mathbf{M}_{12} \\ \mathbf{M}_{13} & \mathbf{M}_{14} \end{bmatrix} \tag{19}$$

where the first subscript tells the transmit quadrant and the second signifies the receive. The coherent data matrix after all transmissions are given by:

$$\mathbf{Vi}_r = \begin{bmatrix} \mathbf{M}_{21} & \mathbf{M}_{11} + \mathbf{M}_{22} & \mathbf{M}_{12} \\ \mathbf{M}_{23} + \mathbf{M}_{31} & \mathbf{M}_{13} + \mathbf{M}_{24} + \mathbf{M}_{31} + \mathbf{M}_{41} & \mathbf{M}_{14} + \mathbf{M}_{42} \\ \mathbf{M}_{333} & \mathbf{M}_{43} + \mathbf{M}_{43} & \mathbf{M}_{44} \end{bmatrix} \tag{20}$$

This matrix includes 5 additional virtual phase centers corresponding to five additional quadrants to make it a total of 9 quadrants. For PWR, each entry in the column is 12λ wide and high, the third row and column would give it an extra 12λ height and width to the receive aperture due to virtual quadrants [23].

Extending our discussion further about MIMO and CS, let's revisit Equation (8) for a sparse scene with L drone targets. The received signal at the q_{th} quadrant after demodulation to baseband for a single frame is in turn given by:

$$\mathbf{x}_q(t) = \sum_{l=0}^{L-1} \sum_{m=0}^{M-1} \alpha_l h(t - \tau_l - p\tau) e^{-jv_l p\tau} e^{-j\beta_{m,q}\varsigma}. \tag{21}$$

where $\varsigma = sine(\theta_l)$ is the azimuth angle of the lth drone target relative to the quadrant θ_l. Also note that, $\beta_{m,q} = (\zeta_q \xi_m)(f_c \lambda/c + 1)$, f_c is the carrier frequency radiated from the quadrant and $\zeta_q, \xi_m \in \mathbf{Vi}_r$. Again $y(n)$ can be written as:

$$y(n) = \sum_{p=0}^{P-1} \psi_p(n)[x_{p,q}(n) + \omega_p(n) + C_p(n)]. \tag{22}$$

and then forming **s**, an sparse coefficient matrix using basis ι as:

$$x(n) = \sum_{p=0}^{P-1} \sum_{q=0}^{M-1} s_{p,q}(n) \iota_{p,q}(n). \tag{23}$$

The optimization problem expressed as l_1 norm (for reconstruction) can now be expressed as:

$$\hat{s} = arg \min_s ||\mathbf{s}||_1 \quad subject\ to\ \Theta\mathbf{s} = \mathbf{y} \tag{24}$$

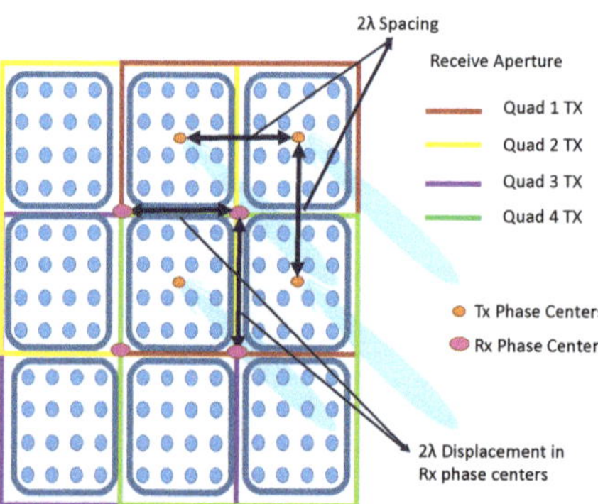

Figure 5. A 8×8 element phased array with multiple transmit phase centers based on the quadrant. The whole array is divided into 4 quadrants.

Theorem 1. *The minimal number of transmit times the number of receive channels required for perfect recovery of L targets in noiseless settings is $\geq 2\,L$ with a minimal number of $\geq 2\,L$ samples per receiver and $\geq 2\,L$ pulses per transmitter [4].*

This is true for Xampling and an FRI framework used in conjunction with CS. In PWR, quadrant MIMO offers four transmit and nine receive channels (four physical and five virtual quadrants), thus $L = 18$ drone targets can be resolved in a CPI or dwell time τ. For this recovery, 36 samples are needed per pulse and 36 pulses are needed per transmitter quadrant for perfect recovery of Doppler for these L targets. This arithmetic is different for a noisy link but then most likely that many drone targets may not be present. Only in the case of drone swarms, that many targets would need to be detected, however, it would be quite a coincidence to get so many of them in a CPI or dwell (one direction) otherwise. This result also implies that many more targets can be perfectly resolved in the DoA sense by using MIMO virtual elements and this framework allows CS theory to be applied for calculating the number of pulses required for a perfect Doppler recovery for all these targets as well.

3.5. An Iterative Approach to Solve the Dwell Time Limitation for a Fast Scanning Drone Radar

The Doppler resolution of the temporal signal can be increased by using the super-resolution algorithms that are frequently used in array processing, such as minimal variance distortionless response (MVDR) and multiple signal classification (MUSIC). To estimate the covariance matrix or carry out eigenanalysis, these algorithms typically need a number of signal snapshots. Some algorithms, like MUSIC, require knowing the number of sources up front as well. However, in surveillance radar, the Doppler processing is carried out over the slow-time samples (over pulses) at each range increment. As a result, there is only one available temporal snapshot. It is also unclear how many target Doppler and micro-Doppler sources there will be. Consequently, it is impossible to use the traditional super-resolution methods. Unlike the conventional MVDR and MUSIC algorithms in which many snapshots are required to estimate the covariance matrix, IAA can work well with only a few or even one snapshot to achieve super-resolution [24].

The formulation of this method is elaborated next. It is similar to the one highlighted in [25]. The basis steering vectors are defined on the grid points that either have the frequency present or do not span the output space of the Doppler processor. Henceforth, the outcome of the Doppler process is written as:

$$\mathbf{y} = \mathbf{A}(f)\mathbf{s} + \omega + \mathbf{C}. \tag{25}$$

where $\omega(n)$ is a zero mean wide-sense stationary random signal with auto-correlation $r_\omega(s) = \sigma^2 \delta(s)$ and C(n) is the clutter component. $A(f) = [\mathbf{a}(f_1)\mathbf{a}(f_2)...\mathbf{a}(f_k)]$ is $P \times K$ dimension where P is the number of pulses and K is the number of finite points in the Doppler grid. $\mathbf{s}$ is a vector of the amplitudes of frequencies at the grid locations $k = 1, 2, ..., K$. The clutter and noise matrix can be defined as:

$$\mathbf{Q}(f_k) = \mathbf{R} - P_k \mathbf{a}(f_k)\mathbf{a}^H(f_k). \tag{26}$$

$\mathbf{R} = \mathbf{A}(f)\mathbf{P}\mathbf{A}^H(f)$ is the auto-correlation matrix of the input and P is a $K \times K$ diagonal matrix, whose diagonals $P_k = |s_k|^2, k = 1, 2, ..., K$ contains the powers at each Doppler frequency on the Doppler grid. The cost function is given by:

$$\Xi = (\mathbf{y} - s_k\mathbf{a}(f_k))^H \mathbf{Q}^{-1}(\mathbf{y} - s_k\mathbf{a}(f_k)). \tag{27}$$

Minimizing the cost function with respect to s_k gives [25]:

$$\hat{s}_k = \frac{\mathbf{a}^H(f_k)\mathbf{R}^{-1}\mathbf{y}}{\mathbf{a}^H(f_k)\mathbf{R}^{-1}\mathbf{a}(f_k)}. \tag{28}$$

Since the iteration requires $\mathbf{R}$, which depends on the unknown powers, it must be implemented as an iterative approach. The initialization can be done by letting $\mathbf{R}$ equal to the identity matrix $\mathbf{I}_P$. The steps are shown in Algorithm 1. Both IAA and CS are capable to enhance the Doppler resolution with fewer pulses, however, CS needs a random pulse transmission within the dwell time. It relies on non-uniform sampling within the dwell time. The pulse time left vacant because a pulse cannot be transmitted in CS can be used for transmitting pulses in other directions though, however, it can make radar operations complex. On the other hand, IAA works with a uniform sampling of the dwell time.

Algorithm 1 An iterative algorithm [25]

$$\hat{P}_k = \frac{1}{\mathbf{a}^H \mathbf{a} \sum_{p=0}^{P-1} |\mathbf{a}^H(f_k)y(n)|^2}$$

while !converge **do**

$\quad \mathbf{R} = \mathbf{A}(f)\mathbf{P}\mathbf{A}^H(f)$

$\quad$ **for** k = 1,2,...,K **do**

$\quad\quad \hat{s}_k = \frac{\mathbf{a}^H(f_k)\mathbf{R}^{-1}\mathbf{y}}{\mathbf{a}^H(f_k)\mathbf{R}^{-1}\mathbf{a}(f_k)} \quad n = 1,2,...,N.$

$\quad\quad \hat{P}_k = 1/N \sum_{n=0}^{N-1} |\hat{s}_k(n)|^2.$

$\quad$ **end for**

end while

---

4. Simulations and Results

In this section, the feasibility and practicality of non-linear methods are discussed and substantiated by simulations using PWR radar parameters and features. A few parameters of PWR that are relevant for simulations and system demonstration are shown in Table 1.

Table 1. PWR Parameters.

Parameter	Value
Frequency	X-band
Pulse Repetition Interval (PRI)	500 us
Scanning	Electronic in elevation

A few micro-Doppler frequencies are simulated making an echo of length 512 samples. This comprises a main Doppler echo from the base motion of the UAV and there are micro-Doppler from the rotary motion of the blade movement modulating the primary echo signal. The signal is corrupted by clutter from the elevation sidelobes of PWR simulated as zero Doppler component being added up to the received echo. The PCA formulation described in an earlier section is used next for the removal of clutter sub-space and reconstruction of the time domain signal for further processing. Current methods including MTI, CLEAN, etc can not realize the real-time removal of ground clutter without suppressing nearby micro-Doppler components. That is why PCA is adopted to remove the clutter components in the echo signal. Figure 6a depicts clutter centered at DC and the micro-Doppler components and the main Doppler signal. If clutter is always a dominant signal in the received echo, then it is pretty easy to remove the highest valued eigenvector from the SVD decomposition, however, if it is not, then we need to figure out the DC power by averaging out the samples and looking at average power. This should be able to give us an estimate of the eigenvector which has similar power levels as the mean power. After that is made certain, we can remove it from our reconstruction process to get rid of clutter. Next, this signal is processed using CS. The time domain and the frequency domain of the echo for CS processing are shown in Figure 7a,b. Random 32 samples (out of 512) are picked from $x(t)$ (Figure 8) and the recovery of the sparse frequency domain is accomplished from these random samples using l_1 minimization.

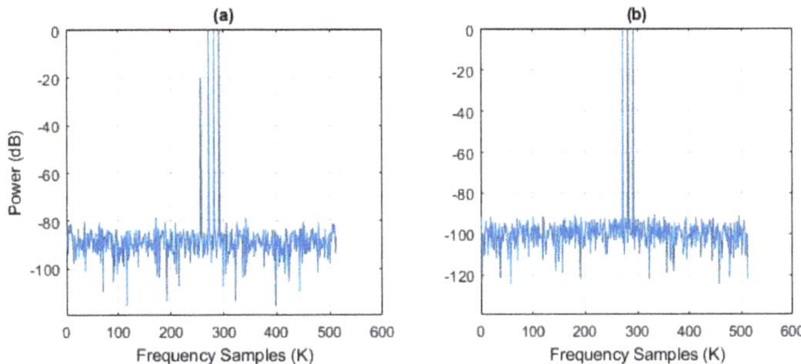

Figure 6. Clutter removal using PCA. (**a**) is with clutter centered at DC and (**b**) clutter removed with no harm to nearby micro-Doppler components.

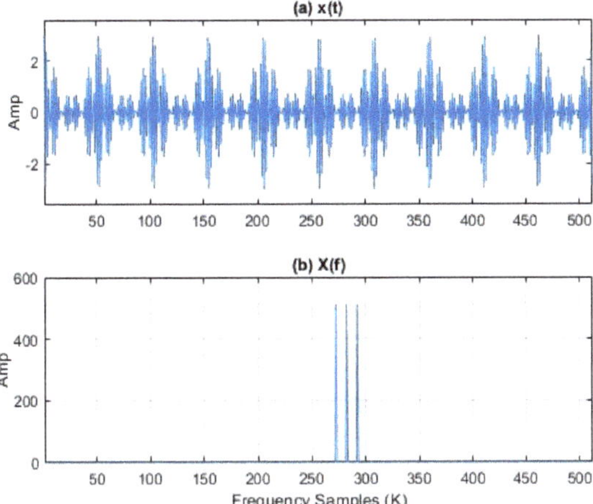

Figure 7. The original signal characteristics for a CS-based micro-Doppler reconstruction.

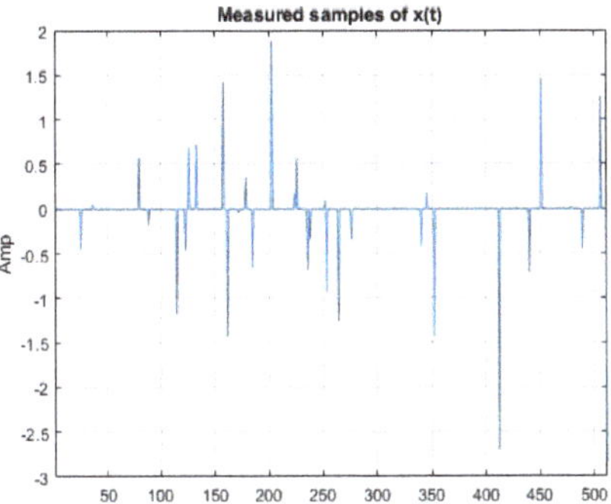

Figure 8. The samples that are picked randomly and CS-based reconstruction is applied.

The recovered high-resolution frequency and time domain samples from the lower dimensional signal are shown in Figure 9a,b.

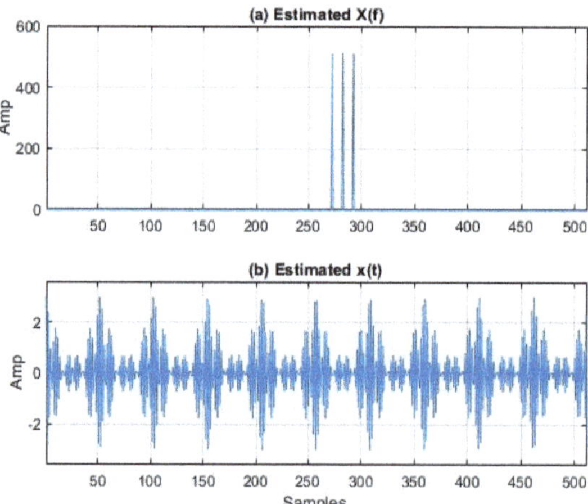

Figure 9. The reconstructed frequency domain and time domain.

Exact recovery of the higher dimensional signal is possible due to sparsity in the frequency domain. Since the frequency domain of a drone echo only consists of a few non-zero components of the base Doppler and micro-Doppler components that are to be recovered, this signal can be considered sparse. However, the performance of CS reconstruction when there are a bunch of other features present in the echo signal, has to be yet demonstrated and can be an analysis for future work. The frequency analysis of the lower dimensional signal is shown in Figure 10 which is the Fourier transform of the first 32 samples from the sequence of the original 512 samples. The loss of resolution is evident and the modulations due to micro-Doppler cannot be observed. This would lead to faulty classification results for the UAV type and detection of UAV based on certain micro-Doppler features.

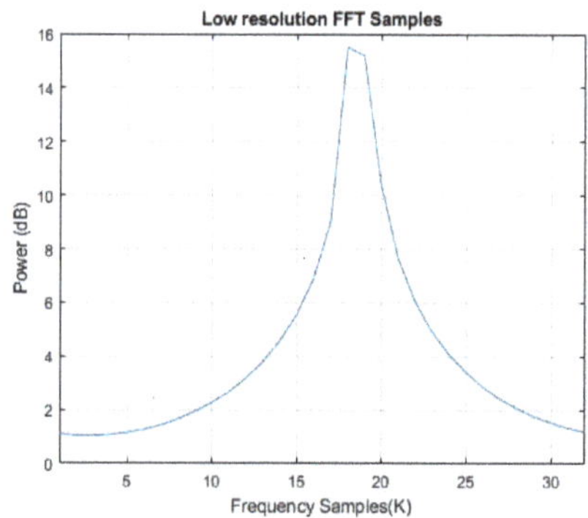

Figure 10. The Fourier transform of the lower-dimensional signal.

It is to be noted that CS needs random $K = 32$ samples from a set of $N = 512$ echo samples. The N can be considered here to be the number of pulses where we reduced it to $K << N$. Thus only K pulses are sufficient to reconstruct the micro-Doppler features of the UAV echo which can easily reduce the dwell time and overall scan time of the radar. These pulses would need to be randomly transmitted in the larger dwell time of N pulses, scanning other elevation states to cover up the total volume in the case of PWR for example. With this, the N pulses would be transmitted at random in different elevation states and then grouped together by the signal processor. This is a little complex and would require beam switching at every pulse instead of every dwell. The IAA however, doesn't rely on non-uniform random sampling and is simulated next.

Figure 11 shows the same setup of micro-Doppler frequencies used before. Instead of CS-based reconstruction, IAA is used for recovery. It is shown that super-resolution can be achieved using a single snapshot of data samples. The IAA iteration was setup simulating three peaks of the micro-Doppler base echo and the modulations from the rotation of the blades of the quadcopter similar to the one for CS and then using $K(32pulses) << N(512pulses)$ that is much fewer than the higher dimensions (512) in the frequency domain required to reconstruct a higher resolution frequency response. It can be observed that the sidelobes are very low for IAA-based reconstruction. To achieve a similar level of sidelobe performance, a very aggressive taper would be needed for FFT-based recovery that would lead to quite a bit of SNR loss. Using IAA can avoid the taper loss in FFT-based Doppler processing and the overall radar detection performance for all targets is also improved. If we compare it with the CS technique, CS works with random non-uniform sampling that is unconventional and as stated earlier about its applicability to PWR, a scheme is required in which different elevation states would be selected at random for transmission of a pulse, and this makes it complex as compared to uniformly sampled IAA. Having said this, it is worth noting that CS can be extended to fast time sampling using Xampling and FRI principles so that lower sampling ADCs are sufficient for below the Nyquist rate sampling of fast time signals. Hence both schemes have their own pros and cons and should be judiciously used. Their relative benefits are shown in Table 2 below:

Table 2. Relative benefits of IAA, CS and traditional FFT approach.

IAA	CS	FFT
Uniform Sampling	Non-uniform Sampling	Uniform Sampling
Peak Sidelobes of the order of -80 dBc	Peak Sidelobes = -40 dBc	Peak Sidelobes = -13 dBc
No Taper required	No taper	Taper required

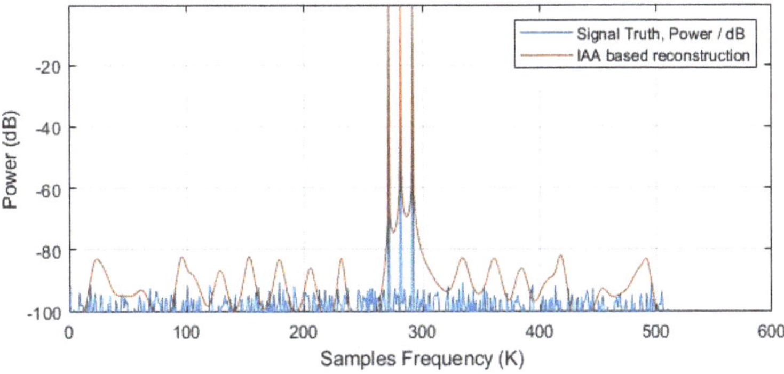

Figure 11. Estimation of frequency response using IAA.

5. Discussion

Both CS and IAA are useful for frequency super-resolution. However, IAA offers distinct advantages in terms of uniform sampling in pulse domain (slow time), hence CS is suggested to be used with fast time (range-time) sampling using a Xampling framework. So as to make it non-confusing, the steps for a fast scanning drone detection radar system could be summarized as:

- Start with PCA-based clutter removal.
- Use CS-based Xampling system for fast time sub-Nyquist sampling. For more details on the Xampling system, please refer to [26].

- IAA to be applied for frequency super-resolution with fewer pulses in slow time.
- Quadrant MIMO to be used for virtual array formation and enhancement of spatial resolution.
- Classification of drones using CNN-based classifier.

6. Conclusions

PCA was simulated for clutter mitigation. CS and IAA were explored for micro-Doppler and spectral retrievals, and MIMO for spatial estimation of drone UAV targets. A unified theoretical framework was developed that stitches together these non-linear methods towards drone micro-Doppler and spatial resolution enhancement for detections from phased array PWR multi-function radar sensors. Both IAA and CS were found to be very useful to recover micro-Doppler drone features so that those targets can be efficiently detected and classified using fewer pulses than the conventional FFT processing. The drawbacks and applicability of each one of these techniques were discussed. Simulations demonstrated that IAA and CS methods can achieve greater Doppler resolutions as compared to FFT-based processing using fewer samples. Additionally, peak sidelobe levels achieved using IAA is substantially lower than the traditional FFT approach. If such levels of sidelobe levels are to be achieved using FFT, significant taper loss has to be encountered in FFT leading to severe SNR degradation. Using PCA, clutter mitigation was demonstrated with benefit in SNR if a few noise eigenvectors were removed in signal reconstruction.

The future investigations in this research might be figuring out the computationally efficient algorithm for realizing the minimization of l_1 norm and comparing it with IAA computations. That will pave the way to a real-time implementation for PWR radar and collection of data samples and performing non-linear processing by the PWR computational server. Also, the computational complexity associated with each of the CS-based and IAA reconstruction algorithms will be evaluated and their feasibility in real-time processors will be demonstrated.

Author Contributions: Both the authors have contributed equally towards conceptualizing and writing the manuscript. All authors have read and agreed to the published version of the manuscript.

Funding: The research has been funded through internal R & D funds of the company.

Data Availability Statement: Data sharing is not applicable to this article.

Conflicts of Interest: The authors declare no conflict of interest.

References

1. Kumar, M.; Chandrasekar, V. Intrapulse Polyphase Coding System for Second Trip Suppression in a Weather Radar. *IEEE Trans. Geosci. Remote Sens.* **2020**, *58*, 3841–3853. [CrossRef]
2. Candes, E.J.; Fernandez-Granda, C. Towards a Mathematical Theory of Super-resolution. *Comm. Pure Appl. Math.* **2014**, *67*, 906–956. [CrossRef]
3. Rani, M.; Dhok, S.B.; Deshmukh, R.B. A Systematic Review of Compressive Sensing: Concepts, Implementations and Applications. *IEEE Access* **2018**, *6*, 4875–4894. [CrossRef]
4. Eldar, Y.C.; Kutyniok, G. *Compressed Sensing: Theory and Applications*; Cambridge University Press: Cambridge, UK, 2012.
5. Mishali, M.; Eldar, Y.C.; Elron, A.J. Xampling: Signal Acquisition and Processing in Union of Subspaces. *IEEE Trans. Signal Process.* **2011**, *59*, 4719–4734. [CrossRef]
6. Zhu, L.; Zhang, S.; Ma, Q.; Zhao, H.; Chen, S.; Wei, D. Classification of UAV-to-Ground Targets Based on Enhanced Micro-Doppler Features Extracted via PCA and Compressed Sensing. *IEEE Sens. J.* **2020**, *20*, 14360–14368. [CrossRef]
7. Gong, J.; Jun, Y.; Deren, L.; Deyong, K. Detection of Micro-Doppler Signals of Drones Using Radar Systems with Different Radar Dwell Times. *Drones* **2022**, *6*, 262. [CrossRef]
8. Bar-Ilan, O.; Eldar, Y.C. Sub-Nyquist Radar via Doppler Focusing. *IEEE Trans. Signal Process.* **2014**, *62*, 1796–1811. [CrossRef]
9. Sira, S.P.; Cochran, D.; Papandreou-Suppappola, A.; Morrell, D.; Moran, W.; Howard, S. A Subspace-Based Approach to Sea Clutter Suppression for Improved Target Detection. In Proceedings of the 2006 Fortieth Asilomar Conference on Signals, Systems and Computers, Pacific Grove, CA, USA, 29 October–1 November 2006; pp. 752–756.
10. Wang, Z.; Wang, Y.; Duan, K.; Xie, W. Subspace-Augmented Clutter Suppression Technique for STAP Radar. *IEEE Geosci. Remote Sens. Lett.* **2016**, *13*, 462–466. [CrossRef]

11. Machidon, A.L.; Pejovic, V. Deep learning for compressive sensing: A ubiquitous systems perspective. *Artif. Intell. Rev.* **2023**, *56*, 3619–3658. [CrossRef]
12. Fan, Z.E.; Lian, F.; Quan, J.N. Global Sensing and Measurements Reuse for Image Compressed Sensing. In Proceedings of the 2022 IEEE/CVF Conference on Computer Vision and Pattern Recognition (CVPR), New Orleans, LA, USA, 18–24 June 2022; pp. 8944–8953.
13. Saideni, W.; Helbert, D.; Courreges, F.; Cances, J. An Overview on Deep Learning Techniques for Video Compressive Sensing. *Appl. Sci.* **2022**, *12*, 2734. [CrossRef]
14. Coluccia, A.; Gianluca, P.; Alessio, F. Detection and Classification of Multirotor Drones in Radar Sensor Networks: A Review. *Sensors* **2020**, *20*, 4172. [CrossRef] [PubMed]
15. Bouzayene, I.; Mabrouk, K.; Gharsallah, A.; Kholodnyak, D. Scan Radar Using an Uniform Rectangular Array for Drone Detection with Low RCS. In Proceedings of the 2019 IEEE 19th Mediterranean Microwave Symposium (MMS), Hammamet, Tunisia, 31 October–2 November 2019; pp. 1–4.
16. Tang, L.; Wang, H.; Feng, Z.; Xu, D.; Wang, Y.; Quan, S.; Xu, W. Small Phased Array Radar Based on AD9361 For UAV Detection. In Proceedings of the IEEE MTT-S International Microwave Biomedical Conference (IMBioC), Nanjing, China, 6–8 May 2019; pp. 1–3.
17. Zhang, P.; Yang, L.; Chen, G.; Li, G. Classification of drones based on micro-Doppler signatures with dual-band radar sensors. In Proceedings of the 2017 Progress in Electromagnetics Research Symposium—Fall (PIERS-FALL), Singapore, 19–22 November 2017; pp. 638–643.
18. Kelly, P.K.; Kumar, M.; McCaskey, R.; Maddocks, E.; Rhodes, J.; Chandrasekar, V.; Radhakrishnan, C.; Kennedy, P. A Novel Portable Phased Array Radar for Meteorological Remote Sensing. In Proceedings of the 2022 IEEE International Symposium on Phased Array Systems and Technology (PAST), Waltham, MA, USA, 11–14 October 2022; pp. 1–8.
19. Kumar, M.; Joshil, S.S.; Chandrasekar, V.; Beauchamp, R.M.; Vega, M.; Zebley, J.W. Performance trade-offs and upgrade of NASA D3R weather radar. In Proceedings of the 2017 IEEE International Geoscience and Remote Sensing Symposium (IGARSS), Fort Worth, TX, USA, 23–28 July 2017; pp. 5260–5263.
20. Kumar, M.; Joshil, S.; Vega, M.; Chandrasekar, V.; Zebley, J.W. Nasa D3R: 2.0, Enhanced Radar with New Data and Control Features. In Proceedings of the IGARSS 2018–2018 IEEE International Geoscience and Remote Sensing Symposium, Valencia, Spain, 22–27 July 2018; pp. 7978–7981.
21. Musa, S.A.; Abdullah, R.S.A.R.; Sali, A.; Ismail, A.; Rashid, N.E.A.; Ibrahim, I.P.; Salah, A.A. A review of copter drone detection using radar systems. *Def. S T Tech. Bull.* **2019**, *12*, 12–16.
22. Wellig, P.; Speirs, P.; Schuepbach, C.; Oechslin, R.; Renker, M.; Boeniger, U.; Pratisto, H. Radar systems and challenges for C-UAV. In Proceedings of the 2018 19th International Radar Symposium (IRS), Bonn, Germany, 20–22 June 2018; pp. 1–8.
23. Kumar, M. Methods and Techniques for a MIMO-based Weather Radar system. *TechRxiv* 2023, *preprint*.
24. Sun, H.; Oh, B.S.; Guo, X.; Lin, Z. Improving the Doppler Resolution of Ground-Based Surveillance Radar for Drone Detection. *IEEE Trans. Aerosp. Electron. Syst.* **2019**, *55*, 3667–3673. [CrossRef]
25. Xue, M. Algorithms and Fast Implementations for Sensing Systems. Ph.D. Thesis, University of Florida, Gainesville, FL, USA, 2011.
26. Mishali, M.; Eldar, Y.C.; Dounaevsky, O.; Shoshan, E. Xampling: Analog to digital at sub-Nyquist rates. *IET Circuits Devices Syst.* **2011**, *5*, 8–20. [CrossRef]

Article

Thermal Image Tracking for Search and Rescue Missions with a Drone

Seokwon Yeom

Department of Artificial Intelligence, Daegu University, Gyeongsan 38453, Republic of Korea; yeom@daegu.ac.kr;
Tel.: +82-53-850-6643

Abstract: Infrared thermal imaging is useful for human body recognition for search and rescue (SAR) missions. This paper discusses thermal object tracking for SAR missions with a drone. The entire process consists of object detection and multiple-target tracking. The You-Only-Look-Once (YOLO) detection model is utilized to detect people in thermal videos. Multiple-target tracking is performed via track initialization, maintenance, and termination. Position measurements in two consecutive frames initialize the track. Tracks are maintained using a Kalman filter. A bounding box gating rule is proposed for the measurement-to-track association. This proposed rule is combined with the statistically nearest neighbor association rule to assign measurements to tracks. The track-to-track association selects the fittest track for a track and fuses them. In the experiments, three videos of three hikers simulating being lost in the mountains were captured using a thermal imaging camera on a drone. Capturing was assumed under difficult conditions; the objects are close or occluded, and the drone flies arbitrarily in horizontal and vertical directions. Robust tracking results were obtained in terms of average total track life and average track purity, whereas the average mean track life was shortened in harsh searching environments.

Keywords: search and rescue missions; thermal image; object detection; target tracking; bounding box gating

Citation: Yeom, S. Thermal Image Tracking for Search and Rescue Missions with a Drone. *Drones* **2024**, *8*, 53. https://doi.org/10.3390/drones8020053

Academic Editor: Giordano Teza

Received: 10 January 2024
Revised: 31 January 2024
Accepted: 3 February 2024
Published: 5 February 2024

1. Introduction

Multirotor drones are widely applied in a variety of fields [1]. The ability to hover and maneuver by the operator or as programmed makes them valuable tools in numerous industries. By capturing images from various angles and heights, drones can obtain cost-effective aerial views covering arbitrary areas.

Thermal imaging cameras detect infrared radiation emitted by objects [2,3]. This radiation is invisible to the human eye, but thermal imaging cameras convert it into a visible image. Thermal imaging requires no illumination or ambient light, penetrating obstacles including smoke, dust, haze, and light foliage. However, the resolution of the thermal image is low, and no texture or color information is provided.

Drones equipped with thermal imaging cameras are highly effective in locating missing people for search and rescue (SAR) missions and surveillance [4–6]. The technology has been also applied to wildlife monitoring and agricultural and industrial inspection [7,8]. People detection with thermal images captured by drones has been studied in [9–15]. Persons and animals were detected using the YOLO detection model [9]. Persons and cars were detected from different observation angles of the drone using the YOLO detection model [10]. The YOLO detection model was adopted to build a human detection system using drones [11]. The spatial gray level co-occurrence matrix estimates temperature differences [12]. In [13], a person is recognized with a two-stage hot-spot detection approach. People and fire detection were studied with high-altitude thermal images obtained by optical and thermal sensors [14]. However, the tracking of people in thermal scenes using drones has been seldom researched [15]. In [15], people tracking with a thermal imaging camera mounted on a small drone was performed for the first time. However, object detection was performed using the *k*-means algorithm, which resulted in degraded performance in complex backgrounds.

Tracking of non-living objects using thermal imaging by fixed-wing drones was studied in [16,17]. The Kalman filter was adopted to track a boat [16], and a colored-noise measurement model was utilized to track a small vessel [17]. Tracking of people and animals using fixed thermal imaging cameras has been studied in [18–24]. Foreground objects were extracted by contour-based background subtraction [18]. The pedestrian detection was performed by local adaptive thresholding [19]. People in the aerial thermal view were detected and tracked with a particle filter [20]. The weighted correlation filter tracked the various targets in the thermal image database [21]. An effective feature representation was introduced based on correlation energy in [22]. An adaptively multi-feature fusion model was proposed to integrate hand-crafted features and a convolutional neural network for thermal object tracking [23]. The Kalman filter was used to detect sports players in thermal videos for real-time tracking [24]. However, the tracking conditions were limited to fixed cameras, and a plain background. In [25], a tracking-by-detection approach was studied in the thermal spectrum. A convolutional neural network, pre-trained with visible images, was transferred to the thermal tracking [26].

A multiple-target tracking system consists of a sensing system that observes the movement of multiple objects in a dynamic environment and a target tracking algorithm that simultaneously estimates the attributes (position, velocity, acceleration, etc.) of multiple objects based on the observation. As a result, a multi-target tracking system forms a trajectory (track) for each target; so, it is essential to establish the identity of the same target while it is present, and this can be achieved by accurately matching the established track with the observation. However, due to high and various maneuvering, close or occluded objects, low object detection, or high false alarms, tracks may be missed, broken, or overlapped on the same target.

In this paper, tracking of people in the mountains is studied using thermal imaging by a drone. The overall process comprises two stages: object detection and multiple-target tracking. For object detection, YOLOv5 is adopted to generate a bounding box of objects. YOLO's deep learning framework has garnered immense popularity for its versatility, ease of use, and high performance [27]. The centroid of the bounding box is considered the measured position (observation) for target tracking.

Multiple-target tracking is performed via three stages: track initialization, maintenance, and termination [28–30]. The track maintenance consists of measurement-to-track association (measurement association), track update, and track-to-track association (track association). Figure 1 shows an entire block diagram of target tracking with infrared thermal videos acquired by drones. Bold fonts inside red bold line boxes include the newly proposed contents in this paper. A scheme of the nearest neighbor measurement association, followed by track association, track termination, and validity testing, has been developed in previous works showing robust performance in ground target tracking from visible images acquired by a drone [29,30].

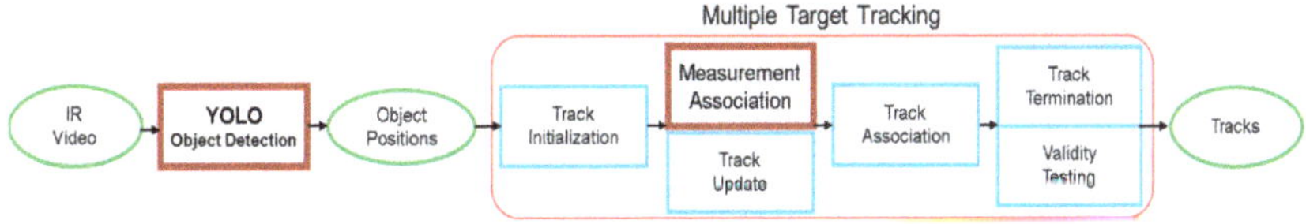

Figure 1. Block diagram of thermal image target tracking.

To the best of my knowledge, ref. [15] was the first study to track people with a thermal imaging camera mounted on a small drone. This paper has been substantially improved from [15]. The contributions of this paper are (1) the YOLO detection mode is applied to extract the position information of the thermal target. The YOLO detection model is pre-trained with visible image datasets, but it was transferred to thermal object detection for multiple-target tracking. (2) A bounding box gating scheme is proposed for the measurement association. This scheme allows track updates if the intersection over union (IoU) between the bounding box of the current frame and the bounding box of the previous frame is sufficiently high. The centroid of the current frame bounding box is the measurement that is statistically nearest to the position prediction at the current frame and the centroid of the previous frame bounding box is the measurement associated with updating the tract at the previous frame. (3) The framework combining the measurement association and the track association with the Kalman filter shows robust tracking performance of the mobility of the platform in complex backgrounds. In the presented scenarios, thermal objects are closely located and heavily occluded, and the sensing platform is allowed to move fast.

In the experiments, the drone flies at a height of 40~60 m. The video shows three hikers simulating being lost in the mountains on a winter night with no ambient light. The drone moves rapidly in horizontal and vertical directions, and the perspective of the camera is arbitrary. People are often occluded by other people or trees and leaves. Figure 2 shows three sample scenes extracted from the three videos, respectively. The experimental results show the average total track lives (TTLs) for the three videos are obtained as 0.987, 0.993, and 0.894, respectively. The corresponding average mean track lives are 0.987, 0.442, and 0.151, respectively. The average track purities (TPs) are obtained as 1, 0.999, and 0.995, respectively, for the three videos. The average MTL is reduced for two videos due to the track breakage in the harsh environments.

Figure 2. Sample frames of three thermal videos capturing people in distress.

The rest of the paper is organized as follows: the YOLOv5 training process is explained in Section 2. Section 3 describes each step of multiple-target tracking. The experimental results are presented in Section 4, and conclusions follow in Section 5.

2. YOLOv5 for Thermal Videos

YOLOv5 is the 5th iteration of the YOLO object detection model [27]. It is designed to provide high-speed, high-accuracy results in real-time. YOLOv5 has several pre-trained models with a visible image dataset; YOLOv5x shows the best performance among them. The output of YOLO is multiple bounding boxes surrounding the object of interest, along with the object's class name and confidence level.

The YOLO pre-trained models are trained with thermal images; the number of training images is 197 and a total of 548 rectangular boxes are manually drawn in them. The training images were obtained from a different video than the ones used for the tracking experiments. Each rectangular box indicates the instance of a person in the scene. The rectangular boxes were manually drawn using the graphical image annotation tool LabelImg [31]. During training, all instances are named with only one class, 'walking'. Three pre-trained models, YOLOv5s, YOLOv5l, and YOLOv5x, were transferred to detect thermal objects. They were trained with 100 epochs. No data augmentation was applied, and no background was used for training. Figure 3 shows five sample images, each image containing three blue rectangular boxes. Supplementary Video S1 shows a movie of all 197 images containing 548 rectangular boxes. The experimental results of detection testing will be shown in Section 4.

Figure 3. Sample training images with thermal object instances.

3. Multiple-Target Tracking

A block diagram of multiple-target tracking is shown in Figure 4. Tracks are initialized with two-point differencing between the measurements in consecutive frames after the initial speed gating. Tracks are maintained by track update, measurement association, and track association. Measurement association is necessary in multiple-target and clutter environments. The track association aims to handle multi-sensor environments [32], but it has been improved to fuse tracks generated with a single sensor [30]. The state of the target is estimated using the Kalman filter. The track is terminated if no measurement is available or track-fusion occurs. The validity of the track is tested with the track-life length after track termination. Each step of the block diagram is described in the following subsections.

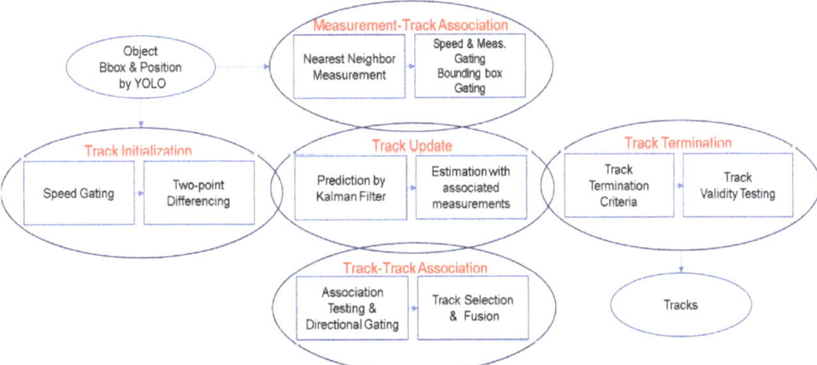

Figure 4. Block diagram of multiple-target tracking.

3.1. System Modeling

For the nearly constant velocity motion, the process noise following a Gaussian distribution imposes uncertainty on the kinematic state of the target. The discrete state equation of a target is

$$x(k+1) = F(\Delta)x(k) + q(\Delta)v(k), \tag{1}$$

$$F(\Delta) = \begin{bmatrix} 1 & \Delta & 0 & 0 \\ 0 & 1 & 0 & 0 \\ 0 & 0 & 1 & \Delta \\ 0 & 0 & 0 & 1 \end{bmatrix}, \; q(\Delta) = \begin{bmatrix} \Delta^2/2 & 0 \\ \Delta & 0 \\ 0 & \Delta^2/2 \\ 0 & \Delta \end{bmatrix}, \tag{2}$$

where $x(k) = \begin{bmatrix} x(k) \; v_x(k) \; y(k) \; v_y(k) \end{bmatrix}^T$ is the state vector of a target at frame k, $x(k)$ and $y(k)$ are positions in the x and y directions, respectively, $v_x(k)$ and $v_y(k)$ are velocities in the x and y directions, respectively, T denotes the matrix transpose, Δ is the sampling time, and $v(k)$ is a process noise vector, which is Gaussian white noise with the covariance matrix

$Q_v = diag\left(\begin{bmatrix} \sigma_x^2 & \sigma_y^2 \end{bmatrix}\right)$. The measurement equation representing the positions in the x and y directions of the target is

$$z(k) = \begin{bmatrix} z_x(k) \\ z_y(k) \end{bmatrix} = Hx(k) + w(k), \tag{3}$$

$$H = \begin{bmatrix} 1000 \\ 0010 \end{bmatrix}, \tag{4}$$

where $w(k)$ is a measurement noise vector, which is Gaussian white noise with the covariance matrix $R = diag\left(\begin{bmatrix} r_x^2 & r_y^2 \end{bmatrix}\right)$.

3.2. Two-Point Initialization

A track is initialized by two positions in consecutive frames if those two positions are close enough to pass the initial speed gating. The initial state and covariance of track t at frame k are estimated as, respectively,

$$\hat{x}_t(k|k) = \begin{bmatrix} \hat{x}_t(k|k) \\ \hat{v}_{tx}(k|k) \\ \hat{y}_t(k|k) \\ \hat{v}_{ty}(k|k) \end{bmatrix} = \begin{bmatrix} z_{tx}(k) \\ \frac{z_{tx}(k)-z_{tx}(k-1)}{\Delta} \\ z_{ty}(k) \\ \frac{z_{ty}(k)-z_{ty}(k-1)}{\Delta} \end{bmatrix}, \quad P_t(k|k) = \begin{bmatrix} r_x^2 & \frac{r_x^2}{\Delta} & 0 & 0 \\ \frac{r_x^2}{\Delta} & \frac{2r_x^2}{\Delta^2} & 0 & 0 \\ 0 & 0 & r_y^2 & \frac{r_y^2}{\Delta} \\ 0 & 0 & \frac{r_y^2}{\Delta} & \frac{2r_y^2}{\Delta^2} \end{bmatrix}, \quad t = 1,\ldots,N(k), \tag{5}$$

where $N(k)$ is the number of established tracks at frame k, which increases by 1 when a track is initialized and decreases by 1 when it terminates. The initial speed is bounded as $\sqrt{[\hat{v}_{tx}(k|k)]^2 + [\hat{v}_{ty}(k|k)]^2} \leq V_{max}$, where V_{max} is the maximum initial speed of the target.

3.3. Prediction and Filter Gain

The state and covariance predictions are iteratively computed as

$$\hat{x}_t(k|k-1) = F\hat{x}_t(k-1|k-1), \tag{6}$$

$$P_t(k|k-1) = FP_t(k-1|k-1)F^T + Q, \tag{7}$$

$$Q = q(\Delta)Q_v q(\Delta)^T, \tag{8}$$

where $\hat{x}_t(k|k-1)$ and $P_t(k|k-1)$, respectively, are the state and the covariance prediction of track t at frame k. The residual covariance $S_t(k)$ and the filter gain $W_t(k)$, respectively, are obtained as

$$S_t(k) = HP_t(k|k-1)H^T + R, \tag{9}$$

$$W_t(k) = P_t(k|k-1)H^T S_t(k)^{-1}. \tag{10}$$

3.4. Measurement-to-Track Association with Bounding Box Gating

The measurement association assigns a suitable measurement to the established track in the multi-target and clutter environments. The nearest-neighbor association rule selects the m_{tk}-th measurement, which has the shortest statistical distance to the position prediction as

$$m_{tk} = \operatorname*{argmin}_{m\,=\,1,\ldots,M(k)} \left\| v_{tm}(k)^T \left[S_t(k) \right]^{-1} v_{tm}(k) \right\|, \tag{11}$$

$$v_{tm}(k) = z_m(k) - H\hat{x}_t(k|k-1), \tag{12}$$

where $z_m(k)$ is the m-th measurement vector at frame k, and $M(k)$ is the number of measurements at frame k. The speed and measurement gating are defined as

$$\frac{\|z_{m_{tk}}(k) - [\hat{x}_t(k|k) \quad \hat{y}_t(k|k)]^T\|}{\Delta} \leq S_{max} \ \& \ v_{tm_{tk}}(k)^T[S_t(k)]^{-1}v_{tm_{tk}}(k) \leq \gamma_f, \tag{13}$$

where S_{max} is the maximum target speed, γ_f is a gating size of the measurement validation region, and '&' is the AND operation. The measurement gating is the chi-square hypothesis testing on Gaussian measurement residuals [32].

The bounding box gating is proposed as

$$\text{IoU}(m_{tk}, \hat{m}_{tk-1}) = \frac{|Bbox(m_{tk}) \ \cap \ Bbox(\hat{m}_{tk-1})|}{|Bbox(m_{tk}) \ \cup \ Bbox(\hat{m}_{tk-1})|} \geq \theta_f, \tag{14}$$

where the $\hat{m}_{tk-1}$-th measurement is already associated with track t at frame $k-1$, Bbox($\cdot$) is the set of pixels in the bounding box obtained by YOLO detection, $|\cdot|$ is the operator that counts the pixels in a set, and θ_f is a threshold for the bounding box gating; if the m_{tk}-th measurement satisfies Equation (14), that is, the IoU of two bounding boxes is equal to or more than the threshold value, the m_{tk}-th measurement is associated with track t at frame k. Figure 4 illustrates the measurement gating and the bounding box gating. In Figure 5a, m_{tk} is located inside the validation region and is associated with track t, whereas in Figure 5b, it is outside the validation region, but its IoU is high enough for the measurement to be associated with track t. As a consequence, the statistically nearest measurement is associated with a track if it passes the speed-measurement gating or the bounding box gating. If the measurements are not associated with any target, they are used for the track initialization at the current or previous frame.

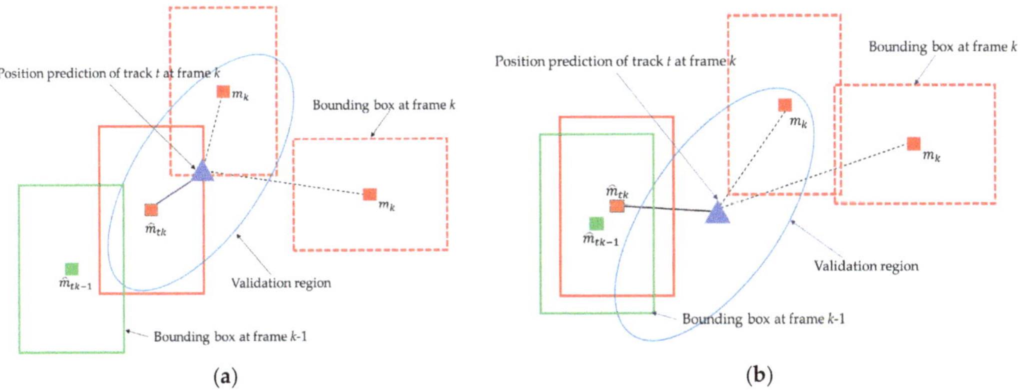

Figure 5. Measurement-target association, (**a**) measurement gating, (**b**) bounding box gating.

3.5. State Estimate and Covariance Update

The state estimate and the covariance are updated after the measurement association as

$$\hat{x}_t(k|k) = \hat{x}_t(k|k-1) + W_t(k)v_{t\hat{m}_{tk}}(k), \tag{15}$$

$$P_t(k|k) = P_t(k|k-1) - W_t(k)S_t(k)W_t(k)^T. \tag{16}$$

If there is no measurement associated, they are the same, with predictions of the state and covariance as

$$\hat{x}_t(k|k) = \hat{x}_t(k|k-1), \tag{17}$$

$$P_t(k|k) = P_t(k|k-1). \tag{18}$$

3.6. Track-to-Track Association

A track fusion method for a multi-sensor environment has been developed to associate redundant tracks [28]. The directional track association was proposed in [29]. The track association procedure was improved to search the fittest track in [30]. The fittest track for track s is

$$\hat{t} = \underset{t=1,\dots,N(k),s\neq t}{argmin} \; [\hat{\boldsymbol{x}}_s(k|k) - \hat{\boldsymbol{x}}_t(k|k)]^T [T_{st}(k)]^{-1} [\hat{\boldsymbol{x}}_s(k|k) - \hat{\boldsymbol{x}}_t(k|k)], \tag{19}$$

$$T_{st}(k) = P_s(k|k) + P_t(k|k) - P_{st}(k|k) - P_{ts}(k|k), \tag{20}$$

$$P_{st}(k|k) = [I - b_s(k)W_s(k)H]\Big[FP_{st}(k-1|k-1)F^T + Q\Big][I - b_t(k)W_t(k)H], \tag{21}$$

where I is the identity matrix, and $b_s(k)$ and $b_t(k)$ become one when track s or t is associated with a measurement, otherwise, they are zero [32]. The fused covariance in Equation (21) is a linear recursion with the initial condition $P_{st}(0|0)$, which is the square zero matrix. The fittest track is also satisfied with the following track and directional gating as

$$[\hat{\boldsymbol{x}}_s(k|k) - \hat{\boldsymbol{x}}_{\hat{t}}(k|k)]^T [T_{s\hat{t}}(k)]^{-1} [\hat{\boldsymbol{x}}_s(k|k) - \hat{\boldsymbol{x}}_{\hat{t}}(k|k)] \leq \gamma_g \; \& $$
$$max\left(\cos^{-1} \frac{|<\hat{\boldsymbol{d}}_{s\hat{t}}(k|k),\, \hat{\boldsymbol{v}}_s(k|k)>|}{\|\hat{\boldsymbol{d}}_{s\hat{t}}(k|k)\|\|\hat{\boldsymbol{v}}_s(k|k)\|},\; \cos^{-1} \frac{|<\hat{\boldsymbol{d}}_{s\hat{t}}(k|k),\, \hat{\boldsymbol{v}}_{\hat{t}}(k|k)>|}{\|\hat{\boldsymbol{d}}_{s\hat{t}}(k|k)\|\|\hat{\boldsymbol{v}}_{\hat{t}}(k|k)\|} \right) \leq \theta_g \tag{22}$$

$$\hat{\boldsymbol{d}}_{s\hat{t}}(k|k) = \begin{bmatrix} \hat{x}_{\hat{t}}(k|k) - \hat{x}_s(k|k) \\ \hat{y}_{\hat{t}}(k|k) - \hat{y}_s(k|k) \end{bmatrix}, \hat{\boldsymbol{v}}_s(k|k) = \begin{bmatrix} \hat{v}_{sx}(k|k) \\ \hat{v}_{sy}(k|k) \end{bmatrix}, \tag{23}$$

where γ_g is a gating size of the track validation region, $< \cdot >$ denotes the inner product operation, and θ_g is a threshold for the directional gating. The track gating is the chi-square hypothesis testing. Tracks s and $\hat{t}$ have the error dependencies on each other if they originated from the same target [32]. The directional gating tests the maximum deviation between the displacement vector and the velocity vector.

After the fittest track is selected, the current state of track s is replaced with a fused estimate and covariance if $|P_s(k|k)| \leq |P_{\hat{t}}(k|k)|$ as

$$\hat{\boldsymbol{x}}_s(k|k) = \hat{\boldsymbol{x}}_s(k|k) + [P_s(k|k) - P_{s\hat{t}}(k|k)][P_s(k|k) + P_{\hat{t}}(k|k) - P_{s\hat{t}}(k|k) - P_{\hat{t}s}(k|k)]^{-1}[\hat{\boldsymbol{x}}_{\hat{t}}(k|k) - \hat{\boldsymbol{x}}_s(k|k)], \tag{24}$$

$$P_s(k|k) = P_s(k|k) - [P_s(k|k) - P_{s\hat{t}}(k|k)][P_s(k|k) + P_{\hat{t}}(k|k) - P_{s\hat{t}}(k|k) - P_{\hat{t}s}(k|k)]^{-1}[P_s(k|k) - P_{\hat{t}s}(k|k)]. \tag{25}$$

A more accurate track has less error covariance; thus, fusion only occurs if the determinant of the covariance matrix of track s is less than the determinant of the selected track $\hat{t}$. After track s becomes a fused track, track $\hat{t}$ becomes a potentially terminated track. The potentially terminated track is still eligible to be associated with other tracks that have not yet been fused. The potentially terminated track is terminated when no track remains for the track association.

3.7. Track Termination and Validation Testing

Tracks are terminated by two criteria. One is if a track is selected as a potentially terminated track but not fused during the track association, then the track is terminated. The other is when the number of frames not updated by the measurement exceeds a threshold, the track is terminated.

After track termination, its validity is tested through the track life length. Track life length is defined as the number of frames between the last frame and the initial frame associated with the measurement. If the track life length is less than the minimum track life length, the track is considered invalid and is removed.

3.8. Performance Evaluation

The tracking performance is evaluated in terms of the number of tracks, total track life (TTL), mean track life (MTL), and track purity (TP) [33]. The TTL, MTL, and TP are defined, respectively, as

$$\text{TTL} = \frac{\text{Sum of lengths of tracks which have the same target ID}}{\text{Target life length} - 1}, \tag{26}$$

$$\text{MTL} = \frac{\text{TTL}}{\text{Number of tracks associated in TTL}}, \tag{27}$$

$$\text{TP} = \frac{\text{Number of measurments with the same target ID in a track}}{\text{Number of measurements in the track}}, \tag{28}$$

where target life length is defined as the number of frames between the last frame and the first frame when a measurement of the target appears, and the target ID of a track is defined as the target with the most measurements associated with the track. It is noted that the track length included in the TTL is only the number of frames for which the measurement associated is the same as the target ID of the track and it also includes predictions between updates. If the track is broken or overlaps with one target, the MTL will be less than the TTL. The TP becomes one if there is only one target associated with a track.

Figure 6 illustrates the TTL, MTL, and TP, where three tracks are generated for two targets. The blue and red circles represent the measurements of Targets t and s, respectively. Mission detections are found at Frames 6 and 8 for Targets t and s, respectively. The squares, triangles, and crosses represent the initialized or updated states of Tracks 1, 2, and 3, respectively. The same color of the target and the track indicates that they are related by the measurement-track association. The empty triangles in Track 2 indicate the predictions between the updated states. The Target ID of Tracks 1 and 2 is t, and Target ID of Track 3 is s. There is a breakage between Tracks 1 and 2 for Target t, and an intersection between Tracks 2 and 3 for Targets t and s. The TTL of Targets t and s are 7/9 and 4/7, respectively. The MTL of Targets t and s are 3.5/9 and 4/7, respectively, since the number of tracks for Targets t and s is 2 and 1, respectively. The TP of Tracks 1, 2, and 3 are 1, 3/5, and 2/3, respectively.

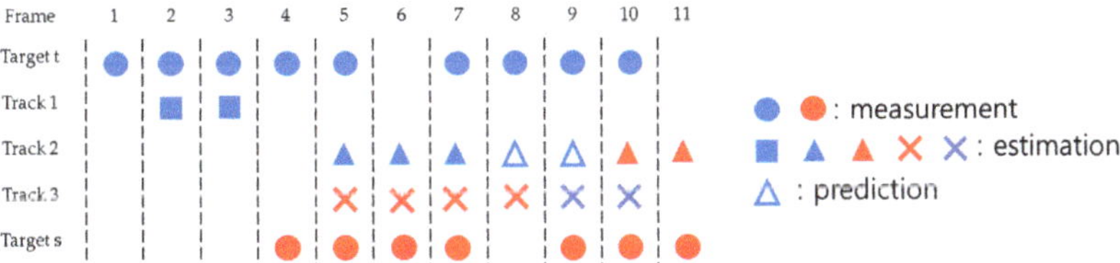

Figure 6. An illustration of three tracks generated for two targets.

4. Results

The experimental results are presented with video description, parameter settings, YOLO object detection, and multiple-target tracking.

4.1. Video Description

An infrared thermal imaging camera captured three videos (Videos 1–3). The camera, FLIR Vue Pro R640 (f = 19 mm, FOV = 32° × 26°, spectrum band: 7.5–13.5 µm, pixel pitch: 17 µm), is mounted on a DJI Inspire 2 drone. The resolution of the image is 640 × 512 pixels, and the frame rate is set to 30 fps. The videos were shot in the mountains on a winter night with no ambient light. In these environments, the visible image is completely dark. The thermal video capturing was assumed to be under harsh circumstances; the drone flies in any direction and altitude slowly or swiftly, and the perspective of the drone is arbitrary. Image characteristics vary from image to image depending on weather conditions and surrounding objects. In Video 1, three hikers walk around the mountain for 60 s. In Videos 2 and 3, three hikers are simulated as missing in the mountains for 120 s. Figure 7a shows the 1st, 501th, 901th, 1301th, and 1701th frame of Video 1, and Figure 7b,c show the 1st, 901th, 1801th, 2701th, 3601th frame of Videos 2 and 3, respectively. In each frame, a target ID, numbered 1 to 3 or 4, was displayed close to the object. It is assumed that there are four people in Video 1 and one of them appears with his legs at the top of the frame.

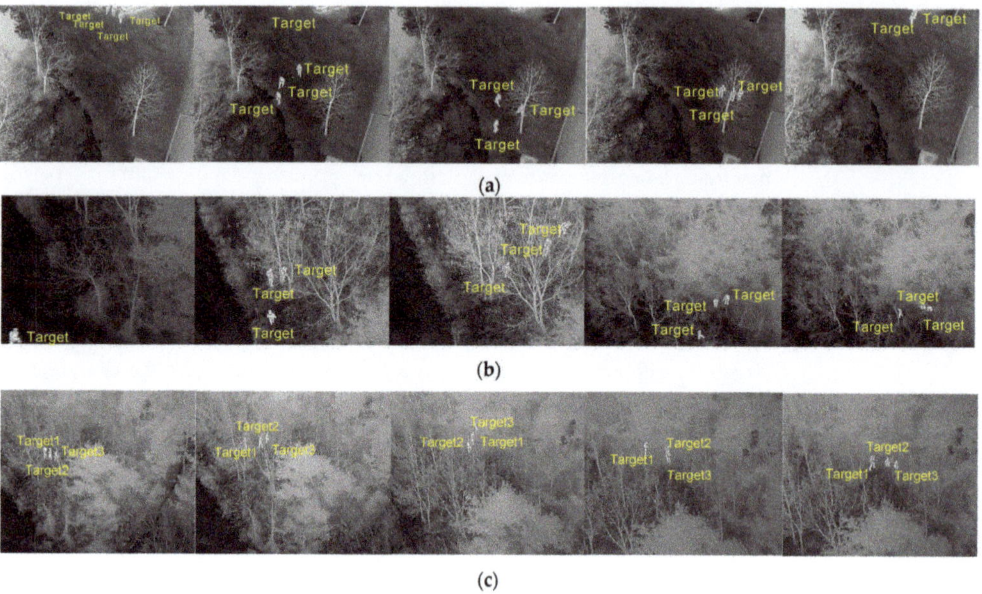

Figure 7. Sample frames with Target ID's of (**a**) Video 1, (**b**) Video 2, (**c**) Video 3.

4.2. People Detection by YOLOv5

YOLOv5 detects people in every frame of Videos 1–3. The detection results are summarized in Table 1. For YOLOv5l, the number of detections in Video 1 is more than the number of true instances, mostly due to false alarms generated on the car at the top of the frame. For YOLOV5x, the number of detections in Video 2 is more than the number of instances. This is because more than one bounding box was generated for one object. The recall of YOLOv5x is calculated lower than YOLOvl for Video 1 because YOLOv5x does not always detect the small part of the legs that appear at the top of the frame.

Table 1. YOLOv5 detection results.

		Video 1	Video 2	Video 3
Descriptions	Num. of frames (duration)	1801 (1 min)	3601 (2 min)	
	Num. of objects (people)	4	3	
	Num. of instances	6000	10,652	10,800
YOLOv5s	Num. of detections	5760	10,492	9143
	Num. of detections over 0.5 conf. level	5176	10,283	8425
	Recall over 0.5 conf. level	0.863	0.965	0.780
YOLOv5l	Num. of detections	6634	10,610	9693
	Num. of detections over 0.5 conf. level	5811	10,474	9248
	Recall over 0.5 conf. level	0.969	0.983	0.856
YOLOv5x	Num. of detections	5734	10,699	9947
	Num. of detections over 0.5 conf. level	5406	10,567	9676
	Recall over 0.5 conf. level	0.901	0.992	0.862

The detection of YOLOv5x with a minimum confidence level of 0.5 was used for the target tracking as it provides the most accurate detections without generating false alarms. For YOLOv5x, the recall of Videos 1 to 3 at a minimum confidence level of 0.5 is 0.901, 0.992, and 0.862, respectively, and the precision is 1 for all videos. In Video 3, harsh conditions, such as close and occluding objects and rapid movement of the drone, degrade the detection performance. Figure 8 shows YOLOv5x detection with a minimum confidence level of 0.5 for the sample frames. The centroids of the object in all frames are shown in Figures 9 and 10 in the first frame and on a white background, respectively. Supplementary Videos S2–S4 show the YOLOv5x object detection results of Videos 1, 2, and 3, respectively.

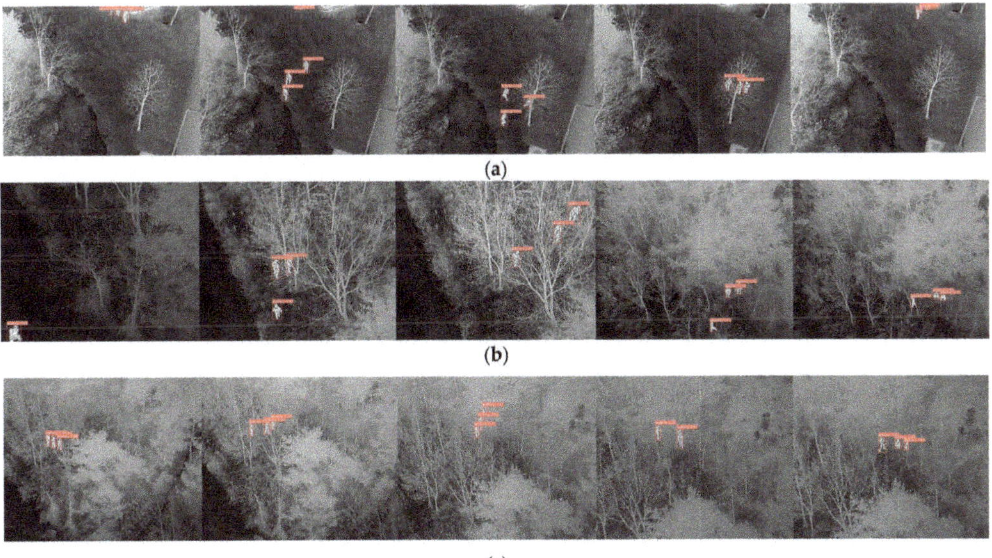

Figure 8. Detections of the sample frames, (**a**) Video 1, (**b**) Video 2, (**c**) Video 3.

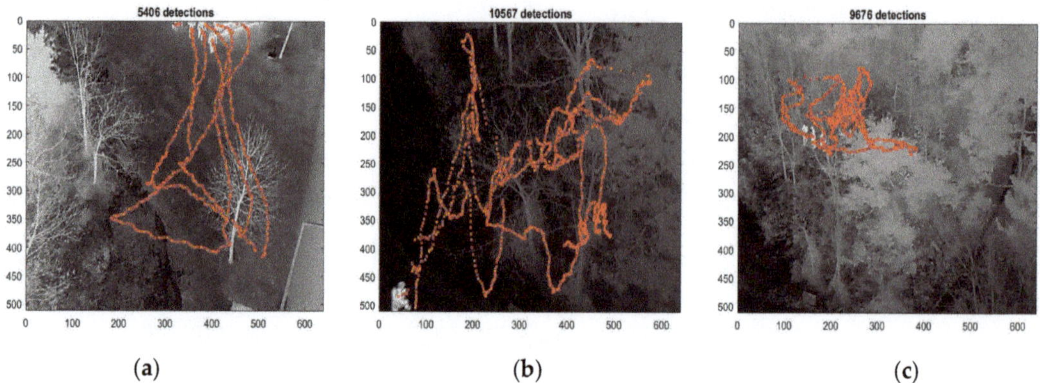

Figure 9. Detections in the 1st frame of (**a**) Video 1: 5406 detections, (**b**) Video 2: 10,567 detections, (**c**) Video 3: 9676 detections.

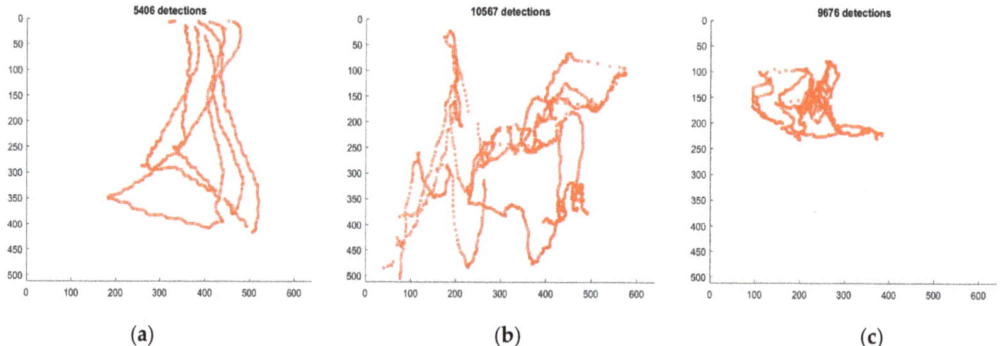

Figure 10. Detections only, (**a**) Video 1: 5406 detections, (**b**) Video 2: 10,567 detections, (**c**) Video 3: 9676 detections.

4.3. Multiple-Target Tracking

4.3.1. Parameter Set-Up

The target-tracking parameters are shown in Table 2. Video 1 was processed every two frames, thus the sampling time is 1/15 s for Video 1, whereas it is 1/30 s for Videos 2 and 3. The one-pixel coordinate is scaled to 0.03 m, 0.04 m, and 0.05 m for Videos 1, 2, and 3, respectively. The maximum initial speed of the target is set to 3 m/s. The process noise standard deviation of the Kalman filter is set to 0.5 m/s^2 for Videos 1 and 2.5 m/s^2 for Videos 2 and 3. The process noise standard deviation is determined considering the acceleration of the target and the rapid movement of the drone. The measurement noise standard deviation is set to 0.5 m for all videos. The maximum target speed for the track maintenance is set to 10 m/s for Videos 1 and 3 and 20 m/s for Video 2. The threshold for measurement gating is set to four and the minimum IoU for the bounding box gating is set to 0.6 for Videos 1 and 2 and 0.4 for Video 3. The gate threshold and angular threshold for the track association are set at 10 and 45° for Videos 1 and 3, respectively, and 20 and 60° for Video 2, respectively. They are set up when better results are produced. For the track termination, the maximum number of searches is set at 20 frames, and tracks shorter than 2 s are removed as invalid.

Table 2. Target tracking parameters.

	Parameters (Unit)	Video 1	Video 2	Video 3
	Sampling Time (second)	1/15	1/30	
	Max. initial target speed, V_{max} (m/s)		3	
Process noise std.	$\sigma_x = \sigma_y$ (m/s^2)	0.5	2.5	
Measurement noise std.	$r_x = r_y$ (m)		0.5	
Measurent association	Max. target speed, S_{max} (m/s)	10	20	10
	Gate threshold, γ_m		4	
	Bbox threshold, b_m	0.6		0.4
Track association	Gate threshold, γ_t	10	20	10
	Angular threshold, θ_t (degree)	45°	60°	45°
Track termination	Maximum searching number (frame)		20	
	Min. track life length for track validity (second)		2	

4.3.2. Target Tracking Results

The target tracking results for Videos 1 to 3, including the number of valid tracks, average TTL, average MTL, and average TP, are shown in Tables 3–5, respectively. Case 1, the first column of the table, is the result of applying both the bounding box gating and the track association. Case 2, the second column, is the result of applying the track association without the bounding box gating, and Case 3, the third column, is the result of applying neither the track association nor the bounding box gating. It is noted that Target 4 at the top of Video 1 was excluded from evaluating the tracking performance because it is a small part of a person's foot that does not move. For Video 1, the tracking results of Case 3 are perfect. For Videos 2 and 3, the results of Case 1 are better than others. For Case 1, the average TTL of Videos 1 to 3 are obtained as 0.987, 0.993, and 0.894, respectively. The corresponding average MTLs are 0.987, 0.442, and 0.151, respectively. The average TPs are obtained as 1, 0.999, and 0.995, respectively, for the three videos. The average MTL is reduced for Videos 2 and 3 due to the track breakage caused by the high maneuvering of the drone or object occlusion.

Table 3. Tracking results of Video 1.

	Case 1	Case 2	Case 3
Num. of Tracks	3	3	3
Avg. TTL	0.987	0.982	1
Avg. MTL	0.987	0.982	1
Avg. TP	1	0.991	1

Table 4. Tracking results of Video 2.

	Case 1	Case 2	Case 3
Num. of Tracks	7	7	18
Avg. TTL	0.993	0.945	0.943
Avg. MTL	0.442	0.417	0.193
Avg. TP	0.999	0.954	0.963

Table 5. Tracking results of Video 3.

	Case 1	Case 2	Case 3
Num. of Tracks	20	22	29
Avg. TTL	0.894	0.878	0.890
Avg. MTL	0.151	0.130	0.097
Avg. TP	0.995	0.995	0.986

Figures 11 and 12 show the tracks for all frames in random colors. The tracks are shown in the first frame in Figure 11 and on a white background in Figure 12. Supplementary Videos S5–S7 show the tracking results for Case 1 for Videos 1, 2, and 3, respectively.

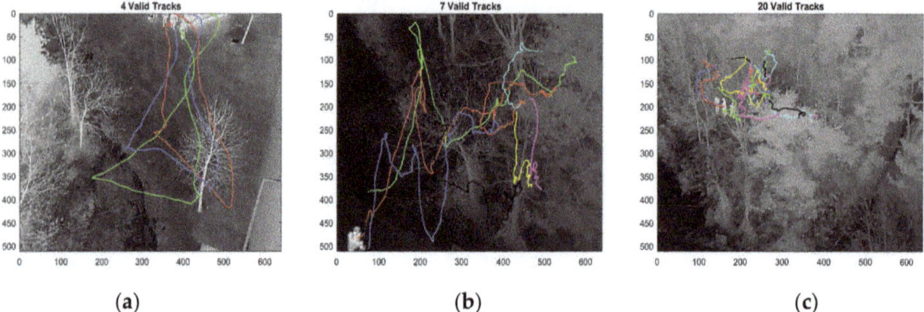

(a)   (b)   (c)

Figure 11. Tracks in the 1st frame, (**a**) Video 1: 4 valid tracks, (**b**) Video 2: 7 valid tracks, (**c**) Video 3: 20 valid tracks.

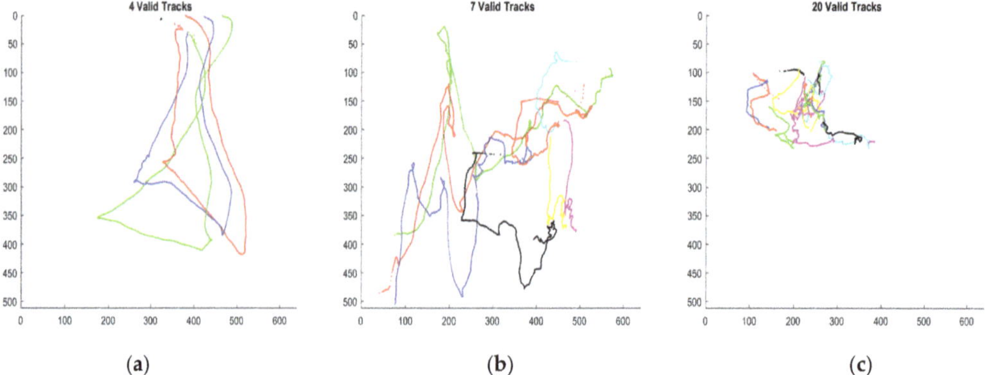

(a)   (b)   (c)

Figure 12. Tracks only, (**a**) Video 1: 4 valid tracks, (**b**) Video 2: 7 valid tracks, (**c**) Video 3: 20 valid tracks.

Seven Supplementary Multimedia Files (MP4 format) are available online. Supplementary Materials Video S1 is a video of YOLO training images. Training instances are indicated by blue rectangles. Supplementary Materials Videos S2–S4 are the YOLOv5x detection results with a minimum confidence level of 0.5 for Videos 1 to 3, respectively. The videos display bounding boxes, including the class name and confidence level. Supplementary Materials Videos S5–S7 are the tracking results from Videos 1 to 3 applying both the bounding box gating and the track association, respectively. The bounding box and its centroid of YOLOv5x are shown as red squares. Position estimates are shown as blue circles. Valid tracks were numbered in the order they were created.

5. Discussion

The thermal videos were captured under extremely challenging conditions, simulating people needing search and rescue missions in non-visible environments. The experimental scenarios included overgrown terrain, in which the warm objects were either partially visible or invisible. The objects are often occluded by other people and natural objects, such as trees, bushes, and foliage. The drone is manually operated, allowing rapid movements.

Since the more accurate the location information, the better the tracking quality, the detections of YOLOv5x with a minimum confidence level of 0.5 were utilized. The detection accuracy is expected to increase with larger training data since less than 200 images were trained in this paper.

TTL and MTL are defined on the target and TP is defined on the track. High average TTLs and average TPs are achieved for all videos. However, the average MTLs were lowered for Videos 2 and 3 as more than one valid track was generated from the target. The number of valid tracks is increased due to the track breakage caused by missing detection. The missing detection is mainly caused by the high maneuvering of the drone or the object occlusion. One track intersection occurs in both Videos 2 and 3, resulting in the slightly lower average TP. The track breakage and intersection both lower the average TTL. The bounding box gating has improved tracking performance for all metrics.

Low-resolution and gray-scaled infrared thermal frames can be transmitted to a ground station with less bandwidth. Adopting a lighter and later version of the YOLO model, such as YOLOv8 [34], is also considered to increase implementation feasibility, as well as detection performance.

6. Conclusions

In the paper, multiple-target tracking using thermal imaging was studied for the purpose of search and rescue missions with a drone. The object-detection multiple-target tracking scheme has been shown to be very powerful for tracking people in thermal videos acquired by drones. The harsh conditions of simulated search and rescue missions, including (1) no ambient lighting environment, (2) complex backgrounds, (3) closely located and heavily occluded targets, and (4) arbitrary moving platforms, can be overcome with the proposed solution.

To evaluate tracking performance, three metrics TTL, MTL, and TP were obtained. TTL and TP provide solid performance, but MTL decreases when tracks are broken. The proposed solution is direct and simple, yet quite effective. It is also suitable for security and surveillance in civil and military fields and wildlife monitoring. It can also be applied to pedestrian tracking in crowds and object tracking in sports analysis. Track segment association to increase the track continuity remains for future studies. Adopting higher iterations of the YOLO model for detection performance and feasible implementation also remains for future work.

Supplementary Materials: The following are available online at https://zenodo.org/records/1046 7489 accessed on 26 January 2024, Video S1: YOLOv5x training data, Video S2: Video 1 YOLOv5x detection, Video S3: Video 2 YOLOv5x detection, Video S4: Video 3 YOLOv5x detection. Videos S2–S4 are the detection results with 0.5 minimum confidence level. Video S5: Video 1 target tracking, Video S6: Video 2 target tracking, Video S7: Video 3 target tracking. Videos S5–S7 are the tracking results with the bounding box gating and the track association.

Funding: This research was supported by Daegu University Research Grant, 2023.

Institutional Review Board Statement: Not applicable.

Informed Consent Statement: Not applicable.

Data Availability Statement: Data are contained within the article and Supplementary Materials.

Conflicts of Interest: The author declares no conflict of interest.

References

1. Alzahrani, B.; Oubbati, O.S.; Barnawi, A.; Atiquzzaman, M.; Alghazzawi, D. UAV assistance paradigm: State-of-the-art in applications and challenges. *J. Netw. Comput. Appl.* **2020**, *166*, 102706. [CrossRef]
2. Vollmer, M.; Mollmann, K.-P. *Infrared Thermal Imaging: Fundamentals, Research and Applications*; Wiley-VCH: Weinheim, Germany, 2010.
3. Kirk, J.; Havens Edward, J. *Sharp, Thermal Imaging Techniques to Survey and Monitor Animals in the Wild*; Academic Press: Cambridge, MA, USA, 2016; ISBN 9780128033845. [CrossRef]
4. Rudol, P.; Doherty, P. Human Body Detection and Geolocalization for UAV Search and Rescue Missions Using Color and Thermal Imagery. In Proceedings of the 2008 IEEE Aerospace Conference, Big Sky, MT, USA, 1–8 March 2008; pp. 1–8. [CrossRef]

5. Jamjoum, M.; Siouf, S.; Alzubi, S.; AbdelSalam, E.; Almomani, F.; Salameh, T.; Al Swailmeen, Y. DRONA: A Novel Design of a Drone for Search and Rescue Operations. In Proceedings of the 2023 Advances in Science and Engineering Technology International Conferences (ASET), Dubai, United Arab Emirates, 20–23 February 2023; pp. 1–5. [CrossRef]
6. Thermal Camera-Equipped UAVs Spot Hard-to-Find Subjects. Available online: https://www.photonics.com/Articles/Thermal_Camera-Equipped_UAVs_Spot_Hard-to-Find/a63435 (accessed on 30 January 2024).
7. Gonzalez, L.F.; Montes, H.G.A.; Puig, E.; Johnson, S.; Mengersen, K.; Gaston, K.J. Unmanned Aerial Vehicles (UAVs) and Artificial Intelligence Revolutionizing Wildlife Monitoring and Conservation. *Sensors* **2016**, *16*, 97. [CrossRef] [PubMed]
8. Messina, G.; Modica, G. Applications of UAV Thermal Imagery in Precision Agriculture: State of the Art and Future Research Outlook. *Remote Sens.* **2020**, *12*, 1491. [CrossRef]
9. Krišto, M.; Ivasic-Kos, M.; Pobar, M. Thermal Object Detection in Difficult Weather Conditions Using YOLO. *IEEE Access* **2020**, *8*, 25459–125476. [CrossRef]
10. Jiang, C.; Ren, H.; Ye, X.; Zhu, J.; Zeng, H.; Nan, Y.; Sun, M.; Ren, X.; Huo, H. Object detection from UAV thermal infrared images and videos using YOLO models. *J. Appl. Earth Obs. Geoinf.* **2022**, *112*, 102912. [CrossRef]
11. Kannadaguli, P. YOLO v4 Based Human Detection System Using Aerial Thermal Imaging for UAV Based Surveillance Applications. In Proceedings of the 2020 International Conference on Decision Aid Sciences and Application (DASA), Sakheer, Bahrain, 8–9 November 2020; pp. 1213–1219. [CrossRef]
12. Levin, E.; Zarnowski, A.; McCarty, J.L.; Bialas, J.; Banaszek, A.; Banaszek, S. Feasibility Study of Inexpensive Thermal Sensor and Small UAS Deployment for Living Human Detection in Rescue Missions Application Scenario. In Proceedings of the International Archives of the Photogrammetry, Remote Sensing and Spatial Information Sciences, Volume XLI-B8, 2016 XXIII ISPRS Congress, Prague, Czech Republic, 12–19 July 2016.
13. Teutsch, M.; Mueller, T.; Huber, M.; Beyerer, J. Low Resolution Person Detection with a Moving Thermal Infrared Camera by Hot Spot Classification. In Proceedings of the 2014 IEEE Conference on Computer Vision and Pattern Recognition Work-Shops, Columbus, OH, USA, 23–28 June 2014; pp. 209–216. [CrossRef]
14. Giitsidis, T.; Karakasis, E.G.; Gasteratos, A.; Sirakoulis, G.C. Human and Fire Detection from High Altitude UAV Images. In Proceedings of the 2015 23rd Euromicro International Conference on Parallel, Distributed, and Network-Based Processing, Turku, Finland, 4–6 March 2015; pp. 309–315. [CrossRef]
15. Yeom, S. Moving People Tracking and False Track Removing with Infrared Thermal Imaging by a Multirotor. *Drones* **2021**, *5*, 65. [CrossRef]
16. Leira, F.S.; Helgensen, H.H.; Johansen, T.A.; Fossen, T.I. Object detection, recognition, and tracking from UAVs using a thermal camera. *J. Field Robot.* **2021**, *38*, 242–267. [CrossRef]
17. Helgesen, H.H.; Leira, F.S.; Johansen, T.A. Colored-Noise Tracking of Floating Objects using UAVs with Thermal Cameras. In Proceedings of the 2019 International Conference on Unmanned Aircraft Systems (ICUAS), Atlanta, GA, USA, 11–14 June 2019; pp. 651–660. [CrossRef]
18. Davis, J.W.; Sharma, V. Background-Subtraction in Thermal Imagery Using Contour Saliency. *Int. J. Comput. Vis.* **2007**, *71*, 161–181. [CrossRef]
19. Soundrapandiyan, R. Adaptive Pedestrian Detection in Infrared Images Using Background Subtraction and Local Thresh-olding. *Procedia Comput. Sci.* **2015**, *58*, 706–713. [CrossRef]
20. Portmann, J.; Lynen, S.; Chli, M.; Siegwart, R. People detection and tracking from aerial thermal views. In Proceedings of the 2014 IEEE International Conference on Robotics and Automation (ICRA), Hong Kong, China, 31 May–7 June 2014; pp. 1794–1800. [CrossRef]
21. He, Y.-J.; Li, M.; Zhang, J.; Yao, J.-P. Infrared target tracking via weighted correlation filter. *Infrared Phys. Technol.* **2015**, *73*, 103–114. [CrossRef]
22. Yu, T.; Mo, B.; Liu, F.; Qi, H.; Liu, Y. Robust thermal infrared object tracking with continuous correlation filters and adaptive feature fusion. *Infrared Phys. Technol.* **2019**, *98*, 69–81. [CrossRef]
23. Yuan, D.; Shu, X.; Liu, Q.; Zhang, X.; He, Z. Robust thermal infrared tracking via an adaptively multi-feature fusion model. *Neural Comput. Appl.* **2022**, *35*, 3423–3434. [CrossRef] [PubMed]
24. Gade, R.; Moeslund, T.B. Thermal Tracking of Sports Players. *Sensors* **2014**, *14*, 13679–13691. [CrossRef] [PubMed]
25. WEl Ahmar, A.; Kolhatkar, D.; Nowruzi, F.E.; AlGhamdi, H.; Hou, J.; Laganiere, R. Multiple Object Detection and Tracking in the Thermal Spectrum. In Proceedings of the 2022 IEEE/CVF Conference on Computer Vision and Pattern Recognition Workshops (CVPRW), New Orleans, LA, USA, 19–24 June 2022; pp. 276–284. [CrossRef]
26. Liu, Q.; Lu, X.; He, Z.; Zhang, C.; Chen, W.-S. Deep convolutional neural networks for thermal infrared object tracking. *Knowl.-Based Syst.* **2017**, *134*, 189–198. [CrossRef]
27. Available online: https://github.com/ultralytics/yolov5 (accessed on 30 January 2024).
28. Yeom, S.; Nam, D.-H. Moving Vehicle Tracking with a Moving Drone Based on Track Association. *Appl. Sci.* **2021**, *11*, 4046. [CrossRef]
29. Yeom, S. Long Distance Moving Vehicle Tracking with a Multirotor Based on IMM-Directional Track Association. *Appl. Sci.* **2021**, *11*, 11234. [CrossRef]
30. Yeom, S. Long Distance Ground Target Tracking with Aerial Image-to-Position Conversion and Improved Track Association. *Drones* **2022**, *6*, 55. [CrossRef]

31. Available online: https://github.com/HumanSignal/labelImg (accessed on 30 January 2024).
32. Bar-Shalom, Y.; Li, X.R. *Multitarget-Multisensor Tracking: Principles and Techniques*; YBS Publishing: Storrs, CT, USA, 1995.
33. Yeom, S.-W.; Kirubarajan, T.; Bar-Shalom, Y. Track segment association, fine-step IMM and initialization with doppler for improved track performance. *IEEE Trans. Aerosp. Electron. Syst.* **2004**, *40*, 293–309. [CrossRef]
34. Available online: https://github.com/ultralytics/ultralytics (accessed on 30 January 2024).

Article

Inverse Airborne Optical Sectioning

Rakesh John Amala Arokia Nathan, Indrajit Kurmi and Oliver Bimber *

Institute of Computer Graphics, Johannes Kepler University Linz, 4040 Linz, Austria
* Correspondence: oliver.bimber@jku.at; Tel.: +43-732-2468-6631

Abstract: We present Inverse Airborne Optical Sectioning (IAOS), an optical analogy to Inverse Synthetic Aperture Radar (ISAR). Moving targets, such as walking people, that are heavily occluded by vegetation can be made visible and tracked with a stationary optical sensor (e.g., a hovering camera drone above forest). We introduce the principles of IAOS (i.e., inverse synthetic aperture imaging), explain how the signal of occluders can be further suppressed by filtering the Radon transform of the image integral, and present how targets' motion parameters can be estimated manually and automatically. Finally, we show that while tracking occluded targets in conventional aerial images is infeasible, it becomes efficiently possible in integral images that result from IAOS.

Keywords: synthetic aperture imaging; through-foliage tracking; occlusion removal

1. Introduction

Higher resolution, wide depth of field, fast framerates, high contrast, or signal-to-noise ratio can often not be achieved with compact imaging systems that apply narrower aperture sensors. Synthetic aperture (SA) sensing is a widely recognized technique to achieve these objectives by acquiring individual signals of multiple or a single moving small-aperture sensor and by computationally combining them to approximate the signal of a physically infeasible, hypothetical wide aperture sensor [1]. This principle has been used in a wide range of applications, such as radar [2–28], telescopes [29,30], microscopes [31], sonar [32–35], ultrasound [36,37], lasers [38,39], and optical imaging [40–47].

In radar, electromagnetic waves are emitted and their backscattered echoes are recorded by an antenna. Electromagnetic waves at typical radar wavelengths (as compared with the visible spectrum) can penetrate scattering media (i.e., clouds, vegetation, and partly soil) and are quite useful for obtaining information in all weather conditions. However, acquiring high spatial resolution images would require an impractically large antenna [2]. Therefore, since its invention in the 1950s [3,4], Synthetic Aperture Radar (SAR) sensors have been placed on space-borne systems, such as satellites [5–8], planes [9–11], and drones [12,13] in different modes of operation, such as strip-map [11,14], spotlight [11,14], and circular [10,14] to observe various sorts of phenomena on Earth's surface. These include crop growth [8], mine detection [12], natural disasters [6], and climate change effects, such as the deforestation [14] or melting of glaciers [7]. Phase differences of multiple SAR recordings (interferometry) have even been used to reconstruct depth information and enables finer resolutions [15].

Analogous to SAR (which utilizes moving radars for synthetic aperture sensing of widely static targets), a technique known as Inverse Synthetic Aperture Radar (ISAR) [16–18] considers the relative motion of moving targets and static radars for SAR sensing. In contrast to SAR (where the radar motion is usually known), ISAR is challenged by the estimation of an unknown target motion. It often requires sophisticated signal processing and is often limited to sensing one target at a time, while SAR can image large areas and monitor multiple (static) targets simultaneously [17,18]. ISAR has been used for non-cooperative target recognition (non-stationary targets) in maritime [19,20], airspace [21,22], near-space [23,24], and overland surveillance applications [25–28]. Recently, spatially distributed systems and

Citation: Amala Arokia Nathan, R.J.; Kurmi, I.; Bimber, O. Inverse Airborne Optical Sectioning. *Drones* **2022**, *6*, 231. https://doi.org/10.3390/drones6090231

Academic Editor: Seokwon Yeom

Received: 27 July 2022
Accepted: 30 August 2022
Published: 2 September 2022

advanced signal processing, such as compressed sensing and machine learning, have been utilized to obtain 3D images of targets, target's reflectivity, and more degrees of freedom for target motion estimation [27,28].

With Airborne Optical Sectioning (AOS) [48–60], we introduced an optical synthetic aperture imaging technique that captures an unstructured light field with an aircraft, such as a drone. We utilized manually, automatically [48–56,58,59], or fully autonomously [57] operated camera drones that sample multispectral (RGB and thermal) images within a certain (synthetic aperture) area above occluding vegetation (such as forest) and combined their signals computationally to remove occlusion. The outcome is a widely occlusion-free integral image of the ground, revealing details of registered targets while unregistered occluders above the ground, such as trunks, branches or leaves disappear in strong defocus. In contrast to SAR, AOS benefits from high spatial resolution, real-time processing rates, and wavelength-independences, making it useful in many domains. So far, AOS has been applied to the visible [48,59] and the far-infrared (thermal) spectrum [51] for various applications, such as archeology [48,49], wildlife observation [52], and search and rescue [55,56]. By employing a randomly distributed statistical model [50,57,60] the limits of AOS and its efficacy with respect to its optimal sampling parameters can be explained. Common image processing tasks, such as classification with deep neural networks [55,56] or color anomaly detection [59] are proven to perform significantly better when applied to AOS integral images compared with conventional aerial images. We also demonstrated the real-time capability of AOS by deploying it on a fully autonomous and classification-driven adaptive search and rescue drone [56]. Yet, the sequential sampling nature of AOS when being used with conventional single-camera drones has limited its applications to recover static targets only. Moving targets lead to motion blur in the AOS integral images, which are nearly impossible to classify or to track.

In [59], we presented a first solution to tracking moving people through densely occluding foliage with parallel synthetic aperture sampling supported by a drone-operated, 10 m wide, 1D camera array (assembling 10 synchronized cameras). Although feasible, such a specialized imaging system is in most cases is impractical as it is bulky and difficult to control.

Being inspired by the principles of ISAR for radar, in this article we present Inverse Airborne Optical Sectioning (IAOS) for detecting and tracking moving targets through occluding foliage (cf. Figure 1b) with a conventional, single-camera drone (cf. Figure 1c). As with ISAR, IAOS relies on the motion of targets being sensed by a static airborne optical sensor (e.g., a drone hovering above forest) over time (cf. Figure 1a) to computationally reconstruct an occlusion-free integral image (cf. Figure 1d). Essential for an efficient reconstruction is the correct estimation of the target's motion.

In this article, we make four main contributions: (1) We introduce the principles of IAOS (i.e., inverse synthetic aperture imaging) in Sections 1 and 2. (2) We explain how the signal of occluders can be further suppressed by filtering the Radon transform of the image integral in Section 2.1 (cf. Figure 1e). (3) We present how a target's motion parameters can be estimated manually and automatically in Sections 2.1 and 2.2. (4) Finally, we show that while tracking occluded targets in conventional aerial images is infeasible, it is efficiently possible in integral images that result from IAOS in Section 3.

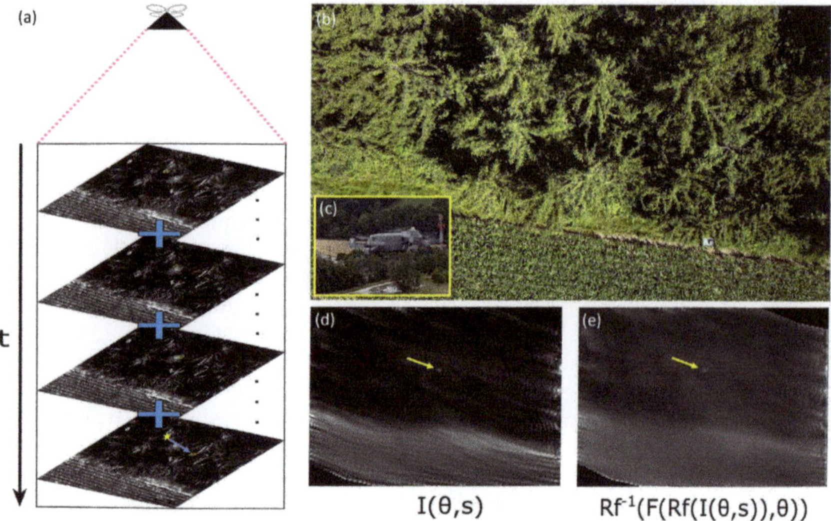

Figure 1. Inverse Airborne Optical Sectioning (IAOS) principle: IAOS relies on the motion of targets being sensed by a static airborne optical sensor (e.g., a drone (**c**) hovering above forest (**b**)) over time (**a**) to computationally reconstruct an occlusion-free integral image I (**d**). Essential for an efficient reconstruction is the correct estimation of the target's motion (direction θ, and speed s). By filtering the Radon transform of I, the signal of occluders can be suppressed further (**e**). Thermal images are shown in (**a,d,e**).

2. Materials and Methods

All field experiments were carried out in compliance with the legal European union Aviation Safety Agency (EASA) flight regulations, using a DJI Mavic 2 Enterprise Advanced, over dense broadleaf, conifer, and mixed forest, and under direct sunlight as well as under cloudy weather conditions. Free flight drone operations were performed using the DJI's standalone smart remote controller with DJI's Pilot application. RGB videos of resolution 1920×1080 (30 fps) and thermal videos of resolution 640×512 (30 fps) were recorded on the drone's internal memory, and were processed offline after landing. For vertical (top-down, as in Figure 3) scans the drone was hovering at an altitude of about 35 m AGL. For horizontal scans (sideways, as in Figure 4) the drone was hovering at a distance of about 10 m away from the vegetation. For quicker processing, we extracted a selection of 1–5 fps from the acquired 30 fps thermal videos using FFmpeg python bindings. Offline processing included intrinsic camera calibration (pre-calibrated transformation matrix computed using MATLAB's camera calibrator application) and image un-distortion/rectification using OpenCV's pinhole camera model (as explained in [48,55]). The undistorted and rectified images were cropped to a field of view of $36°$ and a resolution of 1024×1024 px. Image integration was achieved by averaging the pre-processed images being registered based on manually or automatically estimated motion parameters, as explained in Sections 2.1 and 2.2. Radon transform filtering [61–63] (also explained in Sections 2.1 and 2.2) was implemented in Mathworks' MATLAB R2022a.

2.1. Manual Motion Estimation

If the target's motion parameters (i.e., direction, θ [°] and speed, s [m/s]) are known and assumed to be constant for all time intervals, the captured images can be registered by shifting them accordingly to θ and s. Thereby, θ can directly be mapped to the image plane, while s must be mapped [m/s] to [px/s] (which is easily possibly after camera calibration and knowing the drone's altitude). By averaging the registered images results in an integral image that shows the target in focus (local motion of the target itself, such as arm movements of a walking person, lead to defocus) while the misregistered occluders vanish in defocus.

Large occluders that are shifted in direction θ while being integrated appear as linear directional blur artifacts in the integral image (cf. Figure 1d). Their signal can be suppressed by filtering (zeroing out) the Radon transform of the integral image I(θ,s) in direction θ (+/− an uncertainty range that considers local motion non-linearities of the occluders, such as movements of branches caused by wind, etc.). The inverse Radon transform (filtered back projection [63]) of the filtered sinogram results in a new integral image with suppressed signal of the directionally blurred occluders (cf. Figure 1e). This process is illustrated in Figure 2, and can be summarized mathematically with:

$$I'(\theta,s) = \mathrm{Rf}^{-1}(\mathrm{F}(\mathrm{Rf}(I(\theta,s)),\theta)), \tag{1}$$

where F is the filter function which zeros out coefficients at angle θ (+/− uncertainty range) in the sinogram.

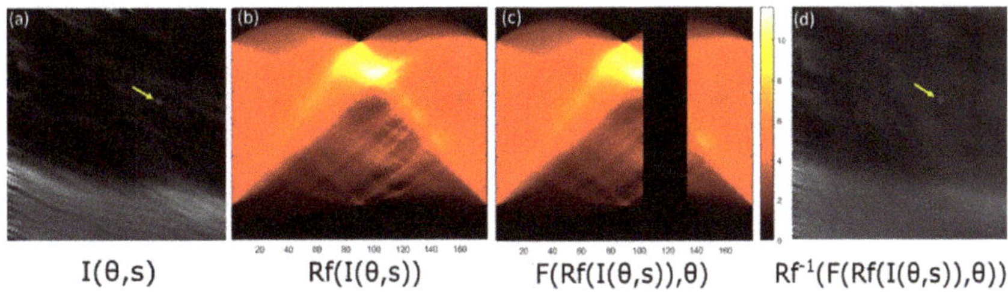

Figure 2. Radon transform filtering: to suppress directional blur artifacts of large occluders integrated in direction θ (**a**), the Radon transform (Rf) of the integral image (**b**) is filtered with function F that zeros out θ, +/− an uncertainty range which takes local motion of the occluders themselves into account (**c**). The inverse Radon transform (Rf^{-1}) of this filtered sinogram suppresses the direction blur artefacts of the occluders (**d**). Note: remaining directional artifacts in orthogonal directions are caused by under-sampling (i.e., the number of images being integrated). They are fluctuating too much to be suppressed in the same manner. In the example above, θ = 118° with +/− 15° (image coordinate system: clockwise, +y-axis = 0°).

One way of estimating the correct motion parameters is by visual search (i.e., θ and s are interactively modified until the target appears best focused in the integral image). Exploring the two-dimensional parameter space within proper bounds is relatively efficient if the motion can be assumed to be constant. Sample results are presented in Section 3. See also Supplementary Video S1 for an example of manual visual search for the motion parameters of results shown in Figure 3k. In case of non-linear motion, the motion parameters must be continuously and automatically estimated. A manual exploration becomes infeasible in this case.

2.2. Automatic Motion Estimation

Automatic estimation of motion parameters requires an error metric which is capable of detecting improvement and degradation in visibility (i.e., focus and occlusion) for different parameters. Here, we utilize simple gray level variance (GLV) [64] as an objective function. We already proved in [53] that, in contrast to traditionally used gradient-, Laplacian , or wavelet based focus metrics [65], GLV does not rely on any image features and is thus invariant to occlusion. In [54] (see also Appendix A), we demonstrated that the variance of an integral image is:

$$Var[I] = \frac{D(1-D)((\mu_o - \mu_s)^2) + D\sigma_o^2 + (1-D)\sigma_s^2}{N} + (1-D)^2(1-\frac{1}{N})\sigma_s^2, \tag{2}$$

where D is the probability of occlusion, while μ_o, σ_o^2 and μ_s, σ_s^2 are the statistical properties of occlusion and the target signal, respectively.

Integrating N individual images with optimal motion parameters results in an occlusion-free view of the target's signal whereas the signal strength of the occluders reduces and disappears in strong defocus. To further suppress occluders, we used Radon filtering [61–63] as described in Section 2.1. However, we now utilize the linearity property of the Radon transform which states that:

$$Rf\left(\sum_i \alpha_i I_i\right) = \sum_i \alpha_i Rf(I_i). \tag{3}$$

Thus, instead of filtering the integral image $I(\theta,s)$, as explained in Equation (1), we apply Radon transform filtering to each single image I_i before integrating it.

For automatic motion parameter estimation, we registered the current integral image I (integrating $I_1' \ldots I_{i-1}'$) to the latest (most recently recorded) inverse Radon transformed filtered image $I_i' = \mathrm{Rf}^{-1}(\mathrm{F}(\mathrm{Rf}(I_i), \theta))$ by maximizing $\mathrm{Var}[I]$ while optimizing for best motion parameters (θ, s). Deterministic-global search, DIRECT [66] (as implemented Nlopt [67]), was applied for optimization. Consequently, we considered each discrete motion component between two recorded images and within the corresponding imaging time (e.g., $1/30$ s for 30 fps) to be piecewise linear. The integration of multiple images, however, can reveal and track a non-linear motion pattern where (θ, s) vary in each recording step. Sample results are presented in Section 3 and in Supplementary Videos S2 and S3.

3. Results

Figure 3 presents results from field studies of IAOS with manual motion estimation, as explained in Section 2.1. Images are recorded top-down, with the drone hovering at a constant position above conifer (Figure 3a–l), broadleaf (Figure 3m–o), and mixed (Figure 3p–r) forest. Estimated motion parameters of hidden walking people were: 118°, 0.5 m/s (Figure 3a–l), 108°, 0.6 m/s (Figure 3m–o), and 90°, 0.6 m/s (Figure 3p–q).

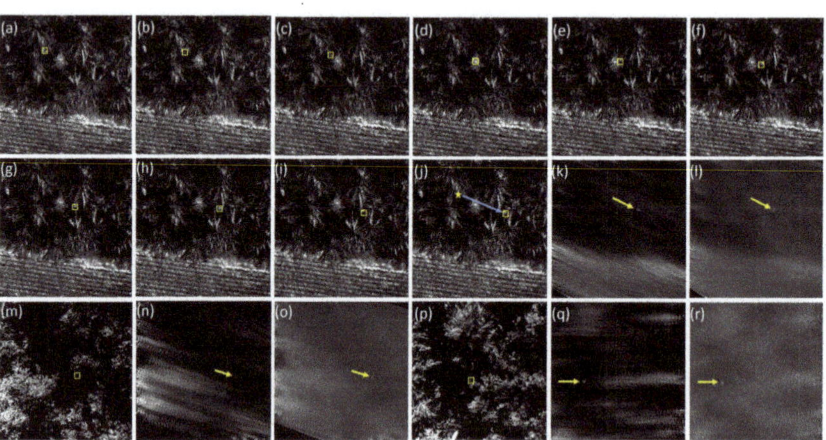

Figure 3. Manual motion estimation (vertical): sequence of single thermal images (**a–j**) with walking persons indicated (yellow box), distance covered by person during capturing time (**j**), computed integral image (**k**), and Radon transform filtered integral image (**l**). Target indicated by yellow arrow. Different forest types: single thermal image example (**m,p**), integral images (**n,q**), and Radon transform filtered integral images (**o,r**).

Figure 4 illustrates an example with the drone hovering at a distance of 10 m in front of dense bushes (at an altitude of 2 m, recording horizontally). The hidden person is walking from right to left at 260° with 0.27 m/s (both manually estimated).

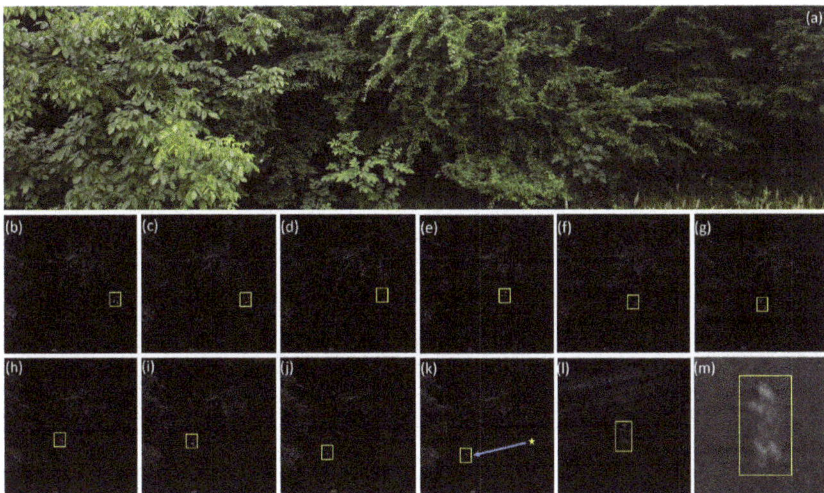

Figure 4. Manual motion estimation (horizontal): Walking person behind dense bushes. RGB image of drone (**a**). Single thermal images with person position indicated with yellow box (**b–k**). Distance covered by person during capturing time (**k**). Integral image (**l**) and close-up (**m**) where the shape of the person can be recognized.

Figure 5 illustrates two examples for automatic motion estimation, as explained in Section 2.2, with the drone hovering at an altitude of 35 m and a hidden person walking through dense forest.

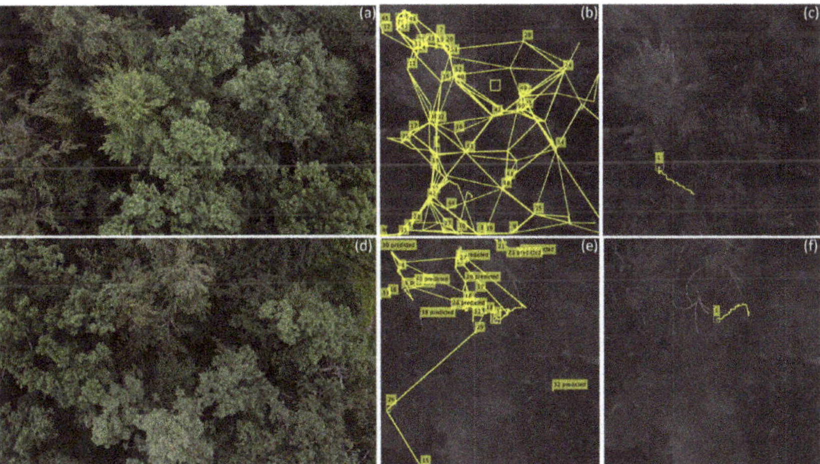

Figure 5. Automatic motion estimation (vertical): Two examples of tracking a moving hidden person within dense forest ((**a**,**d**) RGB images of drone) in either single thermal images (**b**,**e**) and IAOS integral images. Note: the tracking results of the integral images were projected back to a single thermal image for better spatial reference (**c**,**f**). Motion paths are indicated by yellow lines. While tracking in single images leads to many false positive detections, tracking in integral images results in clear track-paths of a single target. See Supplementary Videos S2 and S3 for dynamic examples of these results.

For tracking, moving targets are first detected by utilizing background subtraction based on Gaussian mixture models [68,69]. The resulting foreground mask is further processed using morphological operations to eliminate noise [70,71]. Subsequently blob analysis [72,73] detects connected pixels corresponding to each moving target. Association of detections in subsequent frames is entirely based on motion where the motion of each detected target is estimated by a Kalman filter. The filter predicts the target location in subsequent frame (based on previous motion and associated motion model) and then determines the likelihood of assigning the detection to the target.

For comparison, we applied the above tracking approach to both: the sequence of captured single thermal images and to the sequence of integral images computed from the single images, as described in Section 2.2. For each case, tracking parameters (such as minimum blob size, max. prediction length, number of training images for background subtraction) were individually optimized to achieve best possible results.

While tracking in single images leads to many false positive detections becoming practically infeasible, tracking in integral images results in clear track-paths of a single target. Estimated mean motion parameters were: 291°, 0.82 m/s (Figure 5a–c), 309°, 0.16 m/s for the first leg, and 241°, 0.41 m/s for the second leg (Figure 5d–f). See Supplementary Videos S2 and S3 for dynamic examples of these results.

4. Discussion and Conclusions

In this article we presented Inverse Airborne Optical Sectioning (IAOS), an optical analogy to Inverse Synthetic Aperture Radar (ISAR). Moving targets, such as walking people, that are heavily occluded by vegetation can be made visible and tracked with a stationary optical sensor (e.g., a hovering camera drone above forest). We introduced the principles of IAOS (i.e., inverse synthetic aperture imaging), explained how the signal of occluders can be further suppressed by filtering the Radon transform of the image integral, and presented how targets' motion parameters can be estimated manually and automatically. Furthermore, we showed that while tracking occluded targets in conventional aerial images is infeasible, it is efficiently possible in integral images that result from IAOS.

IAOS has several limitations: We assume that local motion of occluders and of the drone (e.g., caused by wind) is smaller than the motion of the target. Small local motion of the target itself, such as individual moving body parts, appear blurred in integral images. Moreover, the field of view of a hovering drone is limited and moving targets might be out of view quickly. In the future, we will investigate how drone movement being adapted to target movement can increase field of view and reduce blur of local target motion. This corresponds to a combination of IAOS (i.e., occlusion removal by registering target motion) and classical AOS (i.e., occlusion removal by registering drone movement). Furthermore, results of Radon transform filtering have artifacts that are due to under-sampling; higher imaging rates can overcome this. The blob-based tracking approach applied for proof-of-concept is very simple; more sophisticated methods achieve superior tracking results. See supplementary Videos S2 and S3 for dynamic examples of these results. However, we believe that tracking in integral images will always outperform tracking in conventional images.

Supplementary Materials: The following supporting information can be downloaded at: https://github.com/JKU-ICG/AOS/, Video S1: Manual visual search for the motion parameters. Video S2: Automatic motion estimation (example 1). Video S3: Automatic motion estimation (example 2).

Author Contributions: Conceptualization, O.B.; methodology, O.B., R.J.A.A.N. and I.K.; software, R.J.A.A.N., I.K.; validation, R.J.A.A.N., I.K. and O.B.; formal analysis, R.J.A.A.N., I.K. and O.B.; investigation, R.J.A.A.N. and I.K.; resources, R.J.A.A.N. and I.K.; data curation, R.J.A.A.N. and I.K.; writing—original draft preparation, O.B., R.J.A.A.N. and I.K.; writing—review and editing,, O.B., R.J.A.A.N. and I.K.; visualization, O.B., R.J.A.A.N. and I.K.; supervision, O.B.; project administration, O.B.; funding acquisition, O.B. All authors have read and agreed to the published version of the manuscript.

Funding: This research was funded by the Austrian Science Fund (FWF) under grant number P 32185-NBL, and by the State of Upper Austria and the Austrian Federal Ministry of Education, Science and Research via the LIT–Linz Institute of Technology under grant number LIT-2019-8-SEE-114.

Institutional Review Board Statement: Not applicable.

Informed Consent Statement: Informed consent was obtained from all subjects involved in the study.

Data Availability Statement: The data, code, and supplementary material are available on GitHub: https://github.com/JKU-ICG/AOS/ (accessed on 1 September 2022).

Acknowledgments: We want to thank the Upper Austrian Fire Brigade Headquarters for providing the DJI Mavic 2 Enterprise Advanced for our experiments. Open Access Funding by the Austrian Science Fund (FWF).

Conflicts of Interest: The authors declare no conflict of interest.

Appendix A

In the following, we present the derivation of an integral image's variance ($Var[I]$). We applied the statistical model described in [50], where the integral image I is composed of N single image recordings I_i and each single image pixel in I_i is either occlusion-free (S) or occluded (O), determined by Z:

$$I_i = Z_i O_i + (1 - Z_i)S.$$

Similar to [50], all variables are independent and identically distributed with Z_i, following a Bernoulli distribution with success parameter D (i.e., $E[Z_i] = E[Z_i^2] = D$; furthermore note that $E[Z_i(1 - Z_i)] = 0$ is true). The random variable S follows a distribution whose properties can be described with mean $E[S] = \mu_s$ and $E[S^2] = (\mu_s^2 + \sigma_s^2)$. Analogously, the occluded variable O_i follows a distribution whose properties can be described with $E[O_i] = \mu_o$ and $E[O_i^2] = (\mu_o^2 + \sigma_o^2)$. We compute the first and second moments of I_i to determine its mean and variance with:

$$E[I_i] = D\mu_o + (1 - D)\mu_s$$

and

$$E\left[I_i^2\right] = D\left(\mu_o^2 + \sigma_o^2\right) + (1 - D)\left(\mu_s^2 + \sigma_s^2\right).$$

Variances of single images I_i can be obtained as:

$$\begin{aligned}
Var[I_i] &= E\left[I_i^2\right] - (E[I_i])^2 \\
&= D(\mu_o^2 + \sigma_o^2) + (1 - D)(\mu_s^2 + \sigma_s^2) \\
&\quad - (D^2\mu_o^2 + (1 - D)^2\mu_s^2 + 2D(1 - D)\mu_o\mu_s \\
&= D(1 - D)\left((\mu_o - \mu_s)^2\right) + D\sigma_o^2 + (1 - D)\,\sigma_s^2.
\end{aligned}$$

Similarly, for I we determine the first and second moments where the first moment of I is given by:

$$E[I] = E\left[\frac{1}{N}\sum_{i=1}^{N} Z_i O_i + (1 - Z_i)S\right] = D\mu_o + (1 - D)\mu_s$$

and the second moment of I is as derived in [50]:

$$E\left[I^2\right] = \frac{1}{N^2}\left(\begin{array}{c} N\left(D(\sigma_o^2 + \mu_o^2) + (1 - D)(\sigma_s^2 + \mu_s^2)\right) \\ + N(N - 1)\left(\begin{array}{c} D^2\mu_o^2 + 2D(1 - D)\mu_o\mu_s \\ + (1 - D)^2(\sigma_s^2 + \mu_s^2) \end{array}\right) \end{array} \right).$$

Consecutively, we calculate the variance of the integral image as:

$$
\begin{aligned}
Var[I] \;&= E\!\left[I^2\right] - \left(E[I]\right)^2 \\
&= \tfrac{1}{N}\!\left(D\!\left(\sigma_o^2 + \mu_o^2\right) + (1-D)\!\left(\sigma_s^2 + \mu_s^2\right)\right) \\
&\quad + \left(D^2\mu_o^2 + 2D(1-D)\mu_o\mu_s + (1-D)^2\!\left(\sigma_s^2 + \mu_s^2\right)\right) \\
&\quad - \tfrac{1}{N}\!\left(D^2\mu_o^2 + 2D(1-D)\mu_o\mu_s + (1-D)^2\!\left(\sigma_s^2 + \mu_s^2\right)\right) \\
&\quad - \left(D^2\mu_o^2 + (1-D)^2\mu_s^2 + 2D(1-D)\mu_o\mu_s\right) \\
&= \tfrac{1}{N}\!\left(D\!\left(\sigma_o^2 + \mu_o^2\right) + (1-D)\!\left(\sigma_s^2 + \mu_s^2\right)\right) + (1-D)^2\sigma_s^2 \\
&\quad - \tfrac{1}{N}\!\left(D^2\mu_o^2 + 2D(1-D)\mu_o\mu_s + (1-D)^2\!\left(\sigma_s^2 + \mu_s^2\right)\right) \\
&= \tfrac{1}{N}\!\left(D(1-D)\!\left((\mu_o - \mu_s)^2\right) + D\sigma_o^2 + (1-D)\sigma_s^2\right) \\
&\quad + (1-D)^2\!\left(1 - \tfrac{1}{N}\right)\sigma_s^2 \,.
\end{aligned}
$$

References

1. Ryle, M.; Vonberg, D.D. Solar radiation on 175 Mc./s. *Nature* **1946**, *158*, 339–340. [CrossRef]
2. Moreira, A.; Prats-Iraola, P.; Younis, M.; Krieger, G.; Hajnsek, I.; Papathanassiou, K.P. A tutorial on synthetic aperture radar. *IEEE Geosci. Remote Sens. Mag.* **2013**, *1*, 6–43. [CrossRef]
3. May, C.A. Pulsed Doppler Radar Methods and Apparatus. U.S. Patent No. 3,196,436, 20 July 1965.
4. Willey, C.A. Synthetic aperture radars: A paradigm for technology evolution. *IRE Trans. Military Electron.* **1985**, *21*, 440–443.
5. Farquharson, G.; Woods, W.; Stringham, C.; Sankarambadi, N.; Riggi, L. The capella synthetic aperture radar constellation. In Proceedings of the 12th European Conference on Synthetic Aperture Radar, Aachen, Germany, 4–7 June 2018. EUSAR 2018; VDE.
6. Chen, F.; Lasaponara, R.; Masini, N. An overview of satellite synthetic aperture radar remote sensing in archaeology: From site detection to monitoring. *J. Cult. Herit.* **2017**, *23*, 5–11. [CrossRef]
7. Zhang, Z.; Lin, H.; Wang, M.; Liu, X.; Chen, Q.; Wang, C.; Zhang, H. A Review of Satellite Synthetic Aperture Radar Interferometry Applications in Permafrost Regions: Current Status, Challenges, and Trends. *IEEE Geosci. Remote Sens. Mag.* **2022**, *1*, 2–23. [CrossRef]
8. Ranjan, A.K.; Parida, B.R. Predicting paddy yield at spatial scale using optical and Synthetic Ap-erture Radar (SAR) based satellite data in conjunction with field-based Crop Cutting Experiment (CCE) data. *Int. J. Remote Sens.* **2021**, *42*, 2046–2071.
9. Reigber, A.; Scheiber, R.; Jager, M.; Prats-Iraola, P.; Hajnsek, I.; Jagdhuber, T.; Papathanassiou, K.P.; Nannini, M.; Aguilera, E.; Baumgartner, S.; et al. Very-high-resolution airborne synthetic aperture radar imaging: Signal processing and applica-tions. *Proc. IEEE* **2021**, *101*, 759–783. [CrossRef]
10. Sumantyo JT, S.; Chua, M.Y.; Santosa, C.E.; Panggabean, G.F.; Watanabe, T.; Setiadi, B.; Sumantyo, F.D.S.; Tsushima, K.; Sasmita, K.; Mardiyanto, A.; et al. Airborne circularly polarized synthetic aperture radar. *IEEE J. Sel. Top. Appl. Earth Obs. Remote Sens.* **2020**, *14*, 1676–1692. [CrossRef]
11. Tsunoda, S.I.; Pace, F.; Stence, J.; Woodring, M.; Hensley, W.H.; Doerry, A.W.; Walker, B.C. Lynx: A high-resolution synthetic aperture radar. In Proceedings of the 2000 IEEE Aerospace Conference. Proceedings (Cat. No.00TH8484), Big Sky, MT, USA, 25 March 2000; Volume 5, pp. 51–58.
12. Fernández, M.G.; López Y, Á.; Arboleya, A.A.; Valdés, B.G.; Vaqueiro, Y.R.; Andrés FL, H.; García, A.P. Synthetic aperture radar imaging system for landmine detection using a ground penetrat-ing radar on board a unmanned aerial vehicle. *IEEE Access* **2018**, *6*, 45100–45112. [CrossRef]
13. Deguchi, T.; Sugiyama, T.; Kishimoto, M. Development of SAR system installable on a drone. In Proceedings of the EUSAR 2021, 13th European Conference on Synthetic Aperture Radar, VDE, Online, 2 July 2021.
14. Mondini, A.C.; Guzzetti, F.; Chang, K.T.; Monserrat, O.; Martha, T.R.; Manconi, A. Landslide failures detection and mapping using Synthetic Aperture Radar: Past, present and future. *Earth-Sci. Rev.* **2021**, *216*, 103574. [CrossRef]
15. Rosen, P.A.; Hensley, S.; Joughin, I.R.; Li, F.K.; Madsen, S.N.; Rodriguez, E.; Goldstein, R.M. Synthetic aperture radar interferometry. *Proc. IEEE* **2000**, *88*, 333–382. [CrossRef]
16. Prickett, M.J.; Chen, C.C. Principles of inverse synthetic aperture radar/ISAR/imaging. In Proceedings of the EASCON'80, Electronics and Aerospace Systems Conference, Arlington, VA, USA, 29 September–1 October 1980.
17. Vehmas, R.; Neuberger, N. Inverse Synthetic Aperture Radar Imaging: A Historical Perspective and State-of-the-Art Survey. *IEEE Access* **2021**, *9*, 113917–113943. [CrossRef]
18. Özdemir, C. *Inverse Synthetic Aperture Radar Imaging with MATLAB®Algorithms*; Wiley-Interscience: Hoboken, NJ, USA, 2021. [CrossRef]
19. Marino, A.; Sanjuan-Ferrer, M.J.; Hajnsek, I.; Ouchi, K. Ship Detection with Spectral Analysis of Synthetic Aperture Radar: A Comparison of New and Well-Known Algorithms. *Remote Sens.* **2015**, *7*, 5416–5439. [CrossRef]
20. Wang, Y.; Chen, X. 3-D Interferometric Inverse Synthetic Aperture Radar Imaging of Ship Target With Complex Motion. *IEEE Trans. Geosci. Remote Sens.* **2018**, *56*, 3693–3708. [CrossRef]

21.   Xu, G.; Zhang, B.; Chen, J.; Wu, F.; Sheng, J.; Hong, W. Sparse Inverse Synthetic Aperture Radar Imaging Using Structured Low-Rank Method. *IEEE Trans. Geosci. Remote Sens.* **2021**, *60*, 1–12. [CrossRef]
22.   Berizzi, F.; Corsini, G. Autofocusing of inverse synthetic aperture radar images using contrast optimiza-tion. *IEEE Transactions on Aerospace and Electronic Systems* **1996**, *32*, 1185–1191. [CrossRef]
23.   Bai, X.; Zhou, F.; Xing, M.; Bao, Z. Scaling the 3-D Image of Spinning Space Debris via Bistatic Inverse Synthetic Aperture Radar. *IEEE Geosci. Remote Sens. Lett.* **2010**, *7*, 430–434. [CrossRef]
24.   Anger, S.; Jirousek, M.; Dill, S.; Peichl, M. Research on advanced space surveillance using the IoSiS radar system. In Proceedings of the EUSAR 2021, 13th European Conference on Synthetic Aperture Radar, Online, 2 July 2021.
25.   Vossiek, M.; Urban, A.; Max, S.; Gulden, P. Inverse Synthetic Aperture Secondary Radar Concept for Precise Wireless Positioning. *IEEE Trans. Microw. Theory Tech.* **2007**, *55*, 2447–2453. [CrossRef]
26.   Jeng, S.L.; Chieng, W.H.; Lu, H.P. Estimating speed using a side-looking single-radar vehicle detec-tor. *IEEE Trans. Intell. Transp. Syst.* **2013**, *15*, 607–614. [CrossRef]
27.   Ye, X.; Zhang, F.; Yang, Y.; Zhu, D.; Pan, S. Photonics-Based High-Resolution 3D Inverse Synthetic Aperture Radar Imaging. *IEEE Access* **2019**, *7*, 79503–79509. [CrossRef]
28.   Pandey, N.; Ram, S.S. Classification of automotive targets using inverse synthetic aperture radar images. *IEEE Trans. Intell. Veh.* **2022**. Available online: https://scholar.google.com/citations?view_op=view_citation&hl=en&user=dHBHt38AAAAJ&citation_for_view=dHBHt38AAAAJ:zYLM7Y9cAGgC (accessed on 20 July 2022).
29.   Levanda, R.; Leshem, A. Synthetic aperture radio telescopes. *IEEE Signal Process. Mag.* **2009**, *27*, 14–29. [CrossRef]
30.   Dravins, D.; Lagadec, T.; Nuñez, P.D. Optical aperture synthesis with electronically connected telescopes. *Nat. Commun.* **2015**, *6*, 1–5. [CrossRef] [PubMed]
31.   Ralston, T.S.; Marks, D.L.; Carney, P.S.; Boppart, S.A. Interferometric synthetic aperture microscopy. *Nat. Phys.* **2007**, *3*, 129–134. [CrossRef]
32.   Edgar, R. Introduction to Synthetic Aperture Sonar. *Sonar Syst.* **2011**. [CrossRef]
33.   Hayes, M.P.; Gough, P.T. Synthetic Aperture Sonar: A Review of Current Status. *IEEE J. Ocean. Eng.* **2009**, *34*, 207–224. [CrossRef]
34.   Hansen, R.E.; Callow, H.J.; Sabo, T.O.; Synnes, S.A.V. Challenges in Seafloor Imaging and Mapping With Synthetic Aperture Sonar. *IEEE Trans. Geosci. Remote Sens.* **2011**, *49*, 3677–3687. [CrossRef]
35.   Bülow, H.; Birk, A. Synthetic Aperture Sonar (SAS) without Navigation: Scan Registration as Basis for Near Field Synthetic Imaging in 2D. *Sensors* **2020**, *20*, 4440. [CrossRef]
36.   Jensen, J.A.; Nikolov, S.I.; Gammelmark, K.L.; Pedersen, M.H. Synthetic aperture ultrasound imaging. *Ultrasonics* **2006**, *44*, e5–e15. [CrossRef]
37.   Zhang, H.K.; Cheng, A.; Bottenus, N.; Guo, X.; Trahey, G.E.; Boctor, E.M. Synthetic tracked aperture ultrasound imaging: Design, simulation, and experimental evaluation. *J. Med. Imaging* **2016**, *3*, 027001. [CrossRef]
38.   Barber, Z.W.; Dahl, J.R. Synthetic aperture ladar imaging demonstrations and information at very low return levels. *Appl. Opt.* **2014**, *53*, 5531–5537. [CrossRef]
39.   Terroux, M.; Bergeron, A.; Turbide, S.; Marchese, L. Synthetic aperture lidar as a future tool for earth observation. *Proc. SPIE* **2017**, *10563*, 105633V. [CrossRef]
40.   Vaish, V.; Wilburn, B.; Joshi, N.; Levoy, M. Using plane+ parallax for calibrating dense camera arrays. In Proceedings of the 2004 IEEE Computer So-ciety Conference on Computer Vision and Pattern Recognition, Washington, DC, USA, 27 June–2 July 2004; CVPR 2004. Volume 1.
41.   Vaish, V.; Levoy, M.; Szeliski, R.; Zitnick, C.L.; Kang, S.B. Reconstructing occluded surfaces using synthetic apertures: Stereo, focus and robust measures. In Proceedings of the 2006 IEEE Computer Society Conference on Computer Vision and Pattern Recognition (CVPR'06), New York, NY, USA, 17–22 June 2006; Volume 2.
42.   Zhang, H.; Jin, X.; Dai, Q. Synthetic Aperture Based on Plenoptic Camera for Seeing Through Occlusions. In *Pacific Rim Conference on Multimedia*; Springer: Cham, Switzerland, 2018; pp. 158–167. [CrossRef]
43.   Yang, T.; Ma, W.; Wang, S.; Li, J.; Yu, J.; Zhang, Y. Kinect based real-time synthetic aperture imaging through occlusion. *Multimed. Tools Appl.* **2015**, *75*, 6925–6943. [CrossRef]
44.   Joshi, N.; Avidan, S.; Matusik, W.; Kriegman, D.J. Synthetic aperture tracking: Tracking through occlusions. In Proceedings of the 2007 IEEE 11th International Conference on Computer Vision, Rio de Janeiro, Brazil, 14–21 October 2007.
45.   Pei, Z.; Li, Y.; Ma, M.; Li, J.; Leng, C.; Zhang, X.; Zhang, Y. Occluded-Object 3D Reconstruction Using Camera Array Synthetic Aperture Imaging. *Sensors* **2019**, *19*, 607. [CrossRef] [PubMed]
46.   Yang, T.; Zhang, Y.; Yu, J.; Li, J.; Ma, W.; Tong, X.; Yu, R.; Ran, L. All-In-Focus Synthetic Aperture Imaging. In *European Conference on Computer Vision*; Springer: Cham, Switzerland, 2014; pp. 1–15. [CrossRef]
47.   Pei, Z.; Zhang, Y.; Chen, X.; Yang, Y.-H. Synthetic aperture imaging using pixel labeling via energy minimization. *Pattern Recognit.* **2013**, *46*, 174–187. [CrossRef]
48.   Kurmi, I.; Schedl, D.C.; Bimber, O. Airborne Optical Sectioning. *J. Imaging* **2018**, *4*, 102. [CrossRef]
49.   Bimber, O.; Kurmi, I.; Schedl, D.C. Schedl Synthetic aperture imaging with drones. *IEEE Comput. Graph. Appl.* **2019**, *39*, 8–15. [CrossRef] [PubMed]
50.   Kurmi, I.; Schedl, D.C.; Bimber, O. A statistical view on synthetic aperture imaging for occlusion removal. *IEEE Sens. J.* **2019**, *19*, 9374–9383. [CrossRef]

51. Kurmi, I.; Schedl, D.C.; Bimber, O. Thermal Airborne Optical Sectioning. *Remote Sens.* **2019**, *11*, 1668. [CrossRef]

52. Schedl, D.C.; Kurmi, I.; Bimber, O. Airborne Optical Sectioning for Nesting Observation. *Sci. Rep.* **2020**, *10*, 1–7. [CrossRef]

53. Kurmi, I.; Schedl, D.C.; Bimber, O. Fast Automatic Visibility Optimization for Thermal Synthetic Aperture Visualization. *IEEE Geosci. Remote Sens. Lett.* **2021**, *18*, 836–840. [CrossRef]

54. Kurmi, I.; Schedl, D.C.; Bimber, O. Schedl, and Oliver Bimber Pose error reduction for focus enhancement in thermal synthetic aperture visualization. *IEEE Geosci. Remote Sens. Lett.* **2021**, *19*, 1–5. [CrossRef]

55. Schedl, D.C.; Kurmi, I.; Bimber, O. Search and rescue with airborne optical sectioning. *Nat. Mach. Intell.* **2020**, *2*, 783–790. [CrossRef]

56. Kurmi, I.; Schedl, D.C.; Bimber, O. Combined person classification with airborne optical sectioning. *Sci. Rep.* **2022**, *12*, 1–11. [CrossRef]

57. Schedl, D.C.; Kurmi, I.; Bimber, O. An autonomous drone for search and rescue in forests using airborne optical sectioning. *Sci. Robot.* **2021**, *6*, eabg1188. [CrossRef] [PubMed]

58. Ortner, R.; Kurmi, I.; Bimber, O. Acceleration-Aware Path Planning with Waypoints. *Drones* **2021**, *5*, 143. [CrossRef]

59. Nathan, R.J.A.A.; Kurmi, I.; Schedl, D.C.; Bimber, O. Through-Foliage Tracking with Airborne Optical Sectioning. *J. Remote Sens.* **2022**, *2022*, 1–10. [CrossRef]

60. Seits, F.; Kurmi, I.; Nathan RJ, A.A.; Ortner, R.; Bimber, O. On the Role of Field of View for Occlusion Removal with Airborne Optical Sectioning. *arXiv* **2022**, arXiv:2204.13371.

61. Bracewell, R.N. *Two-Dimensional Imaging*; Prentice-Hall: Englewood Cliffs, NJ, USA, 1995.

62. Lim, J.S. *Two-Dimensional Signal and Image Processing*; Prentice-Hall: Englewood Cliffs, NJ, USA, 1990.

63. Kak, A.C.; Slaney, M.; Wang, G. Principles of Computerized Tomographic Imaging. *Am. Assoc. Phys. Med.* **2002**, *29*, 107. [CrossRef]

64. Firestone, L.; Cook, K.; Culp, K.; Talsania, N.; Preston, K., Jr. Comparison of autofocus methods for automated microscopy. *Cytom. J.-Ternational Soc. Anal. Cytol.* **1991**, *12*, 195–206. [CrossRef]

65. Pertuz, S.; Puig, D.; Garcia, M.A. Analysis of focus measure operators for shape-from-focus. *Pattern Recognit.* **2012**, *46*, 1415–1432. [CrossRef]

66. Jones, D.R.; Perttunen, C.D.; Stuckman, B.E. Lipschitzian optimization without the Lipschitz constant. *J. Optim. Theory Appl.* **1993**, *79*, 157–181. [CrossRef]

67. Johnson, S.G. The NLopt Nonlinear-Optimization Package. Available online: http://github.com/stevengj/nlopt (accessed on 20 July 2022).

68. KaewTraKulPong, P.; Bowden, R. An Improved Adaptive Background Mixture Model for Real-time Tracking with Shadow Detection. In *Video-Based Surveillance Systems*; Springer: Boston, MA, USA, 2002; pp. 135–144. [CrossRef]

69. Stauffer, C.; Grimson, W.E.L. Grimson Adaptive background mixture models for real-time tracking. In Proceedings of the 1999 IEEE Computer Society Conference on Computer Vision and Pattern Recognition (Cat. No PR00149), Fort Collins, CO, USA, 23–25 June 1999; Volume 2.

70. Soille, P. *Morphological Image Analysis: Principles and Applications*; Springer: Berlin, Germany, 1999; Volume 2.

71. Dougherty, E.R.; Lotufo, R.A. *Hands-on Morphological Image Processing*; SPIE Press: Washington, DC, USA, 2003. [CrossRef]

72. Dillencourt, M.B.; Samet, H.; Tamminen, M. A general approach to connected-component labeling for arbitrary image representations. *J. ACM* **1992**, *39*, 253–280. [CrossRef]

73. Shapiro, L.G.; Stockman, G.C. *Computer Vision*; Prentice Hall: Englewood Cliffs, NJ, USA, 2001; Volume 3.

Article

Improved Image Synthesis with Attention Mechanism for Virtual Scenes via UAV Imagery

Lufeng Mo [1,2], **Yanbin Zhu** [1], **Guoying Wang** [1,*], **Xiaomei Yi** [1], **Xiaoping Wu** [3] **and Peng Wu** [1]

[1] College of Mathematics and Computer Science, Zhejiang A&F University, Hangzhou 311300, China
[2] Information and Education Technology Center, Zhejiang A&F University, Hangzhou 311300, China
[3] School of Information Engineering, Huzhou University, Huzhou 313000, China
* Correspondence: wgy@zafu.edu.cn

Abstract: Benefiting from the development of unmanned aerial vehicles (UAVs), the types and number of datasets available for image synthesis have greatly increased. Based on such abundant datasets, many types of virtual scenes can be created and visualized using image synthesis technology before they are implemented in the real world, which can then be used in different applications. To achieve a convenient and fast image synthesis model, there are some common issues such as the blurred semantic information in the normalized layer and the local spatial information of the feature map used only in the generation of images. To solve such problems, an improved image synthesis model, SYGAN, is proposed in this paper, which imports a spatial adaptive normalization module (SPADE) and a sparse attention mechanism YLG on the basis of generative adversarial network (GAN). In the proposed model SYGAN, the utilization of the normalization module SPADE can improve the imaging quality by adjusting the normalization layer with spatially adaptively learned transformations, while the sparsified attention mechanism YLG improves the receptive field of the model and has less computational complexity which saves training time. The experimental results show that the Fréchet Inception Distance (FID) of SYGAN for natural scenes and street scenes are 22.1, 31.2; the Mean Intersection over Union (MIoU) for them are 56.6, 51.4; and the Pixel Accuracy (PA) for them are 86.1, 81.3, respectively. Compared with other models such as CRN, SIMS, pix2pixHD and GauGAN, the proposed image synthesis model SYGAN has better performance and improves computational efficiency.

Keywords: deep learning; unmanned aerial vehicle; image synthesis; generative adversarial network; attention mechanism

Citation: Mo, L.; Zhu, Y.; Wang, G.; Yi, X.; Wu, X.; Wu, P. Improved Image Synthesis with Attention Mechanism for Virtual Scenes via UAV Imagery. *Drones* **2023**, *7*, 160. https://doi.org/10.3390/drones7030160

Academic Editor: Seokwon Yeom

Received: 7 February 2023
Revised: 22 February 2023
Accepted: 23 February 2023
Published: 25 February 2023

1. Introduction

The simulation of image scenes has developed rapidly and is one of the current research hotspots [1]. More and more places need to use image synthesis technology, such as interior design, street design, park landscape preview maps, and so on. A real and reasonable image can improve people's impression of the project, and can also make people feel more intuitively about how the project will look on completion. However, there are few angles available for manual image acquisition and it is more time consuming and laborious. The popularity of unmanned aerial vehicles (UAV) makes the collection of remote sensing image data simpler and more convenient. UAV can obtain images from a wider range with more angles, which greatly expands the source of image synthesis datasets. Compared with artificial image acquisition, that derived from UAV has lower costs and a broad application prospect. Similarly, image synthesis based on deep learning is better than artificial image synthesis [2].

At present, image synthesis methods based on deep learning are mainly based on Generative Adversarial Networks (GAN) [3]. Pix2pixHD is one of the most widely used models at present, and it is a supervised learning model. By inputting the semantic labels and the ground truth, realistic composite images can be generated in the model [4]. Chen et al. [5] proposed a Cascaded Refinement Network (CRN), which can repeatedly refine the output from low resolution to high resolution, resulting in high-quality images. Qi et al. [6] proposed SIMS, which first divides semantic labels into each plate, identifies patterns similar to the plate in the material library to supplement, and then refines the connections of each plate.

Although deep learning has made some progress in the field of image synthesis in recent years, some aspects need improvement [7]. For example, part of the structure can be optimized, and the receptive field of the model is inadequate [8]. Park et al. [9] showed that the traditional network architecture, which is a superposition of convolution, normalization, and nonlinear layers, is not optimal because their normalization layers tend to reduce the information contained in the input semantic mask. Transposed convolutional layers are a type of basic constituent layer that can capture the spatial properties of natural images, which are important for generating high-quality images. However, it has a major limitation in that it cannot model complex geometries and long-distance dependencies [10]. To compensate for this limitation and expand the receptive field of the model, some have introduced an attention mechanism into the model. This method was first proposed by SAGAN [11]. However, this mechanism also has the following limitations: first, the calculation cost of the standard dense attention mechanism is relatively high; second, when the attention mechanism is calculated, the spatial characteristics of the image are lost in the step of expanding the two-dimensional spatial structure into a one-dimensional vector [12].

To solve the above problems, an image synthesis model SYGAN is proposed in this paper. It is based on adjusted GAN and a spatially adaptive normalization module SPADE [9] and a sparsified attention mechanism YLG [13] which are imported. Using the SPADE module, both the normalization function and the initial semantic information are well retained. The attention mechanism YLG not only effectively improves the reading of feature point information and expands the receptive field of the model, but also reduces the computational complexity, which decreases the requirements of hardware equipment and improves the computational speed of the model.

The main contributions of this paper are as follows: (1) A new image synthesis model SYGAN is proposed. Compared with other models, the model SYGAN adopts a spatially adaptive normalization module and a sparsified attention mechanism to achieve good performance and low complexity. (2) Image synthesis of two kinds of scenes – natural and street scenes – are examined, and the reasons for the difference between the performance for the two scenes are analyzed. (3) Experiments for the comparison of performance of SYGAN and other models such as CRN, SIMS, pix2pixHD, and GauGAN and ablation experiments are conducted to verify the performance of SYGAN.

2. Materials and Methods

2.1. Main Idea

SYGAN, an image synthesis model based on deep learning whose overall structure is shown in Figure 1, is proposed in this paper.

In SYGAN, the image encoder first encodes the real images and then generates the mean and variance vectors, which are used for the noise input of the generator. In addition to these data, the generator also accepts the label images as input to the SPADE block, and then generates the output images. The output images and the real images are used as the input of the discriminator. Finally, the discriminator makes the judgment classification and outputs the attention map to the generator to help it focus on the regions with higher discrimination in the image.

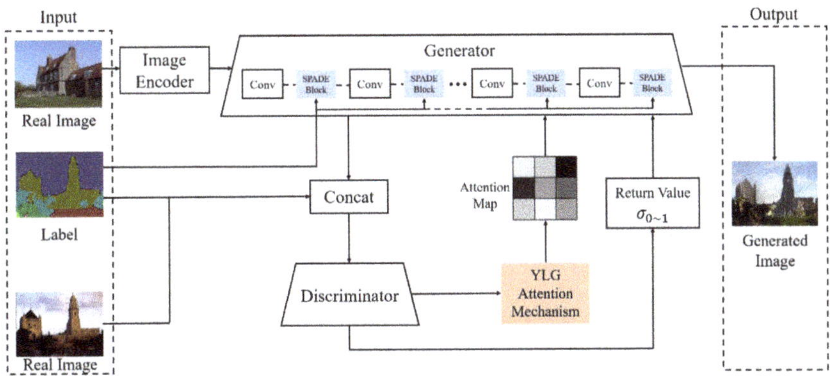

Figure 1. Overall structure of the SYGAN model.

As shown in Figure 1, the main idea of SYGAN includes the following aspects:

(1)   Adjusting GAN as a main framework

The main framework of SYGAN is based on GAN which uses generators and discriminators against each other to obtain a reasonable output. As a generative model, it deals well with the problem of data generation. The neural network structure used in this model can fit the high-dimensional representation of various types of data. GAN uses two neural networks against each other and end-to-end optimization, which can effectively improve the training efficiency [14]. The image encoder is mainly composed of a convolutional layer and a linear layer. Real images are encoded as input to generate vector data as input to the generator. The discriminator adopted by SYGAN refers to the classical design of some other models and is mainly composed of convolutional layers. It takes the label image, the output of the generator, and the real image as inputs and judges them.

(2)   Importing spatially adaptive normalization module SPADE into the generator

In the past, deep learning-based methods often sent semantic images directly to the neural network in the generator for learning. Although these methods have some impact, they are not conducive to generating high-quality images, because the normalization layer in ordinary neural networks will unconsciously reduce the semantic information. In order to solve this problem, in this study a spatially adaptive normalization module SPADE is imported to replace the ordinary normalization layer, use the layout of input semantic information to activate regulation through spatially adaptively learned transformations, and effectively propagate semantic information throughout the network.

(3)   Adding attention mechanism YLG

By modeling the relationships between pixels, the attention mechanism can effectively handle complex geometric shapes and capture long-distance dependencies to further improve network performance [15]. However, common attention also has some of the limitations described above. In view of these, the sparsified attention mechanism YLG is added into SYGAN, which introduces the local sparse attention layer, reducing both the computational complexity and the loss of spatial characteristics when the two-dimensional spatial structure tensor is expanded into one-dimensional spatial structure, and can support good information flow. Compared with other attention mechanisms, the performance and training time have been optimized to a certain extent.

2.2. SYGAN Model

2.2.1. Adjusting GAN as Main Framework

The basic framework for GAN is shown in Figure 2. A set of random noise vectors z satisfying a specified distribution is given as input. The generator G will generate a sample x, and then the discriminator D will make a binary classification decision, resulting in a value $\sigma_{0\sim1}$ (if $\sigma_{0\sim1}$ is 1, it means that the discriminator considers the sample to be a real sample; otherwise, it is a false sample, which means that the sample is generated). There are two types of inputs to discriminator D: generated sample x_f and real sample x_t. In the process of optimizing model parameters through adversarial training, the generator G fits the latent distribution of the real data, so that it is able to synthesize samples that approximate the latent distribution of the real data using the random noise vector z. Then the generated sample x_f and the real data x_t are sent to the discriminator D, which then tries to distinguish the real and fake input samples as much as possible. Meanwhile, the generator G tries to generate samples that are indistinguishable from the real data in order to make the discriminator D judge that the generated samples are true. In the process of confrontation between the generator and the discriminator, both are optimized, and their respective performances are also improved. When the discriminator cannot distinguish the source of the sample data, the optimization ends, and the mathematical expression of the optimization process is shown in Equation (1):

$$\min_{G}\max_{D}V(D,G) = E_{x\sim P_{data}}[logD(x)] + E_{z\sim P_z}[log(1 - D(G(z)))] \tag{1}$$

where z and x represent the random noise vector and the true sample, respectively. x can be generated by randomly sampling from the true data distribution P_{data}, and z can be generated by sampling from the specified prior distribution P_z. In the process of optimizing this adversarial generative model, the generator attempts to minimize $V(D,G)$, while the discriminator maximizes $V(D,G)$. In the optimization process, an alternate iterative updating method is adopted. First, the generator G is fixed to maximize $V(D,G)$ to solve D, and then the discriminator D is fixed to minimize $V(D,G)$ to solve G.

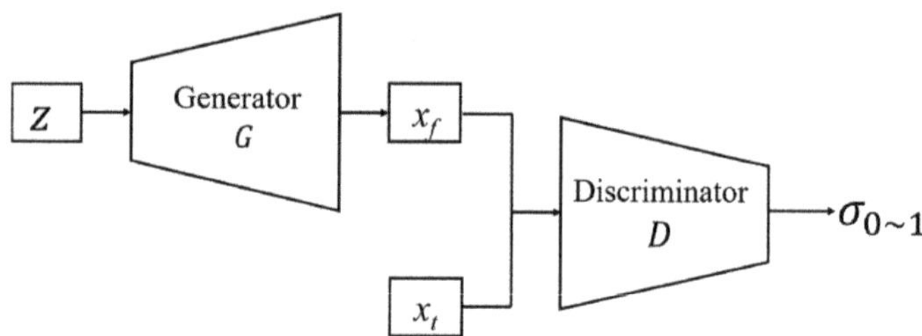

Figure 2. Structure of GAN.

In image synthesis applications, the function of generator G is to process vector data generated by image encoder as input to generate image x_f. The role of the discriminator D is to determine whether the received input samples are generated images x_f or real images x_t. The training goal of generator G is to make its output fool discriminator D, and the goal of discriminator D is to identify which image samples come from discriminator G.

As shown in Figure 3, the encoder consists of six convolutional layers with a step size of 2 and two linear layers to output a mean μ and a variance σ, which are used as the input of the generator. It uses the LReLU activation function [16] and Instance Norm (IN) [17]. LReLU is easy to compute, fast in convergence, and solves the problem of vanishing positive interval gradients. Compared with ReLU [18], it solves the problem that some neurons cannot be activated.

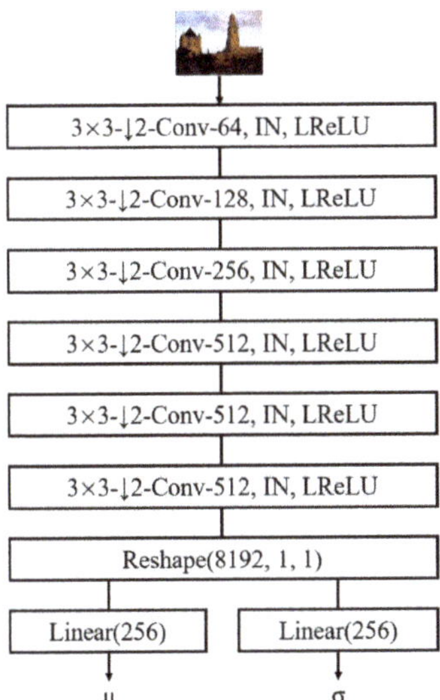

Figure 3. Image encoder.

The discriminator in SYGAN refers to the design of pix2pixHD and Patch-GAN to some extent [19] whose input is segmented images and the connection between the generator output and the real image, uses the LReLU activation function and IN, and takes the convolution layer as the last layer. The output of the discriminator will be received by the attention mechanism YLG (Section 2.2.3) to generate an attention map, which is then input to the generator to assist in its focusing on areas of higher discrimination in the image. Its structure is shown in Figure 4.

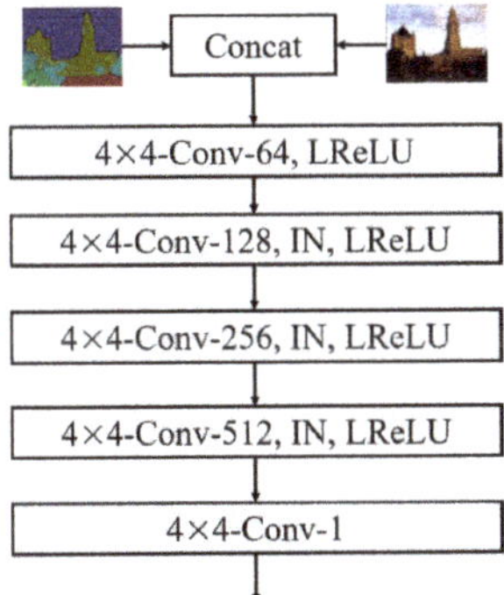

Figure 4. Structure of discriminator.

2.2.2. Importing Spatially Adaptive Normalization Module SPADE into Generator

The structure of the spatially adaptive normalization module SPADE is shown in Figure 5. The label image is first projected onto the embedding space and then convolved to produce the modulation parameters γ and β. Unlike the previous conditional normalization method, γ and β here are not vectors, but tensors with spatial dimensions. The generated γ and β are processed in the next step, similar to batch normalization (BN) [20] It is also regularized in the channel and modulated with the learned scale and bias. The input of SPADE is a segmented image with different colors representing different labels. First, a unified convolution is performed, and then two different convolutions are performed to generate γ and β with the same number and size as the current number of channels. Next, γ is multiplied by the layer that has just been normalized, and β is added. It is equivalent in that each pixel point of each channel in a layer is normalized separately. In contrast to BN, it depends on the input label image and varies depending on the location. With SPADE, there is no need to input semantic images at the first level of the generator, because the learned modulation parameters already encode enough information about the label layout.

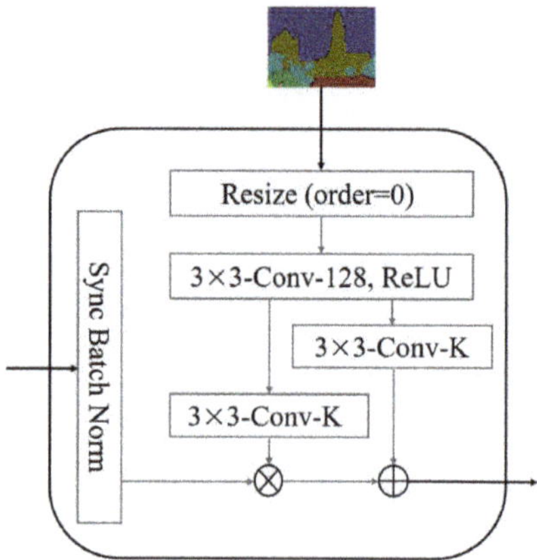

Figure 5. Structure of SPADE.

The SPADE structure is shown in Equations (2)–(4), where h^i represents the activation of the *i*th layer of the deep convolutional network for a batch of N samples. c^i is the number of channels in the layer, H^i and w^i are the height and width of the activation map in the layer. $h^i_{n,c,y,x}$ denotes activation before normalization, μ^i_c and σ^i_c are the mean and variance in channel c. normally, N is set to 1.

$$\gamma^i_{c,\,y,\,x}(m)\frac{h^i_{n,c,y,x} - \mu^i_c}{\sigma^i_c} + \beta^i_{c,y,x}(m), \tag{2}$$

$$\mu^i_c = \frac{1}{NH^iW^i}\sum_{n,y,x} h^i_{n,c,y,x}, \tag{3}$$

$$\sigma^i_c = \sqrt{\frac{1}{NH^iW^i}\sum_{n,y,x}\left(\left(h^i_{n,c,y,x}\right)^2 - \left(\mu^i_c\right)^2\right)} \tag{4}$$

The SPADE is combined with the activation function and convolution to form a SPADE block, refer Mescheder et al. [21] and Miyato et al. [22], the SPADE block replaces the commonly used "convolution $\rightarrow$ activation $\rightarrow$ normalization" module with "SPADE $\rightarrow$ activation $\rightarrow$ convolution". This module can be seen as using the image semantic information to guide the feature map for normalization processing. The structure is shown in Figure 6. In order to solve the problem that the number of channels before and after the residual block is different, a skip connection is added to the structure [23]. That is the portion within the dashed box in Figure 6.

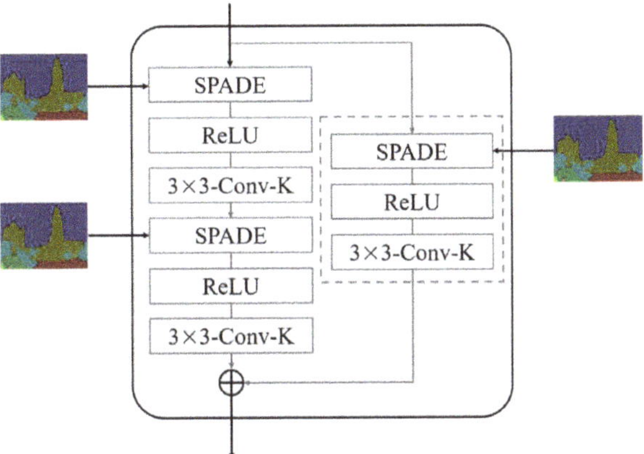

Figure 6. SPADE Block.

Since the learned modulation parameters already encode enough information about the label layout, there is no need to feed the segmented images back to the first layer of the generator, whereby the encoder part of the generator may not be used, which could make the network more lightweight. Figure 7 shows the structure of the generator of SYGAN, which is composed of a series of SPADE blocks and convolutions. The whole network structure is formed by learning the data distribution in a row, and then stacking the SPADE blocks layer by layer. The size of the feature map is from small to large, and the number of channels is from large to small to generate the final real image. In each layer of SPADE block, semantic segmentation images are continuously added to intervene, so that the network can learn multi-scale semantic information in each layer.

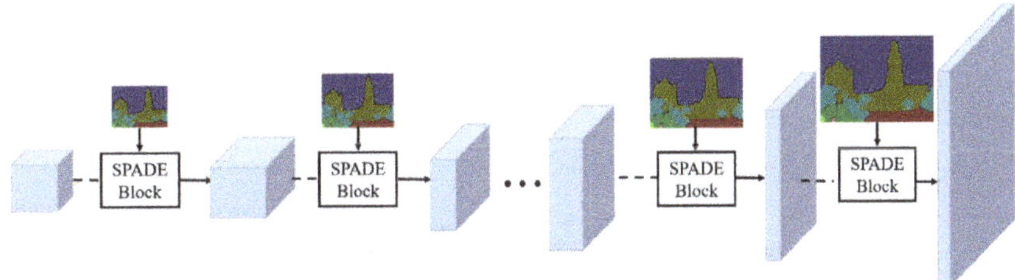

Figure 7. Importing the SPADE block into the SYGAN generator.

In Figure 7, the SPADE block takes the previously output low-resolution image and the different-sized label image of the input image as the input of the next block to generate a higher-resolution image. The growing blue squares are images of increasing size.

2.2.3. Adding Attention Mechanism YLG

The YLG attention mechanism is a sparse attention mechanism, which can improve the computational efficiency of the module. It divides the attention into multiple steps for computation instead of concentrating the computation together. The second-order complexity of the input attention can be expressed by a matrix $A_{X,Y} = X_Q \cdot Y_K^T$.

X, Y is an intermediate representation that associates several matrices with the input. At each step i, attention is directed to a subset of the input locations, which are determined by the binary mask M^i, as shown in Equation (5).

$$A^i_{X,Y}[a,b] = \begin{cases} A_{X,Y}[a,b], & M^i[a,b] = 1 \\ -\infty, & M^i[a,b] = 0 \end{cases} \tag{5}$$

$-\infty$ means that after the function is activated, the value of this position will be cleared, and the calculation will no longer be transferred, so it has no effect on it. Therefore, the design of mask M^i is very important, which is related to the complexity of the data involved in the calculation of attention. The mechanism is designed to solve this problem by using a kind of attention mask that specifies which points have a calculation relationship with points and which points are not settled. The mechanism also refers to the method of Rewon Child et al. [24], which allows individual attention heads to operate on different matrices in parallel, and then connect them in series along the feature dimension. This attention mask also has two modes, which are Left to Right (LTR) in Figure 8a and Right to Left (RTL) in Figure 8b. RTL is the transposed version of LTR. The related information flow diagram is shown in Figure 4. These two modes only allow attention to some areas, which can significantly reduce the quadratic complexity of attention. The mask is actually a superposition of the connected graphs of the two calculations, in which dark blue represents the position of both calculations, light blue represents the position of the first calculation, and green represents the location of the second calculation. The remaining yellow squares represent the positions that are not involved twice, from which the sparsity of the attention mechanism can be reflected.

(a)

(b)

Figure 8. Attention masks. (**a**) Left to right (LTR). (**b**) Right to left (RTL).

2.3. Datasets

Experiments were conducted using the following three datasets:

COCO-Stuff [25]: From the COCO dataset. It has 118,000 training images and 5000 test images from different scenes, containing 182 semantic categories.

ADE20K [26]: Consists of 20,210 training images and 2000 test images. Similar to COCO-Stuff, the dataset contains 150 semantic categories.

UAVid [27]: An image segmentation dataset of urban scenes captured by UAVs, with a total of 3296 images containing 8 semantic categories.

2.4. Design of Experiments

2.4.1. Hardware and Software Configuration

The deep learning framework PyTorch was used to implement the SYGAN model and the experiments. The hardware and software configurations are shown in Table 1.

Table 1. Software and hardware configuration.

Item	Detail
CPU	AMD Ryzen 7 3900X 12-Core processor
GPU	NVIDIA GeForce RTX 3090
RAM	32GB
Operating system	64-bit Windows 11
CUDA	CUDA11.3
Data processing	Python 3.7

2.4.2. Evaluation Indicators

In order to evaluate the accuracy of the model, Pixel Accuracy (PA), Mean Intersection over Union (MIoU) and Fréchet Inception Distance (FID) [28] were used in this paper to measure the gap between the synthetic image distribution and the ground truth distribution.

Pixel Accuracy (PA) is an evaluation criterion for predicting the accuracy of pixels. PA = number of correctly predicted pixels/total number of predicted pixels, as shown in Equation (6):

$$PA = \frac{\sum_{i=0}^{k} p_{ii}}{\sum_{i=0}^{k} \sum_{j=0}^{k} p_{ij}} \tag{6}$$

The definition of MIoU is given in Equation (7). Where $k + 1$ is the number of classes (including null classes), i is the true value, j is the predicted value, and p_{ji} is the number of true values i and predicted values j.

$$MIoU = \frac{1}{k+1} \sum_{i=0}^{k} \frac{p_{ii}}{\sum_{j=0}^{k} p_{ij} + \sum_{j=0}^{k} p_{ji} - p_{ii}} \tag{7}$$

FID is an index commonly used to evaluate GAN. Its idea is to send the samples generated by the generator and those generated by the discriminator to the classifier respectively, extract the abstract features of the middle layer of the classifier, assume that the abstract features conform to the multivariate Gaussian distribution, and estimate the mean value of the Gaussian distribution of the generated samples' μ_g, variance $\sum g$, training samples μ_{data}, and variance $\sum data$ to calculate the Fréchet distance between two gaussian distributions. In addition, tr represents trace. This distance value is the FID, as shown in Equation (8).

$$FID = \|\mu_{data} - \mu_g\|^2 + tr\left(\sum data + \sum g - 2(\sum data \sum g)^{\frac{1}{2}}\right) \tag{8}$$

2.4.3. Parameters of Experiments

(1) Loss function.

The loss function is the combination of ordinary cross entropy loss (Cross Entropy Loss) and Dice Loss. Dice coefficient is an aggregate similarity measure function, which is used to calculate the similarity between two samples. The value is usually between 0 and 1, and the lower the loss value, the better the fitting effect and robustness of the synthetic model.

(2) Training parameters.

The learning rates of the generator and discriminator are set to 0.0001 and 0.0004 respectively, and the setting of the learning rates is referred to Heusel et al. [29] The first 200 epochs are performed, and the learning rate is linearly attenuated to 0.00005 over the course of 150 to 200 epochs. The test found that the loss value reached the lowest value of 0.15 after 110 times of training, and then there was almost no change, so the epoch = 120 was determined after comprehensive consideration. Due to the limitation of GPU memory, when the batch size is greater than 16, it is likely to stop training due to insufficient memory, so it is determined as batch size = 16. The loss function diagram is shown in Figure 9 and the hyperparameter setting is shown in Table 2.

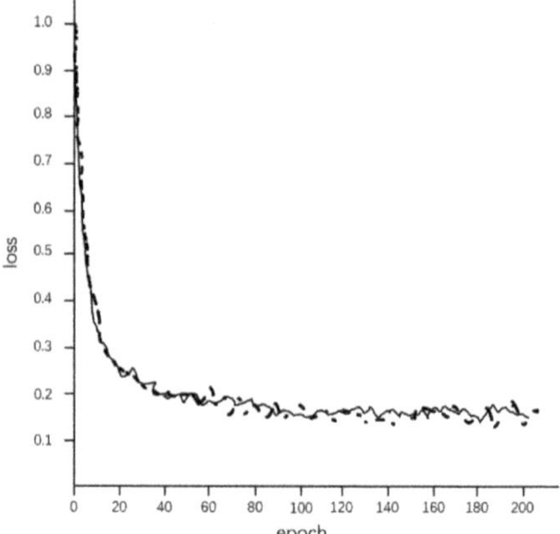

Figure 9. Loss function.

Table 2. Hyperparameter setting.

Item	Value
epoch	120
Batch size	16
$Lr(G)$	0.0001
$Lr(D)$	0.0004
Image size	512×512

2.4.4. Schemes of Experiments

(1)   Comparative experiments.

This part includes two experimental subjects: natural scene and street scene. On the basis of the three datasets, COCO-Stuff, ADE20K, and UAVid, images were selected and classified, and then divided into two new datasets – natural scene and street scene – for training and testing. These are the two most commonly used image scenes, and they have different styles. The difficulty of model training is also different, so it is better to carry out comparative experiments. The training set for each of the two new datasets consists of 10,000 images. The test set for each of the two new datasets consists of 1000 images. The image size used is 512×512. Four other models, CRN, SIMS, pix2pixHD, and GauGAN, were used to conduct the comparison experiments.

(2)   Computational complexity experiments.

COCO-Stuff were used in this experiment. We counted the number of epochs that reach the highest FID and the time it took each epoch. These data are used to calculate the total time required for training for comparison, so as to compare the complexity between SYGAN and SAGAN [11]. In contrast to SYGAN, other models such as CRN, SIMS, pix2pixHD, and GauGAN do not incorporate attention mechanics, so we don't compare the complexity of SYGAN with that of these models.

(3)   Ablation experiments.

The ablation experiment is one of the key factors to assess the quality of the model. The three datasets, COCO-Stuff, ADE20K and UAVid, were used for the experiments to verify the necessity of the corresponding improvement features.

3. Results and Discussion

3.1. Comparative Experiments

In the experiments, the proposed model SYGAN was compared with several image synthesis models: CRN, SIMS, pix2pixHD, GauGAN. CRN uses a deep learning network to repeatedly refine the output from low resolution to high resolution; SIMS uses a semi-parametric method to synthesize real segments from the training set and refine the boundary; pix2pixHD is a conditional image synthesis model based on GAN. A higher value of MIoU and PA indicates better performance, while a lower value of FID indicates better performance. Because the generated image does not need to be completely consistent with the real image, such as vegetation and sky, the image synthesis only needs to be subjectively reasonable to the naked eye, and does not need every tree and cloud to be the same as ground truth, so the MioU index in the above experimental results will be relatively low. However, as it can reflect the coincidence of the generated image and the label image, it can also show the quality of the model to a certain extent.

3.1.1. Natural Scene

Experiments were conducted using natural scene images. MioU, PA, and FID were used as indicators, where the higher the values of MioU and PA, the better the performance, and the lower the value of FID, the better the performance. The results are shown in Figure 10 and Table 3.

| (a) | (b) | (c) | (d) | (e) | (f) | (g) |

Figure 10. Visual comparison of natural scene image synthesis results. (**a**) Label. (**b**) Ground truth. (**c**) CRN. (**d**) SIMS. (**e**) pix2pixHD. (**f**) GauGAN. (**g**) SYGAN (ours).

Table 3. Results of the comparison of natural scene images.

Model	PA (%)	MIoU (%)	FID
CRN	68.4	45.3	48.6
SIMS	63.6	38.6	43.6
pix2pixHD	73.9	46.3	39.8
GauGAN	83.9	54.8	22.6
SYGAN(ours)	86.1	56.6	22.1

It can be seen from Figure 10 that SYGAN, the model proposed in this paper, successfully synthesizes the real details of semantic labels, and the generated images are significantly improved compared with other models, making the generated images closer to human subjective feelings, more natural in various performances, smoother and more natural in the edges of different generated categories. Various indicators also show that the performance of SYGAN is better than the comparative methods.

As is shown in Table 3, the FID of SYGAN was 22.1, which was 26.5, 21.5, 17.7, and 0.5 lower than that of CRN, SIMS, pix2pixHD, and GauGAN, respectively. The MioU of

SYGAN was 56.6%, which was 11.3%, 16%, 10.3%, and 1.8% higher than that of CRN, SIMS, pix2pixHD, and GauGAN, respectively. The PA of SYGAN was 86.1%, which was 17.7%, 22.5%, 12.2%, and 2.2% higher than that of CRN, SIMS, pix2pixHD, and GauGAN, respectively.

3.1.2. Street Scene

The results of experiments for street scenes using SYGAN and the four comparative models are shown in Figure 11 and Table 4.

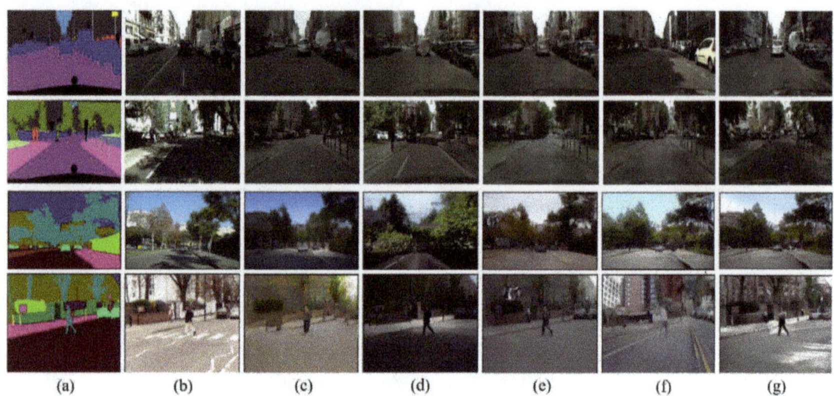

Figure 11. Visual comparison of street scene image synthesis results. (**a**) Label. (**b**) Ground truth. (**c**) CRN. (**d**) SIMS. (**e**) pix2pixHD. (**f**) GauGAN. (**g**) SYGAN (ours).

Table 4. Results of the comparison of street scene images.

Model	PA (%)	MIoU (%)	FID
CRN	67.5	43.5	58.2
SIMS	73.1	34.2	61.3
pix2pixHD	68.9	41.4	47.6
GauGAN	78.8	49.6	33.8
SYGAN(ours)	81.3	51.4	31.2

It can be seen from Figure 11 that the effect of CRN is not good in complex street scenes. Although SIMS looks good, it often deviates from the input label image. Pix2pixHD also has the same problem; the output will be deviated. On the whole, the results of our model SYGAN can achieve more detail than others, which can better generate the semantic information contained in the tags, and the indicators also show that SYGAN has better performance.

As is shown in Table 4, the FID of SYGAN was 31.2, which was 27, 30.1, 16.4, and 2.6 lower than that of CRN, SIMS, pix2pixHD, and GauGAN, respectively. The MIoU of SYGAN was 51.4%, which was 7.9%, 17.2%, 10%, and 1.8% higher than that of CRN, SIMS, pix2pixHD, and GauGAN, respectively. The PA of SYGAN was 81.3%, which was 13.8%, 8.2%, 12.4%, and 2.5% higher than that of CRN, SIMS, pix2pixHD, and GauGAN, respectively.

3.1.3. Comparison of the Two Scenes

According to the indicators in Tables 3 and 4, the performance of all the mentioned methods for natural scenes is better than that for street scenes. As to SYGAN, its PA and MIoU for natural scenes are 86.1 and 56.6 which are 5.90% and 10.12% higher than those for street scenes, respectively, and its FID for natural scenes is 22.1 which is 29.17% lower than that for street scenes.

The reason for the above conclusion is that street scenes are usually more complex than natural scenes. Street scenes usually include many relatively small elements, and there are many complex boundaries between different elements. Conversely, natural scenes tend to have few and large elements, and the boundaries between different elements are relatively long and obvious.

In natural scenes, there are usually four or five elements, and the existence of sky is very frequent. This element often makes up a large proportion of the entire image, ranging from 10% to 70%. Other elements that appear in high proportions are mountains, trees, and water. The distribution of these elements is concentrated, and they usually have long, smooth boundaries. In a street scene, there are usually groups of seven or eight elements. The components are fixed, like buildings, cars, trees. The different elements are scattered and cover each other. Trees usually appear alone, so they have uneven boundaries.

3.2. Computational Complexity Experiments

Compared with CRN, SIMS, pix2pixHD, and GauGAN, SYGAN has a relatively high complexity due to the addition of attention mechanism, but it achieves better synthesis quality. Therefore, the complexity analysis in this paper does not consider the comparison with the above four methods, but only with SAGAN in terms of complexity. SAGAN also introduces the attention mechanism in the network, which solves the limitation of the receptive field size caused by the convolutional structure, and also enables the network to learn different areas that to which attention should be paid in the process of generating images. However, the dense attention mechanism also brings some problems, such as high computational cost. Compared with the comparison method, SYGAN uses the YLG attention mechanism. In terms of ensuring accuracy, it can also reduce the overhead brought by the attention mechanism. The experimental results are shown in Figure 12. It can be seen that the proposed model SYGAN has reached the best FID at about epoch = 110, with an average of 21 min each time, while SAGAN needs about 140 times to reach the best FID, with an average of 19 min each time. The overall time of SYGAN has an advantage over SAGAN and the FID performance is better.

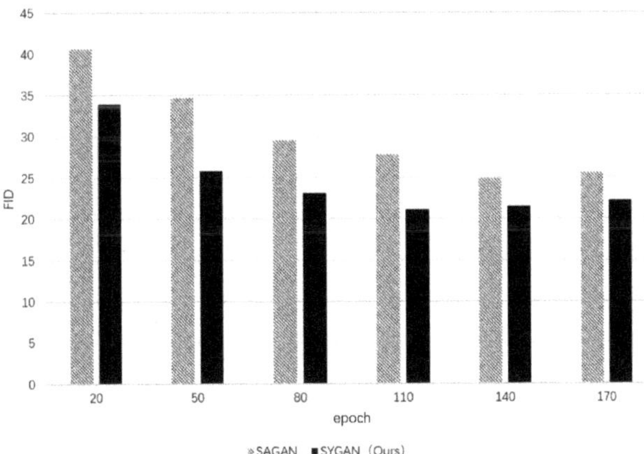

Figure 12. Relationship between epoch and FID.

3.3. Ablation Experiments

Ablation experiments were performed using public datasets with the same hyper-parameters. The results of ablation experiments are shown in Tables 5–7 where SGAN represents the model without YLG, YGAN represents the model without SPADE, GAN represents the models without SPADE and YLG. Higher MIoU and PA values in the table indicate better performance, and lower FID values indicate better performance.

Table 5. Results of ablation experiments on the COCO-stuff.

Model	PA (%)	MIoU (%)	FID
SYGAN	69.5	48.2	22.3
SGAN	66.3	46.1	25.3
YGAN	55.4	38.6	36.5
GAN	33.4	30.6	68.2

Table 6. Results of ablation experiments on the ADE20K.

Model	PA (%)	MIoU (%)	FID
SYGAN	81.4	51.3	37.8
SGAN	78.6	48.1	42.3
YGAN	68.2	41.8	51.2
GAN	44.3	25.6	71.5

Table 7. Results of ablation experiments on the UAVid.

Model	PA (%)	MIoU (%)	FID
SYGAN	86.3	57.1	32.3
SGAN	82.9	54.3	36.2
YGAN	71.5	46.3	46.1
GAN	49.6	29.8	70.3

It can be seen from Tables 5–7 that the FID of SGAN and YGAN has a high improvement compared with that of GAN, indicating that SPADE and YLG have a very good improvement on the performance of the model. The FID of SGAN is improved by about 10 compared with that of YGAN, indicating that the performance improvement of SPADE is greater than that of YLG. SYGAN, when combined with SPADE and YLG, has about 4 and 14 improvements, respectively, compared with SGAN and YGAN. The YLG attention mechanism combined with Figure 12 shows that compared with the usual intensive attention mechanism, it can significantly reduce the computational complexity and improve the training speed.

4. Conclusions

An image synthesis model SYGAN is proposed in this paper, which imports a spatial adaptive normalization module SPADE and an attention mechanism YLG on the basis of GAN. These improvements ensure the model has good performance, increases the accuracy of image synthesis, reduces the generation of false features, expands the receptive field of the model, and shortens the training time. The PA of the model SYGAN is 86.1% in the natural scene dataset, and 81.3% in the street scene dataset. The MIoU of the model SYGAN is 56.6% in the natural scene dataset, and 51.4% in the street scene dataset. The FID score of the model is 22.1 in the natural scene dataset, and 31.2 in the street scene dataset. SYGAN has a better performance in the natural than the street scene. Compared with other models in the experiment, the synthesis effect is better in both datasets. In the computational complexity experiments, the training time of SYGAN is shorter and the FID lower than that of SAGAN with the addition of typical attention mechanisms. From the experimental results, we can see that the model has a good performance as it generates a virtual image through the label image, which can easily preview engineering tasks. This has a very positive significance for the construction of smart cities.

Although SYGAN can complete the task of image synthesis well, it generates some problem images in complex environments, edge generation, and shadow display, which does not conform to the subjective impression of human beings. This will be studied and solved in our future study and work.

Author Contributions: Conceptualization, Y.Z. and L.M.; methodology, Y.Z., G.W., X.Y., X.W. and P.W.; software, Y.Z.; validation, Y.Z.; formal analysis, Y.Z., X.Y., X.W. and P.W.; investigation, Y.Z., G.W., X.Y., X.W. and P.W.; data curation, Y.Z.; writing—original draft. Y.Z.; resources, G.W. and L.M.; writing—review & editing, G.W. and L.M.; visualization, G.W., X.Y. and X.W.; supervision, G.W. and L.M.; project administration. G.W.; funding acquisition, L.M. and P.W. All authors have read and agreed to the published version of the manuscript.

Funding: This study was supported by the National Natural Science Foundation of China (Grant number: U1809208) and the Key Research and Development Program of Zhejiang Province (Grant number: 2021C02005).

Institutional Review Board Statement: Not applicable.

Informed Consent Statement: Not applicable.

Data Availability Statement: Not applicable.

Conflicts of Interest: The authors declare no conflict of interest.

References

1. Botín-Sanabria, D.M.; Mihaita, A.-S.; Peimbert-García, R.E.; Ramírez-Moreno, M.A.; Ramírez-Mendoza, R.A.; Lozoya-Santos, J.D.J. Digital twin technology challenges and applications: A comprehensive review. *Remote Sens.* **2022**, *14*, 1335. [CrossRef]
2. Karras, T.; Laine, S.; Aila, T. A Style-Based Generator Architecture for Generative Adversarial Networks. In Proceedings of the IEEE/CVF Conference on Computer Vision and Pattern Recognition, Long Beach, CA, USA, 16–17 June 2019; pp. 4401–4410.
3. Goodfellow, I.; Pouget-Abadie, J.; Mirza, M.; Xu, B.; Warde-Farley, D.; Ozair, S.; Courville, A.; Bengio, Y. Generative adversarial networks. *Commun. ACM* **2020**, *63*, 139–144. [CrossRef]
4. Wang, T.-C.; Liu, M.-Y.; Zhu, J.-Y.; Tao, A.; Kautz, J.; Catanzaro, B. High-Resolution Image Synthesis and Semantic Manipulation with Conditional Gans. In Proceedings of the IEEE Conference on Computer Vision and Pattern Recognition, Salt Lake City, UT, USA, 18–22 June 2018; pp. 8798–8807.
5. Chen, Q.; Koltun, V. Photographic Image Synthesis with Cascaded Refinement Networks. In Proceedings of the IEEE International Conference on Computer Vision, Venice, Italy, 22–29 October 2017; pp. 1511–1520.
6. Qi, X.; Chen, Q.; Jia, J.; Koltun, V. Semi-Parametric Image Synthesis. In Proceedings of the IEEE Conference on Computer Vision and Pattern Recognition, Salt Lake City, UT, USA, 18–22 June 2018; pp. 8808–8816.
7. Bai, G.; Xi, W.; Hong, X.; Liu, X.; Yue, Y.; Zhao, S. Robust and Rotation-Equivariant Contrastive Learning. *IEEE Trans. Neural Netw. Learn. Syst.* **2023**, 1–14. [CrossRef]
8. Wang, H.; Zhang, Y.; Yu, X. An overview of image caption generation methods. *Comput. Intell. Neurosci.* **2020**, *2020*, 3062706. [CrossRef] [PubMed]
9. Park, T.; Liu, M.-Y.; Wang, T.-C.; Zhu, J.-Y. Semantic Image Synthesis with Spatially-Adaptive Normalization. In Proceedings of the IEEE/CVF Conference on Computer Vision and Pattern Recognition, Long Beach, CA, USA, 16–17 June 2019; pp. 2337–2346.
10. Albawi, S.; Mohammed, T.A.; Al-Zawi, S. Understanding of a Convolutional Neural Network. In Proceedings of the 2017 International Conference on Engineering and Technology (ICET), Antalya, Turkey, 21–23 August 2017; pp. 1–6.
11. Zhang, H.; Goodfellow, I.; Metaxas, D.; Odena, A. Self-Attention Generative Adversarial Networks. In Proceedings of the International Conference on Machine Learning, Long Beach, CA, USA, 9–15 June 2019; pp. 7354–7363.
12. Niu, Z.; Zhong, G.; Yu, H. A review on the attention mechanism of deep learning. *Neurocomputing* **2021**, *452*, 48–62. [CrossRef]
13. Daras, G.; Odena, A.; Zhang, H.; Dimakis, A.G. Your local GAN: Designing Two Dimensional Local Attention Mechanisms for Generative Models. In Proceedings of the IEEE/CVF Conference on Computer Vision and Pattern Recognition, Seattle, WA, USA, 14–19 June 2020; pp. 14531–14539.
14. Creswell, A.; White, T.; Dumoulin, V.; Arulkumaran, K.; Sengupta, B.; Bharath, A.A. Generative adversarial networks: An overview. *IEEE Signal Process. Mag.* **2018**, *35*, 53–65. [CrossRef]
15. Fukui, H.; Hirakawa, T.; Yamashita, T.; Fujiyoshi, H. Attention Branch Network: Learning of Attention Mechanism for Visual Explanation. In Proceedings of the IEEE/CVF Conference on Computer Vision and Pattern Recognition, Long Beach, CA, USA, 16–17 June 2019; pp. 10705–10714.
16. Xu, J.; Li, Z.; Du, B.; Zhang, M.; Liu, J. Reluplex Made More Practical: Leaky ReLU. In Proceedings of the 2020 IEEE Symposium on Computers and communications (ISCC), Rennes, France, 7–10 July 2020; pp. 1–7.
17. Cai, T.; Luo, S.; Xu, K.; He, D.; Liu, T.-Y.; Wang, L. Graphnorm: A Principled Approach to Accelerating Graph Neural Network Training. In Proceedings of the International Conference on Machine Learning, Virtual, 18–24 July 2021; pp. 1204–1215. [CrossRef]
18. Hara, K.; Saito, D.; Shouno, H. Analysis of Function of Rectified Linear Unit Used in Deep Learning. In Proceedings of the 2015 International Joint Conference on Neural Networks (IJCNN), Killarney, Ireland, 12–17 July 2015; pp. 1–8.
19. Isola, P.; Zhu, J.-Y.; Zhou, T.; Efros, A.A. Image-to-Image Translation with Conditional Adversarial Networks. In Proceedings of the IEEE Conference on Computer Vision and Pattern Recognition, Honolulu, HI, USA, 21–26 July 2017; pp. 1125–1134.

20. Ioffe, S.; Szegedy, C. Batch Normalization: Accelerating Deep Network Training by Reducing Internal Covariate Shift. In Proceedings of the International Conference on Machine Learning, Lille, France, 6–11 July 2015; pp. 448–456.
21. Mescheder, L.; Geiger, A.; Nowozin, S. Which Training Methods for GANs do Actually Converge? In Proceedings of the International Conference on Machine Learning, Stockholm, Sweden, 10–15 July 2018; pp. 3481–3490.
22. Miyato, T.; Koyama, M. cGANs with Projection Discriminator. In Proceedings of the International Conference on Learning Representations, Vancouver, BC, Canada, 30 April–3 May 2018.
23. Mazaheri, G.; Mithun, N.C.; Bappy, J.H.; Roy-Chowdhury, A.K. A Skip Connection Architecture for Localization of Image Manipulations. In Proceedings of the CVPR Workshops, Long Beach, CA, USA, 16–20 June 2019; pp. 119–129.
24. Child, R.; Gray, S.; Radford, A.; Sutskever, I. Generating long sequences with sparse transformers. *arXiv* **2019**, arXiv:1904.10509.
25. Caesar, H.; Uijlings, J.; Ferrari, V. Coco-stuff: Thing and stuff classes in context. In Proceedings of the IEEE Conference on Computer Vision and Pattern Recognition, Salt Lake City, UT, USA, 18–22 June 2018; pp. 1209–1218.
26. Zhou, B.; Zhao, H.; Puig, X.; Fidler, S.; Barriuso, A.; Torralba, A. Scene Parsing through ade20k Dataset. In Proceedings of the IEEE Conference on Computer Vision and Pattern Recognition, Honolulu, HI, USA, 21–26 July 2017; pp. 633–641.
27. Pedamonti, D. Comparison of non-linear activation functions for deep neural networks on MNIST classification task. *arXiv* **2018**, arXiv:1804.02763.
28. Obukhov, A.; Krasnyanskiy, M. Quality Assessment Method for GAN Based on Modified Metrics Inception Score and Fréchet Inception Distance. In Proceedings of the Computational Methods in Systems and Software, Online, 14–17 October 2020; pp. 102–114. Available online: https://link.springer.com/chapter/10.1007/978-3-030-63322-6_8 (accessed on 6 February 2023).
29. Heusel, M.; Ramsauer, H.; Unterthiner, T.; Nessler, B.; Hochreiter, S. Gans trained by a two time-scale update rule converge to a local nash equilibrium. *Adv. Neural Inf. Process. Syst.* **2017**, *30*, 1–12.

Review

A Comprehensive Survey of Transformers for Computer Vision

Sonain Jamil [1], Md. Jalil Piran [2,*] and Oh-Jin Kwon [1,*]

[1] Department of Electronics Engineering, Sejong University, Seoul 05006, Republic of Korea
[2] Department of Computer Engineering, Sejong University, Seoul 05006, Republic of Korea
* Correspondence: piran@sejong.ac.kr (M.J.P.); ojkwon@sejong.ac.kr (O.-J.K.)

Abstract: As a special type of transformer, vision transformers (ViTs) can be used for various computer vision (CV) applications. Convolutional neural networks (CNNs) have several potential problems that can be resolved with ViTs. For image coding tasks such as compression, super-resolution, segmentation, and denoising, different variants of ViTs are used. In our survey, we determined the many CV applications to which ViTs are applicable. CV applications reviewed included image classification, object detection, image segmentation, image compression, image super-resolution, image denoising, anomaly detection, and drone imagery. We reviewed the state of the-art and compiled a list of available models and discussed the pros and cons of each model.

Keywords: vision transformers; computer vision; deep learning; image coding; drone imagery; drone surveillance

1. Introduction

Vision transformers (ViTs) are designed for tasks related to vision, including image recognition [1]. Originally, transformers were used to process natural language (NLP). Bidirectional encoder representations from transformers (BERT) [2] and generative pretrained transformer 3 (GPT-3) [3] were the pioneers of transformer models for natural language processing. In contrast, classical image processing systems use convolutional neural networks (CNNs) for different computer vision (CV) tasks. The most common CNN models are AlexNet [4,5], ResNet [6], VGG [7], GoogleNet [8], Xception [9], Inception [10,11], DenseNet [12], and EfficientNet [13].

To track attention links between two input tokens, transformers are used. With an increasing number of tokens, the cost rises inexorably. The pixel is the most basic unit of measurement in photography, but calculating every pixel relationship in a normal image would be time-consuming; memory-intensive [14]. ViTs, however, take several steps to do this, as described below:

- ViTs divide the full image into a grid of small image patches.
- ViTs apply linear projection to embed each patch.
- Then, each embedded patch becomes a token, and the resulting sequence of embedded patches is passed to the transformer encoder (TE).
- Then, TE encodes the input patches, and the output is given to the multilayer perceptron (MLP) head, with the output of the MLP head being the input class.

Figure 1 shows the primary illustration of ViTs. In the beginning, the input image is divided into smaller patches. Each patch is then embedded using linear projection. Tokens are created from embedded patches that are given to the TE as inputs. Multihead attention and normalization are used by the TE to encode the information embedded in patches. The TE output is given to the MLP head, and the MLP head output is the input image class.

For image classification, the most popular architecture uses the TE to convert multiple input tokens. However, the transformer's decoder can also be used for other purposes. As described in 2017, transformers have rapidly spread across NLP, becoming one of the most widely used and promising designs [15].

Citation: Jamil, S.; Jalil Piran, M.; Kwon, O.-J. A Comprehensive Survey of Transformers for Computer Vision. *Drones* **2023**, *7*, 287. https://doi.org/10.3390/drones7050287

Academic Editor: Seokwon Yeom

Received: 24 March 2023
Revised: 19 April 2023
Accepted: 23 April 2023
Published: 25 April 2023

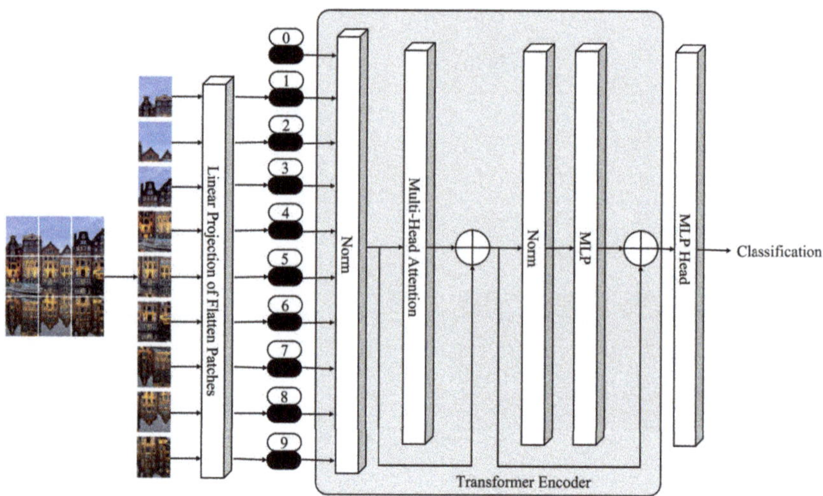

Figure 1. ViT for Image Classification.

For CV tasks, ViTs were applied in 2020 [16]. The aim was to construct a sequence of patches that, once reconstructed into vectors, are interpreted as words by a standard transformer. Imagine that the attention mechanism of NLP transformers was designed to capture the relationships between different words within the text. In this case, the CV takes into account how the different patches of the image relate to one another.

In 2020, a pure transformer outperformed CNNs in image classification [16]. Later, a transformer backend was added to the conventional ResNet, drastically lowering costs while enhancing accuracy [17,18].

In the same year, several key ViT versions were released. These variants were more efficient, accurate, or applicable to specific regions. Swin transformers are the most prominent variants [19]. Using a multistage approach and altering the attention mechanism, the Swin transformer achieved cutting-edge performance on object detection datasets. There is also the TimeSformer, which was proposed for video comprehension issues and may capture spatial and temporal information through divided space–time attention [20].

ViT performance is influenced by decisions such as optimizers, dataset-specific hyperparameters, and network depth. Optimizing a CNN is significantly easier. Even when trained on data quantities that are not as large as those required by ViTs, CNNs perform admirably. Apparently, CNNs exhibit this distinct behavior because of some inductive biases that they can use to comprehend the particularities of images more rapidly, even if they end up restricting them, making it more difficult for them to recognize global connections. ViTs, on the other hand, are devoid of these biases, allowing them to capture a broader and more global set of relationships at the expense of more difficult data training [21].

ViTs are also more resistant to input visual distortions such as hostile patches and permutations [22]. Conversely, preferring one architecture over another may not be the best choice. The combination of convolutional layers with ViTs has been shown to yield excellent results in numerous CV tasks [23–25].

To train these models, alternate approaches were developed due to the massive amount of data required. It is feasible to train a neural network virtually autonomously, allowing it to infer the characteristics of a given issue without requiring a large dataset or precise labeling. It might be the ability to train ViTs without a massive vision dataset that makes this novel architecture so appealing.

ViTs have been employed in numerous CV jobs with outstanding and, in some cases, cutting-edge outcomes. The following are some of the important application areas:

- Image classification;

- Anomaly detection;
- Object detection;
- Image compression;
- Image segmentation;
- Video deepfake detection;
- Cluster analysis.

Figure 2 shows that the percentage of the application of ViTs for image classification, object detection, image segmentation, image compression, image super-resolution, image denoising, and anomaly detection is 50%, 40%, 3%, less than 1%, less than 1%, 2%, and 3% respectively.

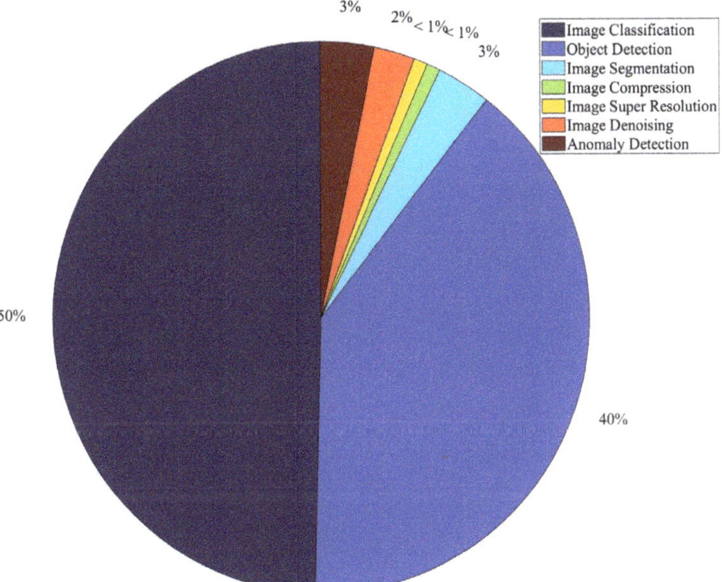

Figure 2. Use of ViTs for CV applications.

ViTs have been widely utilized in CV tasks. ViTs can solve the problems faced by CNNs. Different variants of ViTs are used for image compression, super-resolution, denoising, and segmentation. With the advancement in the ViTs for CV applications, a state-of-the-art survey is needed to demonstrate the performance advantage of ViTs over current CV application approaches.

Our approach was to classify CV applications where ViTs are used, such as in image classification, object detection, image segmentation, image compression, image super-resolution, image denoising, and anomaly detection. Then we surveyed the state-of-the-art in each CV application and tabulated the existing ViT-based models, discussing the pros and cons of each model and lessons learned for each model. We also analyzed the advanced transformers and summarized open-source ViTs, briefly discussing drone imagery using ViTs.

The remainder of this paper is structured as follows: In Section 2, we discuss related work, while in Section 3, we present the application of ViTs in CV. In Section 4, we analyze advanced transformers. In Section 5, we summarize the open-source ViTs and their CV applications. In Section 6, we discuss ViT applications in drone imagery. In Section 7, we examine open research challenges and future directions. In Section 8, we conclude our survey with final thoughts on ViTs in place of CV applications and the results of our survey. A complete organization of the survey is shown in Figure 3.

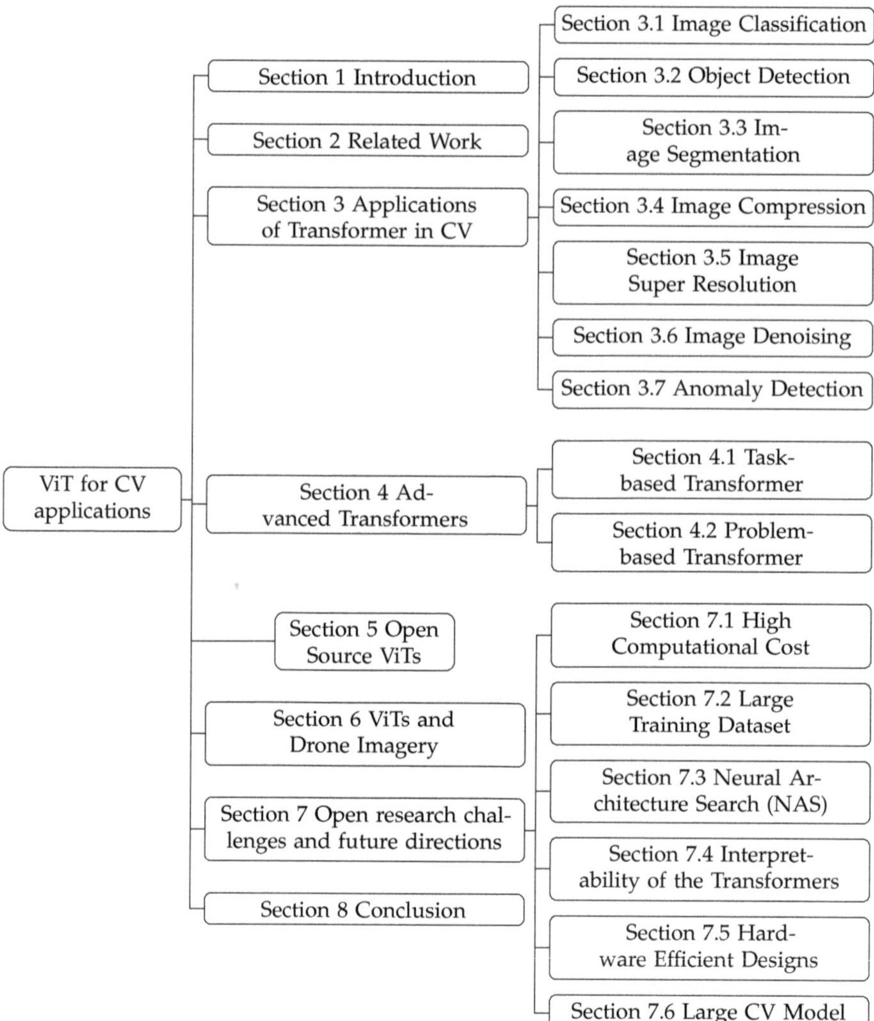

Figure 3. Organization of the survey.

2. Related Work

A number of surveys have been conducted on ViTs in the literature. The authors of [26] review the theoretical concepts, foundation, and applications of the transformer for memory efficiency. They also discussed the applications of efficient transformers in NLP. CV tasks, however, were not included. A similar study, ref. [27], examined the theoretical aspects of the ViTs, the foundations of transformers, the role of multihead attention in transformers, and the applications of transformers in image classification, segmentation, super-resolution, and object detection. The survey did not include applications of transformers for image denoising or compression.

In [28], the authors describe the architectures of transformers for segmenting, classifying, and detecting objects in images. Their survey did not include tasks such as image super-resolution, denoising, or compression associated with CV and image processing.

Lin et al. in [29] summarized different architectures of NLP. Their survey, however, did not include any applications of transformers for CV tasks. In [30], the authors discuss different architectures of transformers for computational visual media, including the application of transformers for low-level vision and generation, such as image colorization, image super-resolution, image generation, and text-to-image conversion. Additionally, the survey focused on high-level vision tasks such as segmentation and object detection. However, the survey did not discuss the transformer for image compression and classification.

Han et al. in [31] surveyed the application of transformers in high-, mid-, and low-level vision and video processing. They also provided a comprehensive discussion of self-attention and the role of transformers in real-device-based applications, and the survey did not discuss the transformer for image compression.

Table 1 summarizes all existing surveys on the ViTs. As a result of an analysis of Table 1, it is evident that the survey is needed to provide insight into the latest developments in ViTs for several image processing and CV tasks, including classification, detection, segmentation, compression, denoising, and super-resolution.

Table 1. Summary of the available surveys on ViTs.

Survey	Year	Scope							Contributions and Limitations
		Class.	Det.	Seg.	Com.	Super Res.	Den.	AD.	
[26]	2020	(P)	✗	✗	✗	✗	✗	✗	• Foundation of transformers • Applications of transformers • History-based survey
[27]	2022	✓	✓	✓	✗	✓	✗	✗	• Basic concepts • Applications of transformers in CV • History-based • Advanced ViTs not explored
[28]	2021	✓	✓	✓	✗	✗	✗	✗	• Different architectures • Future perspectives • Limited to classification, detection, and segmentation models
[30]	2022	✗	✓	✓	✗	✓	✗	✗	• Transformers in computational visual media • Limited to detection, segmentation, and super-resolution models
[31]	2022	✗	✓	✓	✗	✓	✗	✓	• Transformers in high, mid, and low-level vision • Comprehensive discussion of self-attention • Role of transformers in real-device-based applications
Our survey	2023	✓	✓	✓	✓	✓	✓	✓	• Applications of transformers in CV • Advanced transformers • ViTs and drone Imagery • New outlook to the open research gaps

Class.—classification; Det.—detection; Seg.—segmentation; Com.—compression; Super Res.—super-resolution; AD.—anomaly detection; ✓— fully explained; (P)—partially explained; ✗—not explained.

2.1. Bibliometric Analysis and Methodology

We used Google Scholar, Web of Science, IEEE Xplore, and ScienceDirect as the databases to select papers.

2.1.1. Bibliometric Analysis

We considered papers published between 2017 and 2023. In 2017, a total of 21 articles were published on transformers in CV, while 24 papers were published in 2018. A total of 44 papers were published in 2019. Additionally, in 2020, 2021, 2022, and 2023, a total of 52, 418, 494, and 8 papers were published, respectively. Figure 4 shows the number of publications per year from 2017 to 2023.

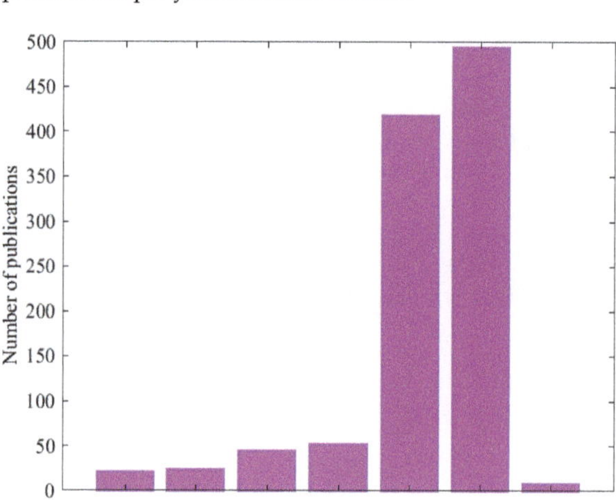

Figure 4. Number of publications in each year from 2017 to 2023 based on Web of Science.

In terms of publishers, the Institute of Electrical and Electronics Engineers (IEEE) has published 527 publications on transformers in CV. Springer Nature, Multidisciplinary Digital Publishing Institute (MDPI), and Elsevier published 177, 104, and 82 papers, respectively. Assoc Computing Machinery (ACM) published the least number of publications on transformers in CV, which is 25. Figure 5 displays the number of publications by different publishers from 2017 to 2023.

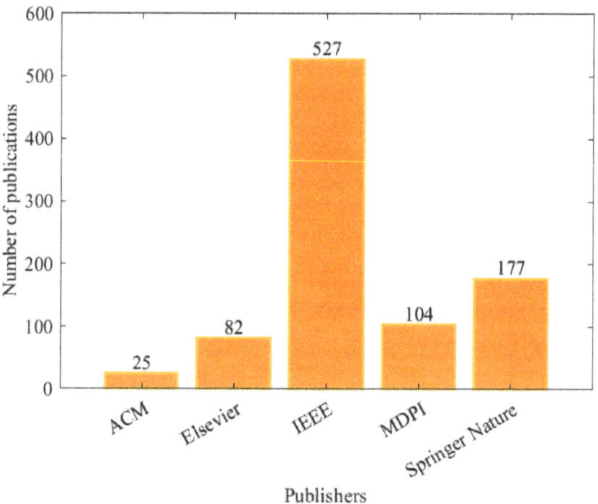

Figure 5. Number of publications by different publishers from 2017 to 2023 based on Web of Science.

In terms of topic popularity in different countries, we present a world map showing the top countries working on transformers in CV. From 2017 to 2023, China published 517 publications on transformers in CV. United States of America (USA), England, and South Korea published 242, 64, and 55 papers, respectively, from 2017 to 2023. Figure 6 shows which countries are leading in learning-driven image compression.

Figure 6. Leading countries working on transformers in CV based on Web of Science.

2.1.2. Methodology

Based on the following criteria as described in Algorithm 1 and Figure 7, 100 papers were selected for analysis:

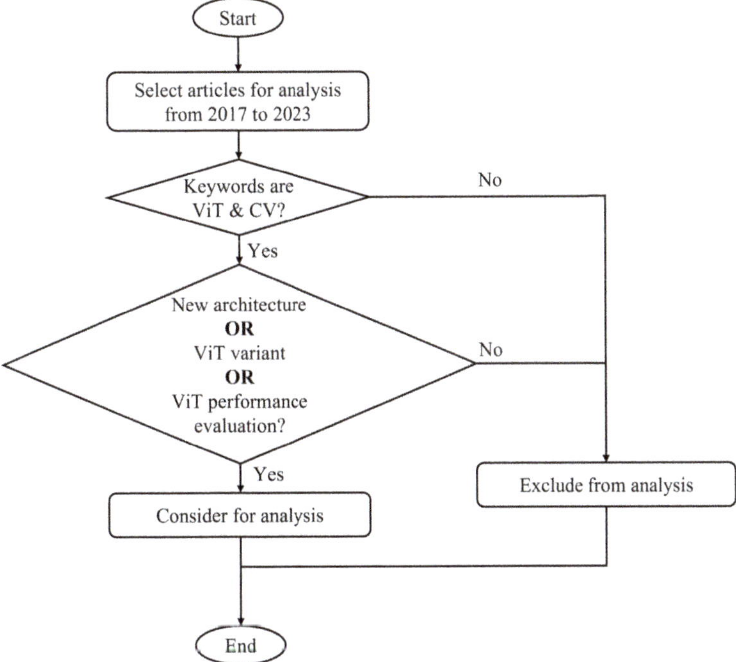

Figure 7. Article Selection Algorithm.

Algorithm 1 Article Selection Criteria

Require: Search on databases
Ensure: Article from 2017 to 2023
    while keyword—transformers in computer vision **do**
       if Discuss new architectures of ViTs | Evaluate performance of ViTs | Variants of ViTs
    then
         Consider for analysis
       else if Does not discuss ViTs architectures in vision **then**
         Exclude from the analysis
       end if
    end while

---

3. Applications of ViTs in CV

In addition to classical ViTs, modified versions of classical ViTs are used for object detection, image segmentation, compression, super-resolution, denoising, and anomaly detection. Figure 8 shows the organization of Section 3.

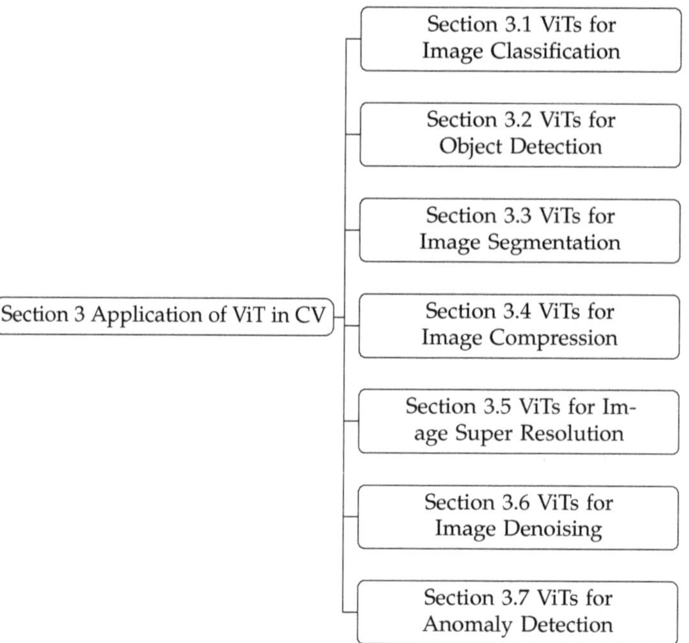

Figure 8. Organization of the Section 3.

3.1. ViTs for Image Classification

In image classification, the image is initially divided into patches; these patches are fed linearly to the transformer encoder, where MLP, normalization, and multihead attention are applied to create embedded patches. Embedded patches are fed to the MLP head, which predicts the output class. These classical ViTs have been used by many researchers to classify visual objects.

In [32], the authors proposed CrossViT-15, CrossViT-18, CrossViT-9†, CrossViT-15†, and CrossViT-18† for image classification. They used the ImageNet1K, CIFAR10, CIFAR100, pet, crop disease, and ChestXRay8 datasets to evaluate the different variants of CrossViT. They achieved 77.1% accuracy on the ImageNet1K dataset by using CrossViT-9†. Similarly, they attained 82.3% and 82.8% accuracy on ImageNet1K dataset using CrossViT-15† and CrossViT-18†, respectively. Similarly, the authors obtained an 99.0% and 99.11% accuracy with CrossViT-15 and CrossViT-18, respectively, on the CIFAR10 dataset. However, they obtained 90.77% and 91.36% accuracy on the CIFAR100 dataset using CrossViT-15 and CrossViT-18, respectively. The authors also used CrossViT for pet classification, crop disease classification, and chest X-ray classification. They observed the highest accuracy of 95.07% with CrossViT-18 for the pet classification. Similarly, they achieved the highest accuracy of 99.97% with CrossViT-15 and CrossViT-18 for the crop diseases classification. Moreover, they achieved the highest accuracy of 55.94% using CrossViT-18 for the chest X-ray classification.

Deng et al. in [33], proposed a combined CNN and ViT model named CTNet for the classification of high-resolution remote sensing images. To evaluate the model, they used the aerial image dataset (AID) and Northwestern Polytechnical University (NWPU)-RESISC45 dataset. CTNet obtained an accuracy of 97.70% and 95.49% using the AID and NWPU-RESISC45 datasets, respectively. Yu et al. in [34], presented multiple instance-enhanced ViT (MIL-ViT) for fundus image classification. They used APTOS 2019 blindness detection and the 2020 retinal fundus multidisease image dataset (RFMiD2020). MIL-ViT yielded an accuracy of 97.9% on the APTOS2019 dataset and 95.9% on the RFMiD2020 dataset. Similarly, Graham et al. in [35] proposed LeViT for fast inference image classification.

In [36], the authors proposed excellent teacher-guiding small networks (ES-GSNet) for the classification of the remote sensing image scenes. They used four datasets: AID, NWPU-RESISC45, UC-Mered Land use dataset (UCM), and OPTIMAL-31. They obtained accuracies of 96.88%, 94.50%, 99.29%, and 96.45% for the AID, NWPU-RESISC45, UCM, and OPTIMAL-31 datasets, respectively.

Xue et al. in [37] proposed deep hierarchical ViT (DHViT) for the hyperspectral and light detection and ranging (LiDAR) data classification. The authors used the Trento, Houston 2013, and Houston 2018 datasets and obtained accuracies of 99.58%, 99.55%, and 96.40%, respectively.

In [38], the authors elaborated on the use of ViT for the satellite imagery multilabel classification and proposed ForestViT. ForestViT demonstrated an accuracy of 94.28% on planet understanding of the Amazon from space (PUAS) dataset. In [39], the researchers put forward the concept of LeViT for pavement image classification. They used Chinese asphalt pavement and German asphalt pavement to evaluate the model's performance. They obtained an accuracy of 91.56% using the Chinese asphalt pavement dataset and 99.17% using the German asphalt pavement dataset.

In [40], the authors used ViT to distinguish malicious drones from airplanes, birds, drones, and helicopters. They demonstrated the efficiency of ViT for the classification over several CNNs such as AlexNet [4], ResNet-50 [41], MobileNet-V2 [42], ShuffleNet [43], SqueezeNet [44], and EfficicentNetb0 [45]. The ViT model achieved 98.3% accuracy on the malicious drones dataset.

Tanzi et al. in [46] applied ViT for the classification of the femur fracture. They used a dataset of real X-rays. The model achieved an accuracy of 83% with 77% precision, 76% recall, and 77% F1-score. In [47], the authors modified the classical ViT and proposed SeedViT for the classification of maize seed quality. They used a custom dataset. The model outperformed CNN and achieved a 96.70% accuracy.

Similarly, in [48], the researchers put forward double output ViT (DOViT) for the classification of air quality and its measurement. They used two datasets named get AQI in one shot-1 (GAOs-1) and get AQI in one shot-2 (GAOs-2). The model achieved a 90.32% accuracy for the GAOs-1 dataset and 92.78% accuracy for the GAOs-2 dataset. In [49], the authors developed a novel multi-instance ViT called MITformer for remote sensing scene classification. They evaluated their model on three different datasets. The model achieved 99.83% accuracy for the UCM dataset, 97.96% accuracy for the AID dataset, and 95.93% accuracy for the NWPU dataset.

Table 2 shows the summary of the application of ViT for image classification.

Table 2. ViT for Image Classification.

Research	Model	Dataset	Objective Classification	Accuracy
[32]	CrossViT-9† [a]	ImageNet1K	Image	77.100%
	CrossViT-15† [a]			82.300%
	CrossViT-18† [a]			82.800%
	CrossViT-15 [a]	CIFAR10	Image	99.000%
		CIFAR100		90.770%
		Pet	Pet classification	94.550%
		Crop Diseases	Crop disease classification	99.970%
		ChestXRay8	Chest X-ray classification	55.890%
	CrossViT-18 [a]	CIFAR10	Image	99.110%
		CIFAR100		91.360%
		Pet	Pet classification	95.070%
		Crop Diseases	Crop diseases	99.970%
		ChestXRay8	Chest X-rays	55.940%
[33]	CTNet	AID	Remote sensing scene	97.700%
		NWPU-RESISC45		95.490%
[34]	MIL-ViT [b]	APTOS2019	Fundus image	97.900%
		RFMiD2020		95.900%
[36]	ET-GSNet	AID	Remote sensing images	96.880%
		NWPU-RESISC45		94.500%
		UCM		99.290%
		OPTIMAL-31		96.450%
[37]	DHViT	Trento	Hyperspectral & LiDAR	99.580%
		Houston 2013		99.550%
		Houston 2018		96.400%
[38]	ForestViT	PUAS	Satellite imagery multilabel	94.280%
[39]	LeViT [c]	Chinese asphalt pavement	Pavement image	91.560%
		German asphalt pavement		99.170%
[40]	ViT	Malicious drone	Malicious drones	98.300%
[46]	ViT	Real X Rays	Femur fracture	83.000%
[47]	SeedViT	Maize seeds	Maize seed quality	96.700%
[48]	DOViT	GAOs-1	Air quality	90.320%
		GAOs-2		92.780%
[49]	MITformer	UCM	Remote sensing scene	99.830%
		AID		97.960%
		NWPU		95.930%

[a] https://github.com/IBM/CrossViT (accessed on 19 April 2023); [b] https://github.com/greentreeys/MIL-VT (accessed on 19 April 2023); [c] https://github.com/facebookresearch/LeViT (accessed on 19 April 2023).

***Key Takeaways**—Transformers in image classification show better performance than do CNNs and use an attention mechanism [50] instead of convolution. However, the major drawback of using transformers in image classification is the requirement of a huge dataset for training [51]. thus, ViT is the best choice for image classification in cases where a huge dataset is easily available.*

3.2. ViTs for Object Detection

The effort to tame pretrained vanilla ViT for object detection has never stopped since the evolution of transformer [15] to CV [16]. Beal et al. [52] were the first to use a faster region-based convolutional neural network (R-CNN) detector with a supervised pretrained ViT for object detection. You only look at one sequence (YOLOS) [53] suggests simply using a pretrained ViT encoder to conduct object detection in a pure sequence-to-sequence manner. Li et al. [54] were the first to complete a large-scale study of vanilla ViT on object detection using sophisticated masked image modeling (MIM) pretrained representations [55,56], confirming vanilla ViT's promising potential and capacity in object-level recognition.

In [57], the authors proposed the unsupervised learning-based technique using ViT for the detection of the manipulation in the satellite images. They used two different datasets for the evaluation of the framework. The ViT model with postprocessing (ViT-PP) achieved an F1-score of 0.354 and a Jaccord index (JI) of 0.275 for dataset 2. The F1-score and JI can be calculated by Equations (1) and (2) respectively.

$$F_1 = \frac{2 * T_P}{2 * T_P + F_P + F_N},\tag{1}$$

$$JI = \frac{T_P}{T_P + F_P + F_N},\tag{2}$$

where T_P, F_P, and F_N denote true positive, false positive, and false negative, respectively.

In [58], the authors proposed a bridged transformer (BrT) for the 3D object detection. The model was applied for the vision and point cloud 3D object detection. They used the ScanNet-V2 [59] and SUN RGB-D [60] datasets to validate their model. The model demonstrated the mean average precision (mAP)@0.5 of 52.8 for the ScanNet-V2 dataset and 55.2 for the SUN RGB-D dataset.

Similarly, in [61], the authors proposed a transformer-based framework for the detection of 3D objects using point cloud data. They used the ScanNet-V2 [59] and SUN RGB-D [60] datasets to validate their model. The model demonstrated a mean average precision (mAP)@0.5 of 52.8 for the ScanNet-V2 dataset and 45.2 for the SUN RGB-D dataset.

Table 3 shows the application of ViT for object detection.

Table 3. ViT for Object Detection.

Research	Model	Dataset	Objective	Perf. Metric	Value
[53]	YOLOS [a]	COCO	Object detection	AP^{box}	42.000
[57]	ViT	Satellite images	Manipulation detection	F1-score	0.354
				JI	0.275
[58]	BrT	ScanNet-V2	3D object detection	mAP@0.5	55.200
		SUN RGB-D			48.100
[61]	ViT [b]	ScanNet-V2	3D object detection	mAP@0.5	52.800
		SUN RGB-D			45.200

[a] https://github.com/hustvl/YOLOS (accessed on 19 April 2023); [b] https://github.com/zeliu98/Group-Free-3D (accessed on 19 April 2023).

***Key Takeaways**—Transformers are used for object detection in three different ways: (a) Feature extraction with transformers and detection with R-CNN as in [62], (b) feature extraction with CNN and detection head based on transformers as in [63,64], and (c) a complete end-to-end pure transformer-based object detection as in [53]. The third method is more feasible and requires more effort to create more end-to-end object detection models using ViTs.*

3.3. ViTs for Image Segmentation

Image segmentation can also be performed using transformers. A combination of ViT and U-Net was used in [65] to segment medical images. The authors replaced the encoder part of the classical U-Net with a transformer. A multi-atlas abdomen-labeling challenge dataset from MICCAI 2015 was used. By using images with a resolution of

224, the TransUNet achieved an average dice score of 77.48%, and while using images of resolution 512, it achieved an average dice score of 84.36%.

In [66], the authors proposed a "ViT for biomedical image segmentation (ViTBIS)" for medical image segmentation. Transformers were used for both encoders and decoders in their transformer-based model. In addition, the MICCAI 2015 multi-atlas abdomen-labeling challenge dataset and the Brain Tumor Segmentation (BraTS 2019) challenge dataset were used. The evaluation metric used was dice score and Hausdorff distance (HD) [67]. According to the MICCAI 2015 dataset, the average dice scores were 80.45%, and the average HDs were 21.24%.

In [68], the authors proposed a novel "language-aware ViT (LAVT)" for image segmentation. They used four different datasets for the evaluation of the model. The datasets were RefCOCO [69], RefCOCO+ [69], G-Ref (UMD partition) [70], and G-Ref (Google partition) [71]. They used intersection over union (IoU) as the performance metric. The value of IoU for the RefCOCO dataset was 72.73%, and for RefCOCO+, the IoU was 62.14%. Similarly, for G-Ref (UMD partition), the IoU was 61.24%, and for G-Ref (Google partition), the IoU was 60.50%. Similarly, another work [72] proposed high-resolution ViT for semantic segmentation. The authors of this work used several branch block co-optimization techniques and achieved good results for the semantic segmentation on the ADE20K and Cityscapes datasets.

Cheng et al. in [73] proposed MaskFormer for image segmentation. This model outperformed state-of-the-art semantic [19,74–78] and panoptic [79] segmentation models.

Hatamizadeh et al. in [80] proposed UNetFormer for medical image segmentation. The model contained a transformer-based encoder, decoder, and bottleneck part. They used the medical segmentation decathlon (MSD) [81] and BraTS 2021 [82] dataset to test UNetFormer. They evaluated dice scores and HD. The dice score using the MSD dataset was 96.03% for the liver and 59.16% for the tumor, whereas the value of HD was 7.21% for the liver and 8.49% for the tumor. Moreover, the average dice score on the BraTS 2021 dataset was 91.54%.

Table 4 shows the application of ViT for image segmentation.

***Key Takeaways**—Transformers use self-attention [31,83,84] for image segmentation. With the help of self-attention, the transformers make rich interactions between pixels. However, the transformers show remarkable performance for image segmentation but require a huge dataset for training.*

3.4. ViTs for Image Compression

In recent years, learning-based image compression has been a major focus of research. For lossy image compression based on learning, different CNN-based architectures have proven effective [85,86]. As ViTs evolved, learning-based image compression has also been performed using transformer-based models. In [87], the authors modified the entropy module of the Ballé 2018 mode [88] with the ViT. Due to the fact that the entropy module used a transformer, this model was called Entroformer. Entroformer effectively captured long-range dependencies in probability distribution estimation. On the Kodak dataset, they demonstrated the performance of the Entroformer. When the model was optimized for the mean squared error (MSE) loss function, the average peak signal-to-noise ratio (PSNR) and multiscale structural similarity (MS-SSIM) were 27.63 dB and 0.90132, respectively. Similarly, in [89], the authors proposed Contextformer and achieved 11% bit savings over Versatile Video Coding (VVC) [90].

***Key Takeaways**—End-to-end transformer-based image compression models outperform other learning-driven image compression models and produce the reconstructed image with better visual quality [91]. However, these models possess high complexity and massive memory utilization [87].*

Table 4. ViT for Image Segmentation.

Research	Model	Dataset	Objective	Performance Metric	Value
[65]	TransUNet [a]	MICCAI 2015	Medical image segmentation	Dice score	77.480%
[66]	ViTBIS	MICCAI 2015	Medical image segmentation	Dice score	80.450%
				HD	21.240%
[68]	LAVT	RefCOCO	Image segmentation	IoU	72.730%
		RefCOCO+			62.140%
		G-Ref (UMD partition)			61.240%
		G-Ref (Google partition)			60.500%
[80]	UNetFormer [b]	MSD	Liver segmentation	Dice score	96.030%
				HD	7.210%
			Tumor segmentation	Dice score	59.160%
				HD	8.490%
		BraTS 2021	Brain tumor segmentation	Dice score	91.540%

[a] https://github.com/Beckschen/TransUNet (accessed on 19 April 2023); [b] https://github.com/Project-MONAI/research-contributions (accessed on 19 April 2023).

3.5. ViTs for Image Super-Resolution

CNN has been used to perform image super-resolution. With ViT's superiority over CNN, image super-resolution can also be achieved by transformers. Spatio-temporal ViT, a transformer-based model for the super-resolution of microscopic images, was developed in [92]. Additionally, the model addressed the problem of video super-resolution. To test the model's performance, the authors used a video dataset. PSNR was calculated for static and dynamic videos. Static, medium, fast, and extreme motions were considered. The PSNR for static was 34.74 dB, whereas the PSNR for medium, fast, and extreme was 30.15 dB, 26.04 dB, and 22.95 dB, respectively.

3.6. ViTs for Image Denoising

Denoising images has also been a challenging problem for researchers. In spite of this, ViT has found a solution. A transformer was used to denoise CT images in [93]. In this work, the authors proposed a model called TED-Net for low-dose CT denoising. The authors used a transformer for both the encoder and decoder parts. Using the AAPM-Mayo clinic LDCT Grand Challenge dataset, they obtained a structural similarity (SSIM) of 0.9144 and a root mean square error (RMSE) of 8.7681.

Luthra et al. in [94] proposed Eformer for medical image denoising. Eformer was based on edge enhancement and incorporated the Sobel operator. They evaluated the model on AAPM-Mayo Clinic Low-Dose CT Grand Challenge Dataset [95]. Eformer achieved a PSNR of 43.487, an RMSE of 0.0067, and an SSIM of 0.9861.

In [96], the authors combined UNet [97] and Swin transformer [19] to propose SUNet for image denoising. The model was evaluated on CBSD68 [98] and Kodak24 [99] dataset. The model achieved a PSNR of 27.85 and SSIM of 0.799 for the CBSD68 dataset when the noise level (σ) was 50. Similarly, the model achieved a PSNR of 29.54 and SSIM of 0.810 for the Kodak24 dataset when the σ was kept at 50.

In [100], the authors proposed DenSformer for image denoising. The DenSformer was composed of three modules, preprocessing, feature extraction, and reconstruction. The model achieved a PSNR of 39.68 and an SSIM of 0.958 on the SIDD dataset [101]. Similarly, the model yielded a PSNR of 39.87 and an SSIM of 0.955 on the Dnd dataset [102].

Xu et al. in [103] proposed the CUR transformer for image denoising. The CUR transformer was deduced from the convolutional unbiased regional transformer. Similarly, in [104–106], the combined transformers and CNN for image denoising and achieved better performance. In [107], the authors proposed Hider, a transformer-based model for image denoising. The model was designed for hyperspectral images. The authors in [108] proposed CSformer for image denoising. The model was based on cross-scale feature fusion.

The model was evaluated on Set12 [109], BSD68 [98] and General100 [110] dataset. The model outperformed DnCNN [109], FDnCNN [109], FFDNet [111], IRCNN [112], DRUnet [113], Uformer [114], and SwinIR [115] in terms of PSNR and SSIM.

***Key Takeaways**—Transformer models can be used for image denoising either in standalone fashion, i.e., end-to-end, or hybrid, i.e., combined with CNN. Standalone models have relatively better performance than do hybrid models.*

3.7. ViTs for Anomaly Detection

Additionally, ViT is used for anomaly detection. A novel ViT network for image anomaly detection and localization (VT-ADL) was developed in [116]. In this study, the authors used a real-world dataset called BTAD. The model was also tested on two publicly available datasets, MNIST and MVTec [117]. For all three datasets, they calculated the model's per region overlap (PRO) score. A mean PRO score of 0.984 was obtained for the MNIST dataset, 0.807 for the MVTec dataset, and 0.89 for the BTAD dataset.

Similarly, in [118], the authors proposed AnoViT for the detection and localization of anomalies. The MNIST, CIFAR, and MVTecAD datasets were used by the authors. Based on the MINST, CIFAR, and MVTecAD datasets, the mean area under the region operating characteristics (AUROC) curve was 92.4, 60.1, and 78, respectively.

Yuan et al. in [119] proposed TransAnomaly, a video ViT and U-Net-based framework for the detection of the anomalies in the videos. They used three datasets, Pred1, Pred2, and Avenue. The calculated area under the curve (AUC) for three datasets, achieving 84.0%, 96.10%, and 85.80%, respectively, without using the sliding windows method (swm). With the swm, the model yielded an AUC of 86.70%, 96.40%, and 87.00% for the Pred1, Pred2, and Avenue datasets, respectively.

Table 5 shows the summary of the ViT for anomaly detection.

Table 5. ViT for Anomaly Detection.

Research	Model	Dataset	Objective	Perf. Metric	Value
[116]	VT-ADL [a]	MNIST	Anomaly detection	PRO	0.984
		MVTec			0.807
		BTAD			0.890
[118]	AnoViT [b]	MNIST	Anomaly detection	AUROC	92.400%
		CIFAR			60.100%
		MVTec			78.000%
[119]	TransAnomaly	Pred1	Anomaly detection	AUC	84.000%
		Pred2			96.100%
		Avenue			85.800%
	TransAnomaly	Pred1			86.700%
		Pred2			96.400%
		Avenue			87.000%

[a] https://github.com/pankajmishra000/VT-ADL (accessed on 19 April 2023); [b] https://github.com/tkdleksms/LG_ES_anomaly (accessed on 19 April 2023).

4. Advanced ViTs

In addition to their promising use in vision, some transformers have been particularly designed to perform a specific task or to solve a particular problem. In this section, we analyze the different advanced transformer models by categorizing them into the following categories:

- Task-based ViTs;
- Problem-based ViTs.

4.1. Task-Based ViTs

In this subsection, we summarize task-based ViTs. Task-based ViTs are those ViTs that are designed for a specific task and perform exceptionally well for that task. Lee et al. in [120] proposed the multipath ViT (MPViT) for dense prediction by embedding features of the same sequence length with the patches of the different scales. The model achieved superior performance for classification, object detection, and segmentation. However, the model is specific to dense prediction.

In [121], the authors proposed the coarse-to-fine ViT (C2FViT) for medical image registration. C2FViT uses convolutional ViT [24,122] an ad multiresolution strategy [123] to learn global affine for image registration. The model was specifically designed for affine medical image registration. Similarly, in [124], the authors proposed TransMorph for medical image registration and achieved state-of-the-art results. However, these models are task-specific, which is why they are categorized as task-based ViTs here.

4.2. Problem-Based ViTs

In this subsection, we present problem-based ViTs. Problem-based ViTs are those ViTs which are proposed to solve a particular problem that cannot be solved by pure ViTs. These types of ViTs are not dependent on tasks but rather on problems. For example, ViTs are not flexible. To make a ViT more flexible and to reduce its complexity, the authors in [125] proposed a messenger (MSG) transformer. They used specialized MSG tokens for each region. By manipulating these tokens, one can flexibly exchange visual information across the regions. This reduces the computational complexity of ViTs.

Similarly, it has been discovered that mixup-based augmentation works well for generalizing models during training, especially for ViTs because they are prone to overfitting. However, the basic presumption of earlier mixup-based approaches is that the linearly interpolated ratio of targets should be maintained constantly with the percentage suggested by input interpolation. As a result, there may occasionally be no valid object in the mixed image due to the random augmentation procedure, but there is still a response in the label space. Chen et al. in [126] proposed TransMix for bridging this gap between the input and label spaces. TransMix blends labels based on the attention maps of ViTs.

In ViTs, global attention is computationally expensive, whereas local attention provides limited interactions between tokens. To solve this problem, the authors in [127] proposed the CSWin transformer based on the cross-shaped window self-attention. This provided efficient computation of self-attention and achieved better results than did the pure ViTs.

5. Open Source ViTs

This section summarizes the available open-source ViTs with potential CV applications. We also provide the links to the source codes of each model discussed. Table 6 presents the comprehensive summary of the open-source ViTs for the different applications of CV such as image classification, object detection, instance segmentation, semantic segmentation, video action classification, and robustness evaluation.

Table 6. Summary of the open-source ViTs present in the literature for different applications of CV.

Research	Year	Model Name	CV Application	Source Code
[1]	2021	PiT [α]	• Img. class. [a] • Object det. [b] • Rob. eval. [e]	https://github.com/naver-ai/pit (accessed on 19 April 2023)
[16]	2020	ViT [*]	• Img. class. [a]	https://github.com/google-research/vision_transformer (accessed on 19 April 2023)
[19]	2021	Swin Transformer	• Img. class. [a] • Object det. [b] • Semantic seg. [c]	https://github.com/microsoft/Swin-Transformer (accessed on 19 April 2023)
[32]	2021	Cross-ViT	• Img. class. [a] • Object det. [b]	https://github.com/IBM/CrossViT (accessed on 19 April 2023)
[122]	2021	CeiT [γ]	• Img. class. [a]	https://github.com/rishikksh20/CeiT-pytorch (accessed on 19 April 2023)
[128]	2022	Swin Transformer V2	• Img. class. [a] • Object det. [b] • Semantic seg. [c] • Vid. act. class. [d]	https://github.com/microsoft/Swin-Transformer (accessed on 19 April 2023)
[129]	2021	DVT [†]	• Img. class. [a]	https://github.com/blackfeather-wang/Dynamic-Vision-Transformer (accessed on 19 April 2023)
[130]	2021	PVT [††]	• Object det. [b] • Instance seg. [c] • Semantic seg. [c]	https://github.com/whai362/PVT (accessed on 19 April 2023)
[131]	2021	Twins	• Img. class. [a] • Dense det. [b] • Seg. [c]	https://github.com/Meituan-AutoML/Twins (accessed on 19 April 2023)
[132]	2021	Mobile-ViT	• Object det. [b]	https://github.com/apple/ml-cvnets (accessed on 19 April 2023)
[133]	2021	Refiner	• Img. class. [a]	https://github.com/zhoudaquan/Refiner_ViT (accessed on 19 April 2023)
[134]	2021	DeepViT [†††]	• Img. class. [b]	https://github.com/zhoudaquan/dvit_repo (accessed on 19 April 2023)
[135]	2021	DeiT [††††]	• Img. class. [a]	https://github.com/facebookresearch/deit (accessed on 19 April 2023)
[136]	2021	Visformer	• Img. class. [a]	https://github.com/danczs/Visformer (accessed on 19 April 2023)

[α] Pooling-based Vision Transformer, [*] Vision transformer, [†] Dynamic vision transformer, [††] Pyramid vision transformer, [†††] Deeper vision transformer, [††††] Data-efficient image Transformer, [γ] Convolution-enhanced image Transformer; [a] Image classification, [b] Detection, [c] Segmentation, [d] Video action classification, [e] Robustness evaluation.

6. ViTs and Drone Imagery

In drone imagery, unmanned aerial vehicles (UAVs) or drones capture images or videos. Images of this type can provide a birds-eye view of a particular area, which can be useful for various applications, such as land surveys, disaster management, agricultural planning, and urban development.

Initially, DL models such as CNNs [137], recurrent neural networks (RNNs) [138], fully convolutional networks (FCNs) [139], and generative adversarial networks (GANs) [140] were widely used for tasks in which drone image processing was involved. CNNs are commonly used for image classification and object detection using drone images. These are particularly useful, as these models can learn to detect features such as buildings, roads, and other objects of interest. Similarly, RNNs are commonly used for processing time-series data, such as drone imagery. These models are able to learn to detect changes in the landscape over time. These are useful for tasks such as crop monitoring and environmental monitoring.

FCNs are mainly used for semantic segmentation tasks, such as identifying different types of vegetation in drone imagery. These can be used to create high-resolution maps of the landscape, which can be useful for various applications.

GANs are commonly used for image synthesis tasks, such as generating high-resolution images of the landscape from low-resolution drone imagery. These can also be used for data augmentation, which can help to improve the performance of other DL models.

When it comes to drone imagery, ViTs can be used for a variety of tasks because of the advantages of ViTs over traditional DL models [26–28,31]. ViTs use a self-attention mechanism that allows these models to focus on relevant parts of the input data [21]. This is particularly useful when processing drone imagery, which may contain a lot of irrelevant information, such as clouds or trees, that can distract traditional DL models. By selectively attending to relevant parts of the image, transformers can improve their accuracy. Similarly, traditional DL models typically process data in a sequential manner, which is slow and inefficient, especially when dealing with large amounts of data. ViTs, on the other hand, can process data in parallel, making them much faster and more efficient. Another advantage of ViTs over traditional DL models is efficient transfer learning ability [141], as ViTs are pretrained on large amounts of data, allowing them to learn general features that can be applied to a wide range of tasks. This means that they can be easily fine-tuned for specific tasks, such as processing drone imagery, with relatively little training data. Moreover, one of the most important advantages is the ability to handle variable-length input as drone imagery can vary in size and shape, making it difficult for traditional DL models to process. ViTs, on the other hand, can handle variable-length input, making them better suited for processing drone imagery.

Similarly, another main advantage of using ViTs for drone imagery analysis is their ability to handle long sequences of inputs. This is particularly useful for drone imagery, where large images or video frames must be processed. Additionally, ViTs can learn complex spatial relationships between different image features, leading to more accurate results than those produced with other DL models.

ViTs are used for object detection, disease detection, prediction, classification, and segmentation using drone imagery. This section briefly summarizes the applications of ViT using drone imagery.

In [142], the authors proposed TransVisDrone, which is a spatio-temporal transformer for the detection of drones in aerial videos. The model obtained state-of-the-art performance on the NPS [143], FLDrones [144], and Airborne Object Tracking (AOT) datasets.

Liu et al. [145] reported the use of ViT for drone crowd counting. The dataset used in the challenge was collected by drones.

In [146], the authors used unmanned aerial vehicle (UAV) images of date palm trees to investigate the reliability and efficiency of various deep-ViTs. They used different ViTs such as Segformer [147], the Segmeter [148], the UperNet-Swin transformer, and dense prediction transformers (DPT) [149]. Based on the comprehensive experimental analysis, Segformer achieved the highest performance.

Zhu et al. [150], proposed TPH-YOLOv5 for which they replaced the original prediction head of YOLOv5 with the transformer prediction head (TPH) to overcome the challenges of objection in the drone-captured images.

In [151], the authors summarized the results of the challenger VisDrone-DET2021 in which the proponents used different transformers, such as Scaled-YOLOv4 with transformer and BiFPN (SOLOER), Swin-transformer (Swin-T), stronger visual information for tiny object detection (VistrongerDet), and EfficientDet for object detection in the drone imagery. Thai et al. [152] demonstrated the use of ViT for cassava leaf disease classification and achieved better performance than did the CNNs. A detailed summary of the existing ViTs for drone imagery data is presented in Table 7.

Table 7. ViTs for drone imagery.

Ref.	Model	Dataset	Objective	Perf. Metric	Value
[142]	TransVisDrone [†]	• NPS • FLDrones • AOT	Drone detection	AP@0.5IoU	• 0.95 • 0.75 • 0.80
[146]	• Segformer • Segmenter • UperNet-Swin • DPT	Date palm trees	Segmentation	mIoU [α]	• ≈86.2% • ≈85.3% • ≈85.8% • ≈85.4%
[146]	• Segformer • Segmenter • UperNet-Swin • DPT	Date palm trees	Segmentation	mF-Score [β]	• ≈92.5% • ≈91.9% • ≈92.3% • ≈92.0%
[146]	• Segformer • Segmenter • UperNet-Swin • DPT	Date palm trees	Segmentation	mAcc [γ]	• ≈92.8% • ≈92.0% • ≈92.4% • ≈91.7%
[150]	TPH-YOLOv5 [††]	VisDrone2021	Object detection	mAP [δ]	• 39.2%
[151]	• DBNet • SOLOer • Swin-T • TPH-YOLOv5 • VistrongerDet • cascade [††] • DNEFS • EfficientDet • DPNet-ensemble • DroneEye2020 • Cascade R-CNN	VisDrone-DET2021	Object detection	AP	• 39.43% • 39.42% • 39.40% • 39.18% • 38.77% • 38.72% • 38.53% • 38.51% • 37.37% • 34.57% • 16.09%

[†] https://github.com/tusharsangam/TransVisDrone (accessed on 19 April 2023), [††] https://github.com/cv516Buaa/tph-yolov5 (accessed on 19 April 2023); [α] mean intersection over union, [β] mean F-Score, [γ] mean accuracy, [δ] mean average precision. red color text: Worst performing model, green color text: best-performing model.

7. Open Research Challenges and Future Directions

Despite showing promising results for different image coding and CV tasks, in addition to high computational costs, large training datasets, neural architecture search, interpretability of transformers, and efficient hardware designs, ViTs implementation still faces challenges. The purpose of this section is to explain the challenges and future directions of ViTs.

7.1. High Computational Cost

There are millions of parameters in ViT-based models. Computers with high computational power are needed to train these models. Due to their high cost, these high-performance computers increase the computational cost of ViTs. In comparison to CNN, ViT performs better; however, its computational cost is much higher. One of the biggest challenges researchers face is reducing the computational cost of ViTs. In [153], the authors proposed the glance-and-gaze ViT to reduce the memory consumption and computational cost of the ViT. However, this critical issue needs more attention and research to make ViTs more effective in terms of computational cost.

7.2. Large Training Dataset

The training of ViTs requires a large amount of data. With a small training dataset, ViTs perform poorly [154]. A ViT trained with the ImageNet1K dataset was found to perform worse than did ResNet, but a ViT trained with ImageNet21K performed better than did ResNet [16]. However, in [51], the authors trained ViT on a small dataset, but the model could not be generalized. Similarly, other authors [155] demonstrated the method to train ViT on 2040 images. Chen et al. in [136] proposed Visformer to reduce the model's complexity and train ViTs on a small dataset. Despite these efforts from researchers, training a ViT with a small dataset to achieve remarkable performance is still challenging.

7.3. Neural Architecture Search (NAS)

There has been a great deal of exploration of NAS for CNNs [156–159]. In contrast, NAS has not yet been fully explored for ViTs. In [160], the authors surveyed several NAS techniques for ViTs. To the best of our knowledge, there are limited studies on the NAS exploration in ViTs [161–166], and more attention is needed in the future. The NAS exploration for ViTs may a new direction for young investigators in the future.

7.4. Interpretability of the Transformers

It is difficult to visualize the relative contribution of input tokens to the final predictions with ViTs since the attention that originates in each layer is intermixed in succeeding layers. In [167,168], the authors demonstrated the interpretability of the transformers to some extent. However, the problem remains unresolved.

7.5. Hardware-Efficient Designs

Power and processing requirements can make large-scale ViTs networks unsuitable for edge devices and resource-constrained contexts such as the Internet of Things (IoT). In [169], the authors proposed a framework for low-bit ViTs. However, the issue is still unresolved.

7.6. Large CV Model

With the advancement of technologies such as green communication [170], digital twins [171], and usage of ViTs, researchers have started to focus on large CV models and large language models [172] with billions of parameters. A large model can be used as a basis for transfer learning and fine-tuning, so there is interest in developing increasingly high-performance models. Google ViT-22B [173] paves the way for new large transformers and to revolutionizing CV. It consists of 22 billion parameters and is trained on four billion images. It can be used for image classification, video classification, semantic segmentation, and depth estimation. Inspired by Google ViT-22B, the authors in [174] proposed the EVA-02 model with 304 M parameters. Despite the remarkable performance of large models, retraining these models on new datasets is a daunting task, but one approach to solving this problem is the usage of a pretrained model. The researchers in [175] demonstrated that continual learning (CL) can help pretrained vision-language models efficiently adapt to new or undertrained data distributions without retraining. Although they achieved good performance, this problem still needs to be explored in future research.

8. Conclusions

It is becoming more common to use ViTs for image coding and CV instead of CNNs. The use of ViTs for classification, detection, segmentation, compression, and image super-resolution has risen dramatically since the introduction of the classical ViT for image classification. This survey presented the existing surveys on ViTs in the literature. This survey highlighted the applications of different variants of ViTs in CV and further examined the use of ViTs for image classification, object detection, image segmentation, image compression, image super-resolution, image denoising, and anomaly detection. We also presented the lessons learned in each category. From the detailed analysis, we observed that ViTs are replacing traditional DL models for CV applications. Significant achievement has been made in image classification and object detection, where ViTs are widely used because of the self-attention mechanism and effective transfer learning. Additionally, we discussed the open research challenges faced by researchers during the implementation of ViTs, which include the high computational costs, large training datasets, interpretability of transformers, and hardware efficiency. By providing future directions, we offer young researchers a new perspective. Recently, large CV models have attracted the focus of the research community as Google's ViT22B has paved the way for future research in this direction.

Author Contributions: Conceptualization, S.J. and M.J.P.; methodology, S.J.; formal analysis, S.J.; investigation, M.J.P.; resources, O.-J.K.; data curation, S.J.; writing—original draft preparation, S.J.; writing—review and editing, M.J.P. and O.-J.K.; visualization, S.J.; supervision, M.J.P.; project administration, O.-J.K.; funding acquisition, O.-J.K. All authors have read and agreed to the published version of the manuscript.

Funding: This research received no external funding.

Data Availability Statement: Not applicable.

Conflicts of Interest: The authors declare no conflicts of interest.

Abbreviations

The following abbreviations are used in this manuscript:

AID	Aerial image dataset
AP	Average precision
AQI	Air quality index
AUC	Area under the curve
AUROC	Area under receiver operating characteristic curve
AP^{box}	Box average precision
BERT	Bidirectional encoder representations from transformers
bpp	Bits per pixel
BrT	Bridged transformer
BTAD	BeanTech anomaly detection
CIFAR	Canadian Institute for Advanced Research
CV	Computer vision
CNN	Convolutional neural network
DHViT	Deep hierarchical ViT
DOViT	Double output ViT
ES-GSNet	Excellent teacher guiding small networks
GAOs-1	Get AQI in one shot-1
GAOs-2	Get AQI in one shot-2
GPT-3	Generative pretrained transformer 3
HD	Hausdorff distance
IoU	Intersection over union
JI	Jaccord index
LiDAR	Light detection and ranging
mAP	Mean average precision
MLP	Multilayer perceptron
MIL-ViT	Multiple instance-enhanced ViT
MIM	Masked image modeling
MITformer	Multi-instance ViT
MSE	Mean squared error
MS-SSIM	Multiscale structural similarity
NLP	Natural language processing
NWPU	Northwestern Polytechnical University
PSNR	Peak signal to noise ratio
PRO	Per region overlap
PUAS	Planet Understanding the Amazon from Space
RFMiD2020	2020 retinal fundus multidisease image dataset
R-CNN	Region-based convolutional neural network
RMSE	Root mean square error
SSIM	Structural similarity
TE	Transformer encoder

UCM	UC-Mered land use dataset
ViTs	Vision transformers
VT-ADL	ViT network for image anomaly detection and localization
ViT-PP	ViT with postprocessing
YOLOS	You only look at one sequence

References

1. Heo, B.; Yun, S.; Han, D.; Chun, S.; Choe, J.; Oh, S.J. Rethinking spatial dimensions of vision transformers. In *Proceedings of the IEEE/CVF International Conference on Computer Vision*; IEEE: New York, NY, USA, 2021; pp. 11936–11945.
2. Tenney, I.; Das, D.; Pavlick, E. BERT rediscovers the classical NLP pipeline. *arXiv* **2019**, arXiv:1905.05950.
3. Floridi, L.; Chiriatti, M. GPT-3: Its nature, scope, limits, and consequences. *Minds Mach.* **2020**, *30*, 681–694. [CrossRef]
4. Krizhevsky, A.; Sutskever, I.; Hinton, G.E. Imagenet classification with deep convolutional neural networks. *Commun. ACM* **2017**, *60*, 84–90. [CrossRef]
5. Jamil, S.; Rahman, M.; Ullah, A.; Badnava, S.; Forsat, M.; Mirjavadi, S.S. Malicious UAV detection using integrated audio and visual features for public safety applications. *Sensors* **2020**, *20*, 3923. [CrossRef]
6. Wu, Z.; Shen, C.; Van Den Hengel, A. Wider or deeper: Revisiting the resnet model for visual recognition. *Pattern Recognit.* **2019**, *90*, 119–133. [CrossRef]
7. Hammad, I.; El-Sankary, K. Impact of approximate multipliers on VGG deep learning network. *IEEE Access* **2018**, *6*, 60438–60444. [CrossRef]
8. Yao, X.; Wang, X.; Karaca, Y.; Xie, J.; Wang, S. Glomerulus classification via an improved GoogLeNet. *IEEE Access* **2020**, *8*, 176916–176923. [CrossRef]
9. Chollet, F. Xception: Deep learning with depthwise separable convolutions. In Proceedings of the IEEE Conference on Computer Vision and Pattern Recognition, Honolulu, HI, USA, 21–26 July 2017; pp. 1251–1258.
10. Wang, C.; Chen, D.; Hao, L.; Liu, X.; Zeng, Y.; Chen, J.; Zhang, G. Pulmonary image classification based on inception-v3 transfer learning model. *IEEE Access* **2019**, *7*, 146533–146541. [CrossRef]
11. Jamil, S.; Fawad; Abbas, M.S.; Habib, F.; Umair, M.; Khan, M.J. Deep learning and computer vision-based a novel framework for himalayan bear, marco polo sheep and snow leopard detection. In Proceedings of the 2020 International Conference on Information Science and Communication Technology (ICISCT), Karachi, Pakistan, 8–9 February 2020; pp. 1–6.
12. Zhang, K.; Guo, Y.; Wang, X.; Yuan, J.; Ding, Q. Multiple feature reweight densenet for image classification. *IEEE Access* **2019**, *7*, 9872–9880. [CrossRef]
13. Wang, J.; Yang, L.; Huo, Z.; He, W.; Luo, J. Multi-label classification of fundus images with efficientnet. *IEEE Access* **2020**, *8*, 212499–212508. [CrossRef]
14. Hossain, M.A.; Nguyen, V.; Huh, E.N. The trade-off between accuracy and the complexity of real-time background subtraction. *IET Image Process.* **2021**, *15*, 350–368. [CrossRef]
15. Vaswani, A.; Shazeer, N.; Parmar, N.; Uszkoreit, J.; Jones, L.; Gomez, A.N.; Kaiser, Ł.; Polosukhin, I. Attention is all you need. *Adv. Neural Inf. Process. Syst.* **2017**, *30*. Available online: https://proceedings.neurips.cc/paper_files/paper/2017/file/3f5ee243 547dee91fbd053c1c4a845aa-Paper.pdf (accessed on 18 April 2023)
16. Dosovitskiy, A.; Beyer, L.; Kolesnikov, A.; Weissenborn, D.; Zhai, X.; Unterthiner, T.; Dehghani, M.; Minderer, M.; Heigold, G.; Gelly, S.; et al. An image is worth 16x16 words: Transformers for image recognition at scale. *arXiv* **2020**, arXiv:2010.11929.
17. Wu, B.; Xu, C.; Dai, X.; Wan, A.; Zhang, P.; Yan, Z.; Tomizuka, M.; Gonzalez, J.; Keutzer, K.; Vajda, P. Visual transformers: Token-based image representation and processing for computer vision. *arXiv* **2020**, arXiv:2006.03677.
18. Xiao, T.; Singh, M.; Mintun, E.; Darrell, T.; Dollár, P.; Girshick, R. Early convolutions help transformers see better. *Adv. Neural Inf. Process. Syst.* **2021**, *34*, 30392–30400.
19. Liu, Z.; Lin, Y.; Cao, Y.; Hu, H.; Wei, Y.; Zhang, Z.; Lin, S.; Guo, B. Swin transformer: Hierarchical vision transformer using shifted windows. In *Proceedings of the IEEE/CVF International Conference on Computer Vision*; IEEE: New York, NY, USA, 2021; pp. 10012–10022.
20. Bertasius, G.; Wang, H.; Torresani, L. Is space-time attention all you need for video understanding? In *ICML*; PMLR; 2021; Volume 2, p. 4. Available online: https://proceedings.mlr.press/v139/bertasius21a.html (accessed on 18 April 2023).
21. Raghu, M.; Unterthiner, T.; Kornblith, S.; Zhang, C.; Dosovitskiy, A. Do vision transformers see like convolutional neural networks? *Adv. Neural Inf. Process. Syst.* **2021**, *34*, 12116–12128.
22. Naseer, M.M.; Ranasinghe, K.; Khan, S.H.; Hayat, M.; Shahbaz Khan, F.; Yang, M.H. Intriguing properties of vision transformers. *Adv. Neural Inf. Process. Syst.* **2021**, *34*, 23296–23308.
23. Dai, Z.; Liu, H.; Le, Q.V.; Tan, M. Coatnet: Marrying convolution and attention for all data sizes. *Adv. Neural Inf. Process. Syst.* **2021**, *34*, 3965–3977.
24. Wu, H.; Xiao, B.; Codella, N.; Liu, M.; Dai, X.; Yuan, L.; Zhang, L. Cvt: Introducing convolutions to vision transformers. In *Proceedings of the IEEE/CVF International Conference on Computer Vision*; IEEE: New York, NY, USA, 2021; pp. 22–31.
25. Coccomini, D.A.; Messina, N.; Gennaro, C.; Falchi, F. Combining efficientnet and vision transformers for video deepfake detection. In *International Conference on Image Analysis and Processing*; Springer: Cham, Switzerland, 2022; pp. 219–229.

26. Tay, Y.; Dehghani, M.; Bahri, D.; Metzler, D. Efficient transformers: A survey. *ACM Comput. Surv. (CSUR)* **2020**, *55*, 1–28. [CrossRef]

27. Khan, S.; Naseer, M.; Hayat, M.; Zamir, S.W.; Khan, F.S.; Shah, M. Transformers in vision: A survey. *ACM Comput. Surv. (CSUR)* **2022**, *54*, 1–41. [CrossRef]

28. Liu, Y.; Zhang, Y.; Wang, Y.; Hou, F.; Yuan, J.; Tian, J.; Zhang, Y.; Shi, Z.; Fan, J.; He, Z. A survey of visual transformers. *arXiv* **2021**, arXiv:2111.06091.

29. Lin, T.; Wang, Y.; Liu, X.; Qiu, X. A survey of transformers. *AI Open* **2022**, *3*, 111–132. [CrossRef]

30. Xu, Y.; Wei, H.; Lin, M.; Deng, Y.; Sheng, K.; Zhang, M.; Tang, F.; Dong, W.; Huang, F.; Xu, C. Transformers in computational visual media: A survey. *Comput. Vis. Media* **2022**, *8*, 33–62. [CrossRef]

31. Han, K.; Wang, Y.; Chen, H.; Chen, X.; Guo, J.; Liu, Z.; Tang, Y.; Xiao, A.; Xu, C.; Xu, Y.; et al. A survey on vision transformer. *IEEE Trans. Pattern Anal. Mach. Intell.* **2022**, *45*, 87–110. [CrossRef] [PubMed]

32. Chen, C.F.R.; Fan, Q.; Panda, R. Crossvit: Cross-attention multi-scale vision transformer for image classification. In *Proceedings of the IEEE/CVF International Conference on Computer Vision*; IEEE: New York, NY, USA, 2021; pp. 357–366.

33. Deng, P.; Xu, K.; Huang, H. When CNNs meet vision transformer: A joint framework for remote sensing scene classification. *IEEE Geosci. Remote. Sens. Lett.* **2021**, *19*, 1–5. [CrossRef]

34. Yu, S.; Ma, K.; Bi, Q.; Bian, C.; Ning, M.; He, N.; Li, Y.; Liu, H.; Zheng, Y. Mil-vt: Multiple instance learning enhanced vision transformer for fundus image classification. In *International Conference on Medical Image Computing and Computer-Assisted Intervention*; Springer: Cham, Switzerland, 2021; pp. 45–54.

35. Graham, B.; El-Nouby, A.; Touvron, H.; Stock, P.; Joulin, A.; Jégou, H.; Douze, M. Levit: A vision transformer in convnet's clothing for faster inference. In *Proceedings of the IEEE/CVF International Conference on Computer Vision*; IEEE: New York, NY, USA, 2021; pp. 12259–12269.

36. Xu, K.; Deng, P.; Huang, H. Vision transformer: An excellent teacher for guiding small networks in remote sensing image scene classification. *IEEE Trans. Geosci. Remote Sens.* **2022**, *60*, 1–15. [CrossRef]

37. Xue, Z.; Tan, X.; Yu, X.; Liu, B.; Yu, A.; Zhang, P. Deep Hierarchical Vision Transformer for Hyperspectral and LiDAR Data Classification. *IEEE Trans. Image Process.* **2022**, *31*, 3095–3110. [CrossRef]

38. Kaselimi, M.; Voulodimos, A.; Daskalopoulos, I.; Doulamis, N.; Doulamis, A. A Vision Transformer Model for Convolution-Free Multilabel Classification of Satellite Imagery in Deforestation Monitoring. *IEEE Trans. Neural Netw. Learn. Syst.* 2022, *Online ahead of print.*

39. Chen, Y.; Gu, X.; Liu, Z.; Liang, J. A Fast Inference Vision Transformer for Automatic Pavement Image Classification and Its Visual Interpretation Method. *Remote Sens.* **2022**, *14*, 1877. [CrossRef]

40. Jamil, S.; Abbas, M.S.; Roy, A.M. Distinguishing Malicious Drones Using Vision Transformer. *AI* **2022**, *3*, 260–273. [CrossRef]

41. He, K.; Zhang, X.; Ren, S.; Sun, J. Deep residual learning for image recognition. In Proceedings of the IEEE Conference on Computer Vision and Pattern Recognition, Las Vegas, NV, USA, 26 June–1 July 2016; pp. 770–778.

42. Sandler, M.; Howard, A.; Zhu, M.; Zhmoginov, A.; Chen, L.C. Mobilenetv2: Inverted residuals and linear bottlenecks. In Proceedings of the IEEE Conference on Computer Vision and Pattern Recognition, Salt Lake City, UT, USA, 18–22 June 2018; pp. 4510–4520.

43. Zhang, X.; Zhou, X.; Lin, M.; Sun, J. Shufflenet: An extremely efficient convolutional neural network for mobile devices. In Proceedings of the IEEE Conference on Computer Vision and Pattern Recognition, Salt Lake City, UT, USA, 18–22 June 2018; pp. 6848–6856.

44. Iandola, F.N.; Han, S.; Moskewicz, M.W.; Ashraf, K.; Dally, W.J.; Keutzer, K. SqueezeNet: AlexNet-level accuracy with 50x fewer parameters and< 0.5 MB model size. *arXiv* **2016**, arXiv:1602.07360.

45. Tan, M.; Le, Q. Efficientnet: Rethinking model scaling for convolutional neural networks. In *International Conference on Machine Learning*; PMLR: Cambridge, MA, USA, 2019; pp. 6105–6114.

46. Tanzi, L.; Audisio, A.; Cirrincione, G.; Aprato, A.; Vezzetti, E. Vision Transformer for femur fracture classification. *Injury* **2022**, *53*, 2625–2634. [CrossRef] [PubMed]

47. Chen, J.; Luo, T.; Wu, J.; Wang, Z.; Zhang, H. A Vision Transformer network SeedViT for classification of maize seeds. *J. Food Process. Eng.* **2022**, *45*, e13998. [CrossRef]

48. Wang, Z.; Yang, Y.; Yue, S. Air Quality Classification and Measurement Based on Double Output Vision Transformer. *IEEE Internet Things J.* **2022**, *9*, 20975–20984. [CrossRef]

49. Sha, Z.; Li, J. MITformer: A Multi-Instance Vision Transformer for Remote Sensing Scene Classification. *IEEE Geosci. Remote Sens. Lett.* **2022**, *19*, 1–5. [CrossRef]

50. Kim, K.; Wu, B.; Dai, X.; Zhang, P.; Yan, Z.; Vajda, P.; Kim, S.J. Rethinking the self-attention in vision transformers. In *Proceedings of the IEEE/CVF International Conference on Computer Vision*; IEEE: New York, NY, USA, 2021; pp. 3071–3075.

51. Lee, S.H.; Lee, S.; Song, B.C. Vision transformer for small-size datasets. *arXiv* **2021**, arXiv:2112.13492.

52. Beal, J.; Kim, E.; Tzeng, E.; Park, D.H.; Zhai, A.; Kislyuk, D. Toward transformer-based object detection. *arXiv* **2020**, arXiv:2012.09958.

53. Fang, Y.; Liao, B.; Wang, X.; Fang, J.; Qi, J.; Wu, R.; Niu, J.; Liu, W. You only look at one sequence: Rethinking transformer in vision through object detection. *Adv. Neural Inf. Process. Syst.* **2021**, *34*, 26183–26197.

54. Li, Y.; Xie, S.; Chen, X.; Dollar, P.; He, K.; Girshick, R. Benchmarking detection transfer learning with vision transformers. *arXiv* **2021**, arXiv:2111.11429.
55. Bao, H.; Dong, L.; Wei, F. Beit: Bert pre-training of image transformers. *arXiv* **2021**, arXiv:2106.08254.
56. He, K.; Chen, X.; Xie, S.; Li, Y.; Dollár, P.; Girshick, R. Masked autoencoders are scalable vision learners. In Proceedings of the IEEE/CVF Conference on Computer Vision and Pattern Recognition, New Orleans, LA, USA, 18–24 June 2022; pp. 16000–16009.
57. Horváth, J.; Baireddy, S.; Hao, H.; Montserrat, D.M.; Delp, E.J. Manipulation detection in satellite images using vision transformer. In Proceedings of the IEEE/CVF Conference on Computer Vision and Pattern Recognition, Nashville, TN, USA, 20–25 June 2021; pp. 1032–1041.
58. Wang, Y.; Ye, T.; Cao, L.; Huang, W.; Sun, F.; He, F.; Tao, D. Bridged Transformer for Vision and Point Cloud 3D Object Detection. In Proceedings of the IEEE/CVF Conference on Computer Vision and Pattern Recognition, New Orleans, LA, USA, 18–24 June 2022; pp. 12114–12123.
59. Dai, A.; Chang, A.X.; Savva, M.; Halber, M.; Funkhouser, T.; Nießner, M. Scannet: Richly-annotated 3d reconstructions of indoor scenes. In Proceedings of the IEEE Conference on Computer Vision and Pattern Recognition, Honolulu, HI, USA, 21–26 July 2017; pp. 5828–5839.
60. Song, S.; Lichtenberg, S.P.; Xiao, J. Sun rgb-d: A rgb-d scene understanding benchmark suite. In Proceedings of the IEEE Conference on Computer Vision and Pattern Recognition, Boston, MA, USA, 7–12 June 2015; pp. 567–576.
61. Liu, Z.; Zhang, Z.; Cao, Y.; Hu, H.; Tong, X. Group-free 3D object detection via transformers. In *Proceedings of the IEEE/CVF International Conference on Computer Vision*; IEEE: New York, NY, USA, 2021; pp. 2949–2958.
62. Prangemeier, T.; Reich, C.; Koeppl, H. Attention-based transformers for instance segmentation of cells in microstructures. In Proceedings of the 2020 IEEE International Conference on Bioinformatics and Biomedicine (BIBM), Seoul, Republic of Korea, 16–19 December 2020; pp. 700–707.
63. Chen, T.; Saxena, S.; Li, L.; Fleet, D.J.; Hinton, G. Pix2seq: A language modeling framework for object detection. *arXiv* **2021**, arXiv:2109.10852.
64. Wang, Y.; Zhang, X.; Yang, T.; Sun, J. Anchor detr: Query design for transformer-based detector. *AAAI Conf. Artif. Intell.* **2022**, *36*, 2567–2575. [CrossRef]
65. Chen, J.; Lu, Y.; Yu, Q.; Luo, X.; Adeli, E.; Wang, Y.; Lu, L.; Yuille, A.L.; Zhou, Y. Transunet: Transformers make strong encoders for medical image segmentation. *arXiv* **2021**, arXiv:2102.04306.
66. Sagar, A. Vitbis: Vision transformer for biomedical image segmentation. In *Clinical Image-Based Procedures, Distributed and Collaborative Learning, Artificial Intelligence for Combating COVID-19 and Secure and Privacy-Preserving Machine Learning*; Springer: Cham, Switzerland, 2021; pp. 34–45.
67. Huttenlocher, D.P.; Klanderman, G.A.; Rucklidge, W.J. Comparing images using the Hausdorff distance. *IEEE Trans. Pattern Anal. Mach. Intell.* **1993**, *15*, 850–863. [CrossRef]
68. Yang, Z.; Wang, J.; Tang, Y.; Chen, K.; Zhao, H.; Torr, P.H. LAVT: Language-Aware Vision Transformer for Referring Image Segmentation. In Proceedings of the IEEE/CVF Conference on Computer Vision and Pattern Recognition, New Orleans, LA, USA, 18–24 June 2022; pp. 18155–18165.
69. Yu, L.; Poirson, P.; Yang, S.; Berg, A.C.; Berg, T.L. Modeling context in referring expressions. In *European Conference on Computer Vision*; Springer: Cham, Switzerland, 2016; pp. 69–85.
70. Nagaraja, V.K.; Morariu, V.I.; Davis, L.S. Modeling context between objects for referring expression understanding. In *European Conference on Computer Vision*; Springer: Cham, Switzerland, 2016; pp. 792–807.
71. Mao, J.; Huang, J.; Toshev, A.; Camburu, O.; Yuille, A.L.; Murphy, K. Generation and comprehension of unambiguous object descriptions. In Proceedings of the IEEE Conference on Computer Vision and Pattern Recognition, Las Vegas, NV, USA, 26 June–1 July 2016; pp. 11–20.
72. Gu, J.; Kwon, H.; Wang, D.; Ye, W.; Li, M.; Chen, Y.H.; Lai, L.; Chandra, V.; Pan, D.Z. Multi-scale high-resolution vision transformer for semantic segmentation. In Proceedings of the IEEE/CVF Conference on Computer Vision and Pattern Recognition, New Orleans, LA, USA, 18–24 June 2022; pp. 12094–12103.
73. Cheng, B.; Schwing, A.; Kirillov, A. Per-pixel classification is not all you need for semantic segmentation. *Adv. Neural Inf. Process. Syst.* **2021**, *34*, 17864–17875.
74. Chen, L.C.; Zhu, Y.; Papandreou, G.; Schroff, F.; Adam, H. Encoder-decoder with atrous separable convolution for semantic image segmentation. In *European Conference on Computer Vision (ECCV)*; Springer: Cham, Switzerland, 2018; pp. 801–818.
75. Xiao, T.; Liu, Y.; Zhou, B.; Jiang, Y.; Sun, J. Unified perceptual parsing for scene understanding. In *European Conference on Computer Vision (ECCV)*; Springer: Cham, Switzerland, 2018; pp. 418–434.
76. Yuan, Y.; Chen, X.; Wang, J. Object-contextual representations for semantic segmentation. In Proceedings of the Computer Vision–ECCV 2020: 16th European Conference, Glasgow, UK, 23–28 August 2020; Proceedings, Part VI 16; Springer: Cham, Switzerland, 2020; pp. 173–190.
77. Yuan, Y.; Huang, L.; Guo, J.; Zhang, C.; Chen, X.; Wang, J. OCNet: Object context for semantic segmentation. *Int. J. Comput. Vis.* **2021**, *129*, 2375–2398. [CrossRef]
78. Zheng, S.; Lu, J.; Zhao, H.; Zhu, X.; Luo, Z.; Wang, Y.; Fu, Y.; Feng, J.; Xiang, T.; Torr, P.H.; et al. Rethinking semantic segmentation from a sequence-to-sequence perspective with transformers. In Proceedings of the IEEE/CVF Conference on Computer Vision and Pattern Recognition, Nashville, TN, USA, 20–25 June 2021; pp. 6881–6890.

79. Wang, H.; Zhu, Y.; Adam, H.; Yuille, A.; Chen, L.C. Max-deeplab: End-to-end panoptic segmentation with mask transformers. In Proceedings of the IEEE/CVF Conference on Computer Vision and Pattern Recognition, Nashville, TN, USA, 20–25 June 2021; pp. 5463–5474.

80. Hatamizadeh, A.; Xu, Z.; Yang, D.; Li, W.; Roth, H.; Xu, D. UNetFormer: A Unified Vision Transformer Model and Pre-Training Framework for 3D Medical Image Segmentation. *arXiv* **2022**, arXiv:2204.00631.

81. Antonelli, M.; Reinke, A.; Bakas, S.; Farahani, K.; Kopp-Schneider, A.; Landman, B.A.; Litjens, G.; Menze, B.; Ronneberger, O.; Summers, R.M.; et al. The medical segmentation decathlon. *Nat. Commun.* **2022**, *13*, 4128. [CrossRef]

82. Baid, U.; Ghodasara, S.; Mohan, S.; Bilello, M.; Calabrese, E.; Colak, E.; Farahani, K.; Kalpathy-Cramer, J.; Kitamura, F.C.; Pati, S.; et al. The rsna-asnr-miccai brats 2021 benchmark on brain tumor segmentation and radiogenomic classification. *arXiv* **2021**, arXiv:2107.02314.

83. Zhao, H.; Jia, J.; Koltun, V. Exploring self-attention for image recognition. In Proceedings of the IEEE/CVF Conference on Computer Vision and Pattern Recognition, Seattle, WA, USA, 13–19 June 2020; pp. 10076–10085.

84. Ramachandran, P.; Parmar, N.; Vaswani, A.; Bello, I.; Levskaya, A.; Shlens, J. Stand-alone self-attention in vision models. *Adv. Neural Inf. Process. Syst.* **2019**, *32*.

85. Mishra, D.; Singh, S.K.; Singh, R.K. Deep architectures for image compression: A critical review. *Signal Process.* **2022**, *191*, 108346. [CrossRef]

86. Jamil, S.; Piran, M. Learning-Driven Lossy Image Compression; A Comprehensive Survey. *arXiv* **2022**, arXiv:2201.09240.

87. Qian, Y.; Lin, M.; Sun, X.; Tan, Z.; Jin, R. Entroformer: A Transformer-based Entropy Model for Learned Image Compression. *arXiv* **2022**, arXiv:2202.05492.

88. Ballé, J.; Minnen, D.; Singh, S.; Hwang, S.J.; Johnston, N. Variational image compression with a scale hyperprior. *arXiv* **2018**, arXiv:1802.01436.

89. Koyuncu, A.B.; Gao, H.; Boev, A.; Gaikov, G.; Alshina, E.; Steinbach, E. Contextformer: A Transformer with Spatio-Channel Attention for Context Modeling in Learned Image Compression. In *European Conference on Computer Vision*; Springer: Cham, Switzerland, 2022; pp. 447–463.

90. Bross, B.; Wang, Y.K.; Ye, Y.; Liu, S.; Chen, J.; Sullivan, G.J.; Ohm, J.R. Overview of the versatile video coding (VVC) standard and its applications. *IEEE Trans. Circuits Syst. Video Technol.* **2021**, *31*, 3736–3764. [CrossRef]

91. Lu, M.; Guo, P.; Shi, H.; Cao, C.; Ma, Z. Transformer-based image compression. *arXiv* **2021**, arXiv:2111.06707.

92. Christensen, C.N.; Lu, M.; Ward, E.N.; Lio, P.; Kaminski, C.F. Spatio-temporal Vision Transformer for Super-resolution Microscopy. *arXiv* **2022**, arXiv:2203.00030.

93. Wang, D.; Wu, Z.; Yu, H. Ted-net: Convolution-free t2t vision transformer-based encoder-decoder dilation network for low-dose ct denoising. In *International Workshop on Machine Learning in Medical Imaging*; Springer: Cham, Switzerland, 2021; pp. 416–425.

94. Luthra, A.; Sulakhe, H.; Mittal, T.; Iyer, A.; Yadav, S. Eformer: Edge enhancement based transformer for medical image denoising. *arXiv* **2021**, arXiv:2109.08044.

95. McCollough, C.H.; Bartley, A.C.; Carter, R.E.; Chen, B.; Drees, T.A.; Edwards, P.; Holmes, D.R., III; Huang, A.E.; Khan, F.; Leng, S.; et al. Low-dose CT for the detection and classification of metastatic liver lesions: Results of the 2016 low dose CT grand challenge. *Med. Phys.* **2017**, *44*, e339–e352. [CrossRef]

96. Fan, C.M.; Liu, T.J.; Liu, K.H. SUNet: Swin Transformer UNet for Image Denoising. *arXiv* **2022**, arXiv:2202.14009.

97. Ronneberger, O.; Fischer, P.; Brox, T. U-net: Convolutional networks for biomedical image segmentation. In *International Conference on Medical Image Computing and Computer-Assisted Intervention*; Springer: Cham, Switzerland, 2015; pp. 234–241.

98. Martin, D.; Fowlkes, C.; Tal, D.; Malik, J. A database of human segmented natural images and its application to evaluating segmentation algorithms and measuring ecological statistics. In Proceedings of the Eighth IEEE International Conference on Computer Vision. ICCV 2001, Vancouver, BC, Canada, 7–14 July 2001; Volume 2; pp. 416–423.

99. Franzen, R. Kodak Lossless True Color Image Suite. 1999; Volume 4. Available online: http://r0k.us/graphics/kodak (accessed on 19 April 2023).

100. Yao, C.; Jin, S.; Liu, M.; Ban, X. Dense residual Transformer for image denoising. *Electronics* **2022**, *11*, 418. [CrossRef]

101. Abdelhamed, A.; Lin, S.; Brown, M.S. A high-quality denoising dataset for smartphone cameras. In Proceedings of the IEEE Conference on Computer Vision and Pattern Recognition, Salt Lake City, UT, USA, 18–22 June 2018; pp. 1692–1700.

102. Plotz, T.; Roth, S. Benchmarking denoising algorithms with real photographs. In Proceedings of the IEEE Conference on Computer Vision and Pattern Recognition, Honolulu, HI, USA, 21–26 July 2017; pp. 1586–1595.

103. Xu, K.; Li, W.; Wang, X.; Wang, X.; Yan, K.; Hu, X.; Dong, X. CUR Transformer: A Convolutional Unbiased Regional Transformer for Image Denoising. *ACM Trans. Multimed. Comput. Commun. Appl. (TOMM)* **2022**, *19*, 1–22. [CrossRef]

104. Xue, T.; Ma, P. TC-net: Transformer combined with cnn for image denoising. *Appl. Intell.* **2022**, *53*, 6753–6762. [CrossRef]

105. Zhao, M.; Cao, G.; Huang, X.; Yang, L. Hybrid Transformer-CNN for Real Image Denoising. *IEEE Signal Process. Lett.* **2022**, *29*, 1252–1256. [CrossRef]

106. Pang, L.; Gu, W.; Cao, X. TRQ3DNet: A 3D Quasi-Recurrent and Transformer Based Network for Hyperspectral Image Denoising. *Remote Sens.* **2022**, *14*, 4598. [CrossRef]

107. Chen, H.; Yang, G.; Zhang, H. Hider: A Hyperspectral Image Denoising Transformer With Spatial–Spectral Constraints for Hybrid Noise Removal. *IEEE Trans. Neural Netw. Learn. Syst.* **2022**, *Early Access*.

108. Yin, H.; Ma, S. CSformer: Cross-Scale Features Fusion Based Transformer for Image Denoising. *IEEE Signal Process. Lett.* **2022**, *29*, 1809–1813. [CrossRef]
109. Zhang, K.; Zuo, W.; Chen, Y.; Meng, D.; Zhang, L. Beyond a gaussian denoiser: Residual learning of deep cnn for image denoising. *IEEE Trans. Image Process.* **2017**, *26*, 3142–3155. [CrossRef] [PubMed]
110. Dong, C.; Loy, C.C.; Tang, X. Accelerating the super-resolution convolutional neural network. In *European Conference on Computer Vision*; Springer: Cham, Switzerland, 2016; pp. 391–407.
111. Zhang, K.; Zuo, W.; Zhang, L. FFDNet: Toward a fast and flexible solution for CNN-based image denoising. *IEEE Trans. Image Process.* **2018**, *27*, 4608–4622. [CrossRef]
112. Zhang, K.; Zuo, W.; Gu, S.; Zhang, L. Learning deep CNN denoiser prior for image restoration. In Proceedings of the IEEE Conference on Computer Vision and Pattern Recognition, Honolulu, HI, USA, 21–26 July 2017; pp. 3929–3938.
113. Zhang, K.; Li, Y.; Zuo, W.; Zhang, L.; Van Gool, L.; Timofte, R. Plug-and-play image restoration with deep denoiser prior. *IEEE Trans. Pattern Anal. Mach. Intell.* **2021**, *44*, 6360–6376. [CrossRef]
114. Wang, Z.; Cun, X.; Bao, J.; Zhou, W.; Liu, J.; Li, H. Uformer: A general u-shaped transformer for image restoration. In Proceedings of the IEEE/CVF Conference on Computer Vision and Pattern Recognition, New Orleans, LA, USA, 18–24 June 2022; pp. 17683–17693.
115. Liang, J.; Cao, J.; Sun, G.; Zhang, K.; Van Gool, L.; Timofte, R. Swinir: Image restoration using swin transformer. In *Proceedings of the IEEE/CVF International Conference on Computer Vision*; IEEE: New York, NY, USA, 2021; pp. 1833–1844.
116. Mishra, P.; Verk, R.; Fornasier, D.; Piciarelli, C.; Foresti, G.L. VT-ADL: A vision transformer network for image anomaly detection and localization. In Proceedings of the 2021 IEEE 30th International Symposium on Industrial Electronics (ISIE), Kyoto, Japan, 20–23 June 2021; pp. 1–6.
117. Bergmann, P.; Fauser, M.; Sattlegger, D.; Steger, C. MVTec AD–A comprehensive real-world dataset for unsupervised anomaly detection. In Proceedings of the IEEE/CVF Conference on Computer Vision and Pattern Recognition, Long Beach, CA, USA, 15–20 June 2019; pp. 9592–9600.
118. Lee, Y.; Kang, P. AnoViT: Unsupervised Anomaly Detection and Localization With Vision Transformer-Based Encoder-Decoder. *IEEE Access* **2022**, *10*, 46717–46724. [CrossRef]
119. Yuan, H.; Cai, Z.; Zhou, H.; Wang, Y.; Chen, X. TransAnomaly: Video Anomaly Detection Using Video Vision Transformer. *IEEE Access* **2021**, *9*, 123977–123986. [CrossRef]
120. Lee, Y.; Kim, J.; Willette, J.; Hwang, S.J. MPViT: Multi-path vision transformer for dense prediction. In Proceedings of the IEEE/CVF Conference on Computer Vision and Pattern Recognition, New Orleans, LA, USA, 18–24 June 2022; pp. 7287–7296.
121. Mok, T.C.; Chung, A. Affine Medical Image Registration with Coarse-to-Fine Vision Transformer. In Proceedings of the IEEE/CVF Conference on Computer Vision and Pattern Recognition, New Orleans, LA, USA, 18–24 June 2022; pp. 20835–20844.
122. Yuan, K.; Guo, S.; Liu, Z.; Zhou, A.; Yu, F.; Wu, W. Incorporating convolution designs into visual transformers. In *Proceedings of the IEEE/CVF International Conference on Computer Vision*; IEEE: New York, NY, USA, 2021; pp. 579–588.
123. Sun, W.; Niessen, W.J.; Klein, S. Hierarchical vs. simultaneous multiresolution strategies for nonrigid image registration. In *International Workshop on Biomedical Image Registration*; Springer: Berlin/Heidelberg, Germany, 2012; pp. 60–69.
124. Chen, J.; Frey, E.C.; He, Y.; Segars, W.P.; Li, Y.; Du, Y. TransMorph: Transformer for unsupervised medical image registration. *Med. Image Anal.* **2022**, *82*, 102615. [CrossRef] [PubMed]
125. Fang, J.; Xie, L.; Wang, X.; Zhang, X.; Liu, W.; Tian, Q. Msg-transformer: Exchanging local spatial information by manipulating messenger tokens. In Proceedings of the IEEE/CVF Conference on Computer Vision and Pattern Recognition, New Orleans, LA, USA, 18–24 June 2022; pp. 12063–12072.
126. Chen, J.N.; Sun, S.; He, J.; Torr, P.H.; Yuille, A.; Bai, S. Transmix: Attend to mix for vision transformers. In Proceedings of the IEEE/CVF Conference on Computer Vision and Pattern Recognition, New Orleans, LA, USA, 18–24 June 2022; pp. 12135–12144.
127. Dong, X.; Bao, J.; Chen, D.; Zhang, W.; Yu, N.; Yuan, L.; Chen, D.; Guo, B. Cswin transformer: A general vision transformer backbone with cross-shaped windows. In Proceedings of the IEEE/CVF Conference on Computer Vision and Pattern Recognition, New Orleans, LA, USA, 18–24 June 2022; pp. 12124–12134.
128. Liu, Z.; Hu, H.; Lin, Y.; Yao, Z.; Xie, Z.; Wei, Y.; Ning, J.; Cao, Y.; Zhang, Z.; Dong, L.; et al. Swin transformer v2: Scaling up capacity and resolution. In Proceedings of the IEEE/CVF Conference on Computer Vision and Pattern Recognition, New Orleans, LA, USA, 18–24 June 2022; pp. 12009–12019.
129. Wang, Y.; Huang, R.; Song, S.; Huang, Z.; Huang, G. Not all images are worth 16x16 words: Dynamic transformers for efficient image recognition. *Adv. Neural Inf. Process. Syst.* **2021**, *34*, 11960–11973.
130. Wang, W.; Xie, E.; Li, X.; Fan, D.P.; Song, K.; Liang, D.; Lu, T.; Luo, P.; Shao, L. Pyramid vision transformer: A versatile backbone for dense prediction without convolutions. In *Proceedings of the IEEE/CVF International Conference on Computer Vision*; IEEE: New York, NY, USA, 2021; pp. 568–578.
131. Chu, X.; Tian, Z.; Wang, Y.; Zhang, B.; Ren, H.; Wei, X.; Xia, H.; Shen, C. Twins: Revisiting the design of spatial attention in vision transformers. *Adv. Neural Inf. Process. Syst.* **2021**, *34*, 9355–9366.
132. Mehta, S.; Rastegari, M. Mobilevit: Light-weight, general-purpose, and mobile-friendly vision transformer. *arXiv* **2021**, arXiv:2110.02178.
133. Zhou, D.; Shi, Y.; Kang, B.; Yu, W.; Jiang, Z.; Li, Y.; Jin, X.; Hou, Q.; Feng, J. Refiner: Refining self-attention for vision transformers. *arXiv* **2021**, arXiv:2106.03714.

134. Zhou, D.; Kang, B.; Jin, X.; Yang, L.; Lian, X.; Jiang, Z.; Hou, Q.; Feng, J. Deepvit: Towards deeper vision transformer. *arXiv* **2021**, arXiv:2103.11886.

135. Touvron, H.; Cord, M.; Douze, M.; Massa, F.; Sablayrolles, A.; Jégou, H. Training data-efficient image transformers & distillation through attention. In *International Conference on Machine Learning*; PMLR: Cambridge, MA, USA, 2021; pp. 10347–10357.

136. Chen, Z.; Xie, L.; Niu, J.; Liu, X.; Wei, L.; Tian, Q. Visformer: The vision-friendly transformer. In *Proceedings of the IEEE/CVF International Conference on Computer Vision*; IEEE: New York, NY, USA, 2021; pp. 589–598.

137. Zhang, L.; Zhang, L.; Du, B. Deep learning for remote sensing data: A technical tutorial on the state of the art. *IEEE Geosci. Remote. Sens. Mag.* **2016**, *4*, 22–40. [CrossRef]

138. Zhu, X.X.; Tuia, D.; Mou, L.; Xia, G.S.; Zhang, L.; Xu, F.; Fraundorfer, F. Deep learning in remote sensing: A comprehensive review and list of resources. *IEEE Geosci. Remote. Sens. Mag.* **2017**, *5*, 8–36. [CrossRef]

139. Long, J.; Shelhamer, E.; Darrell, T. Fully convolutional networks for semantic segmentation. In Proceedings of the IEEE Conference on Computer Vision and Pattern Recognition, Boston, MA, USA, 7–12 June 2015; pp. 3431–3440.

140. Jozdani, S.; Chen, D.; Pouliot, D.; Johnson, B.A. A review and meta-analysis of generative adversarial networks and their applications in remote sensing. *Int. J. Appl. Earth Obs. Geoinf.* **2022**, *108*, 102734. [CrossRef]

141. Zhou, H.Y.; Lu, C.; Yang, S.; Yu, Y. ConvNets vs. Transformers: Whose visual representations are more transferable? In *Proceedings of the IEEE/CVF International Conference on Computer Vision*; IEEE: New York, NY, USA, 2021; pp. 2230–2238.

142. Sangam, T.; Dave, I.R.; Sultani, W.; Shah, M. Transvisdrone: Spatio-temporal transformer for vision-based drone-to-drone detection in aerial videos. *arXiv* **2022**, arXiv:2210.08423.

143. Li, J.; Ye, D.H.; Chung, T.; Kolsch, M.; Wachs, J.; Bouman, C. Multi-target detection and tracking from a single camera in Unmanned Aerial Vehicles (UAVs). In Proceedings of the 2016 IEEE/RSJ International Conference on Intelligent Robots and Systems (IROS), Daejeon, Republic of Korea, 9–14 October 2016; pp. 4992–4997.

144. Rozantsev, A.; Lepetit, V.; Fua, P. Detecting flying objects using a single moving camera. *IEEE Trans. Pattern Anal. Mach. Intell.* **2016**, *39*, 879–892. [CrossRef] [PubMed]

145. Liu, Z.; He, Z.; Wang, L.; Wang, W.; Yuan, Y.; Zhang, D.; Zhang, J.; Zhu, P.; Van Gool, L.; Han, J.; et al. VisDrone-CC2021: The vision meets drone crowd counting challenge results. In *Proceedings of the IEEE/CVF International Conference on Computer Vision*; IEEE: New York, NY, USA, 2021; pp. 2830–2838.

146. Gibril, M.B.A.; Shafri, H.Z.M.; Al-Ruzouq, R.; Shanableh, A.; Nahas, F.; Al Mansoori, S. Large-Scale Date Palm Tree Segmentation from Multiscale UAV-Based and Aerial Images Using Deep Vision Transformers. *Drones* **2023**, *7*, 93. [CrossRef]

147. Xie, E.; Wang, W.; Yu, Z.; Anandkumar, A.; Alvarez, J.M.; Luo, P. SegFormer: Simple and efficient design for semantic segmentation with transformers. *Adv. Neural Inf. Process. Syst.* **2021**, *34*, 12077–12090.

148. Strudel, R.; Garcia, R.; Laptev, I.; Schmid, C. Segmenter: Transformer for semantic segmentation. In *Proceedings of the IEEE/CVF International Conference on Computer Vision*; IEEE: New York, NY, USA, 2021; pp. 7262–7272.

149. Ranftl, R.; Bochkovskiy, A.; Koltun, V. Vision transformers for dense prediction. In *Proceedings of the IEEE/CVF International Conference on Computer Vision*; IEEE: New York, NY, USA, 2021; pp. 12179–12188.

150. Zhu, X.; Lyu, S.; Wang, X.; Zhao, Q. TPH-YOLOv5: Improved YOLOv5 based on transformer prediction head for object detection on drone-captured scenarios. In *Proceedings of the IEEE/CVF International Conference on Computer Vision*; IEEE: New York, NY, USA, 2021; pp. 2778–2788.

151. Cao, Y.; He, Z.; Wang, L.; Wang, W.; Yuan, Y.; Zhang, D.; Zhang, J.; Zhu, P.; Van Gool, L.; Han, J.; et al. VisDrone-DET2021: The vision meets drone object detection challenge results. In *Proceedings of the IEEE/CVF International Conference on Computer Vision*; IEEE: New York, NY, USA, 2021; pp. 2847–2854.

152. Thai, H.T.; Tran-Van, N.Y.; Le, K.H. Artificial cognition for early leaf disease detection using vision transformers. In Proceedings of the 2021 International Conference on Advanced Technologies for Communications (ATC), Ho Chi Minh City, Vietnam, 14–16 October 2021; pp. 33–38.

153. Yu, Q.; Xia, Y.; Bai, Y.; Lu, Y.; Yuille, A.L.; Shen, W. Glance-and-gaze vision transformer. *Adv. Neural Inf. Process. Syst.* **2021**, *34*, 12992–13003.

154. Bai, J.; Yuan, L.; Xia, S.T.; Yan, S.; Li, Z.; Liu, W. Improving Vision Transformers by Revisiting High-frequency Components. *arXiv* **2022**, arXiv:2204.00993.

155. Cao, Y.H.; Yu, H.; Wu, J. Training vision transformers with only 2040 images. *arXiv* **2022**, arXiv:2201.10728.

156. Liu, Y.; Sun, Y.; Xue, B.; Zhang, M.; Yen, G.G.; Tan, K.C. A survey on evolutionary neural architecture search. *IEEE Trans. Neural Netw. Learn. Syst.* **2021**, *34*, 550–570. [CrossRef]

157. Ren, P.; Xiao, Y.; Chang, X.; Huang, P.Y.; Li, Z.; Chen, X.; Wang, X. A comprehensive survey of neural architecture search: Challenges and solutions. *ACM Comput. Surv. (CSUR)* **2021**, *54*, 1–34. [CrossRef]

158. Hu, Y.Q.; Yu, Y. A technical view on neural architecture search. *Int. J. Mach. Learn. Cybern.* **2020**, *11*, 795–811. [CrossRef]

159. Elsken, T.; Metzen, J.H.; Hutter, F. Neural architecture search: A survey. *J. Mach. Learn. Res.* **2019**, *20*, 1997–2017.

160. Chitty-Venkata, K.T.; Emani, M.; Vishwanath, V.; Somani, A.K. Neural Architecture Search for Transformers: A Survey. *IEEE Access* **2022**, *10*, 108374–108412. [CrossRef]

161. Chen, B.; Li, P.; Li, C.; Li, B.; Bai, L.; Lin, C.; Sun, M.; Yan, J.; Ouyang, W. Glit: Neural architecture search for global and local image transformer. In *Proceedings of the IEEE/CVF International Conference on Computer Vision*; IEEE: New York, NY, USA, 2021; pp. 12–21.

162. Li, C.; Tang, T.; Wang, G.; Peng, J.; Wang, B.; Liang, X.; Chang, X. Bossnas: Exploring hybrid cnn-transformers with block-wisely self-supervised neural architecture search. In *Proceedings of the IEEE/CVF International Conference on Computer Vision*; IEEE: New York, NY, USA, 2021; pp. 12281–12291.

163. Gong, C.; Wang, D.; Li, M.; Chen, X.; Yan, Z.; Tian, Y.; Chandra, V. NASViT: Neural Architecture Search for Efficient Vision Transformers with Gradient Conflict aware Supernet Training. In Proceedings of the International Conference on Learning Representations, Vienna, Austria, 4–8 May 2021.

164. Liu, Z.; Li, D.; Lu, K.; Qin, Z.; Sun, W.; Xu, J.; Zhong, Y. Neural architecture search on efficient transformers and beyond. *arXiv* **2022**, arXiv:2207.13955.

165. Wu, B.; Li, C.; Zhang, H.; Dai, X.; Zhang, P.; Yu, M.; Wang, J.; Lin, Y.; Vajda, P. Fbnetv5: Neural architecture search for multiple tasks in one run. *arXiv* **2021**, arXiv:2111.10007.

166. Ni, B.; Meng, G.; Xiang, S.; Pan, C. NASformer: Neural Architecture Search for Vision Transformer. In *Asian Conference on Pattern Recognition*; Springer: Cham, Switzerland, 2022; pp. 47–61.

167. Chefer, H.; Gur, S.; Wolf, L. Transformer interpretability beyond attention visualization. In Proceedings of the IEEE/CVF Conference on Computer Vision and Pattern Recognition, Nashville, TN, USA, 20–25 June 2021; pp. 782–791.

168. Rigotti, M.; Miksovic, C.; Giurgiu, I.; Gschwind, T.; Scotton, P. Attention-based Interpretability with Concept Transformers. In Proceedings of the International Conference on Learning Representations, Vienna, Austria, 4–8 May 2021. Available online: https://openreview.net/forum?id=Qaw16njk6L (accessed on 18 April 2023).

169. Sun, M.; Ma, H.; Kang, G.; Jiang, Y.; Chen, T.; Ma, X.; Wang, Z.; Wang, Y. VAQF: Fully Automatic Software-hardware Co-design Framework for Low-bit Vision Transformer. *arXiv* **2022**, arXiv:2201.06618.

170. Jamil, S.; Abbas, M.S.; Umair, M.; Hussain, Y. A review of techniques and challenges in green communication. In Proceedings of the 2020 International Conference on Information Science and Communication Technology (ICISCT), Karachi, Pakistan, 8–9 February 2020; pp. 1–6.

171. Jamil, S.; Rahman, M. A Comprehensive Survey of Digital Twins and Federated Learning for Industrial Internet of Things (IIoT), Internet of Vehicles (IoV) and Internet of Drones (IoD). *Appl. Syst. Innov.* **2022**, *5*, 56. [CrossRef]

172. Hu, E.J.; Shen, Y.; Wallis, P.; Allen-Zhu, Z.; Li, Y.; Wang, S.; Wang, L.; Chen, W. Lora: Low-rank adaptation of large language models. *arXiv* **2021**, arXiv:2106.09685.

173. Dehghani, M.; Djolonga, J.; Mustafa, B.; Padlewski, P.; Heek, J.; Gilmer, J.; Steiner, A.; Caron, M.; Geirhos, R.; Alabdulmohsin, I.; et al. Scaling vision transformers to 22 billion parameters. *arXiv* **2023**, arXiv:2302.05442.

174. Fang, Y.; Sun, Q.; Wang, X.; Huang, T.; Wang, X.; Cao, Y. EVA-02: A Visual Representation for Neon Genesis. *arXiv* **2023**, arXiv:2303.11331.

175. Zheng, Z.; Ma, M.; Wang, K.; Qin, Z.; Yue, X.; You, Y. Preventing Zero-Shot Transfer Degradation in Continual Learning of Vision-Language Models. *arXiv* **2023**, arXiv:2303.06628.

MDPI
St. Alban-Anlage 66
4052 Basel
Switzerland
www.mdpi.com

Drones Editorial Office
E-mail: drones@mdpi.com
www.mdpi.com/journal/drones